IMPORTANT

MW00745637

HERE IS YOUR REGISTRATION CODE TO AC

YOUR PREMIUM McGRAW-HILL ONLINE RESOURCES.

For key premium online resources you need THIS CODE to gain access. Once the code is entered, you will be able to use the Web resources for the length of your course.

If your course is using **WebCT** or **Blackboard**, you'll be able to use this code to access the McGraw-Hill content within your instructor's online course.

Access is provided if you have purchased a new book. If the registration code is missing from this book, the registration screen on our Website, and within your WebCT or Blackboard course, will tell you how to obtain your new code.

Registering for McGraw-Hill Online Resources

To gain access to your McGraw-Hill web resources simply follow the steps below:

(1) USE YOUR WEB BROWSER TO GO TO: **www.mhhe.com/tillery**

(2) CLICK ON **FIRST TIME USER**.

(3) ENTER THE REGISTRATION CODE* PRINTED ON THE TEAR-OFF BOOKMARK ON THE RIGHT.

(4) AFTER YOU HAVE ENTERED YOUR REGISTRATION CODE, CLICK **REGISTER**.

(5) FOLLOW THE INSTRUCTIONS TO SET-UP YOUR PERSONAL UserID AND PASSWORD.

(6) WRITE YOUR UserID AND PASSWORD DOWN FOR FUTURE REFERENCE. KEEP IT IN A SAFE PLACE.

TO GAIN ACCESS to the McGraw-Hill content in your instructor's **WebCT** or **Blackboard** course simply log in to the course with the UserID and Password provided by your instructor. Enter the registration code exactly as it appears in the box to the right when prompted by the system. You will only need to use the code the first time you click on McGraw-Hill content.

Thank you, and welcome to your McGraw-Hill online Resources!

VGMG-3XVP-9SU5-1Q84-P2TR

REGISTRATION CODE

* YOUR REGISTRATION CODE CAN BE USED ONLY ONCE TO ESTABLISH ACCESS. IT IS NOT TRANSFERABLE.

0-07-291915-9 T/A TILLERY: PHYSICAL SCIENCE, 6/E

THE POWER TO CREATE

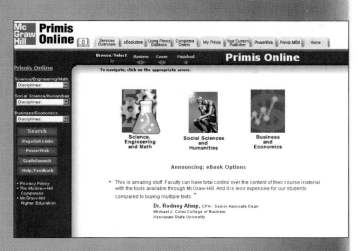

- Customize your course online
- Create a custom printed book (or eBook)
- Mix and match from millions of resources
- Create a textbook or lab manual
- It's easy to do—just vist the website

PRIMIS ONLINE

CREATE THE PERFECT TEXT FOR YOUR STUDENTS!

WWW.MHHE.COM/PRIMIS/ONLINE

THE TOOLS TO LEARN

- Receive a free copy in 7-10 days
- Review your custom book
- Adopt or modify your custom creation
- Students buy a book tailored to match your course

POWERWEB
PHYSICAL & INTEGRATED SCIENCE
Academic Editor: **James Agnew, Elon University**

Harness the power of the web for your course—**McGraw-Hill's PowerWeb** allows you to supplement your students' learning experience by providing Internet-based course material to supplement their learning experience.

PowerWeb provides current articles, curriculum-based materials, weekly updates with assessment, informative and timely world news, referred web links, research tools, student study tools, interactive exercises, and much more!

Instead of unorganized material provided by common search engines, all the content within **McGraw-Hill's PowerWeb** has been refereed by content experts and educational professionals—ensuring that you and your students only receive the most pertinent and topical information for your course.

UNIT 1. The Nature of Science

Part A. Scientific Method & Interdisciplinary Nature of Science
Part B. Science, Nonscience, and Pseudoscience
Part C. Fundamental Constants

UNIT 2. Physics

Part A. Motion and Gravity
Part B. Heat and Energy
Part C. Sound and Light

UNIT 3. Chemistry

Part A. Atomic Structure
Part B. Periodic Table

UNIT 4. The Universe

Part A. Beyond Earth
Part B. Stellar Evolution
Part C. Solar System
Part D. The Earth and Moon in Space

UNIT 5. Earth

Part A. Life on Earth
Part B. Natural Disasters
Part C. Weather and Climate
Part D. Water on Earth

UNIT 6. Biology

Part A. Life on Earth
Part B. Origin and Evolution of Life
Part C. Ecology and Environment
Part D. Health
Part E. Genetics
Part F. Ethical Concerns

Visit the web for a complete list of readings and abstracts at:

www.dushkin.com/powerweb

Conversion Factors

Length
1 in = 2.54 cm
1 cm = 0.394 in
1 ft = 30.5 cm
1 m = 39.4 in = 3.281 ft
1 km = 0.621 mi
1 mi = 5280 ft = 1.609 km
1 light-year = 9.461×10^{15} m

Mass
1 lb = 453.6 g (where g = 9.8 m/s^2)
1 kg = 2.205 lb (where g = 9.8 m/s^2)
1 atomic mass unit u = 1.66061×10^{-27} kg

Volume
1 liter = 1.057 quarts
1 in^3 = 16.39 cm^3
1 gallon = 3.786 liter
1 ft^3 = 0.02832 m^3

Energy
1 cal = 4.184 J
1 J = 0.738 ft·lb = 0.239 cal
1 ft·lb = 1.356 J
1 Btu = 252 cal = 778 ft·lb
1 kWh = 3.60×10^6 J = 860 kcal
1 hp = 550 ft·lb/s = 746 W
1 W = 0.738 ft·lb/s
1 Btu/h = 0.293 W
Absolute zero (0K) = −273.15°C
1 J = 6.24×10^{18} eV
1 eV = 1.6022×10^{-19} J

Speed
1 km/h = 0.2778 m/s = 0.6214 mi/h
1 m/s = 3.60 km/h = 2.237 mi/h = 3.281 ft/s
1 mi/h = 1.61 km/h = 0.447 m/s = 1.47 ft/s
1 ft/s = 0.3048 m/s = 0.6818 mi/h

Force
1 N = 0.2248 lb
1 lb = 4.448 N

Pressure
1 atm = 1.013 bar = 1.013×10^5 N/m^2 = 14.7 lb/in^2
1 lb/in^2 = 6.90×10^3 N/m^2

Metric Prefixes

Prefix	Symbol	Meaning	Unit Multiplier
exa-	E	quintillion	10^{18}
peta-	P	quadrillion	10^{15}
tera-	T	trillion	10^{12}
giga-	G	billion	10^{9}
mega-	M	million	10^{6}
kilo-	k	thousand	10^{3}
hecto-	h	hundred	10^{2}
deka-	da	ten	10^{1}
unit			
deci-	d	one-tenth	10^{-1}
centi-	c	one-hundredth	10^{-2}
milli-	m	one-thousandth	10^{-3}
micro-	μ	one-millionth	10^{-6}
nano-	n	one-billionth	10^{-9}
pico-	p	one-trillionth	10^{-12}
femto-	f	one-quadrillionth	10^{-15}
atto-	a	one-quintillionth	10^{-18}

Physical Constants

Quantity	Approximate Value
Gravity (earth)	g = 9.8 m/s^2
Gravitational law constant	G = 6.67×10^{-11} N·m^2/kg^2
Earth radius (mean)	6.38×10^6 m
Earth mass	5.97×10^{24} kg
Earth-sun distance (mean)	1.50×10^{11} m
Earth-moon distance (mean)	3.84×10^8 m
Fundamental charge	1.60×10^{-19} C
Coulomb law constant	k = 9.00×10^9 N·m^2/C^2
Electron rest mass	9.11×10^{-31} kg
Proton rest mass	1.6726×10^{-27} kg
Neutron rest mass	1.6750×10^{-27} kg
Bohr radius	5.29×10^{-11} m
Avogadro's number	6.022045×10^{23}/mol
Planck's constant	6.62×10^{-34} J·s
Speed of light (vacuum)	3.00×10^8 m/s
Pi	π = 3.1415926536

Periodic Table of the Elements

Group headers

| Group | IA (1) | IIA (2) | IIIB (3) | IVB (4) | VB (5) | VIB (6) | VIIB (7) | VIIIB (8) | VIIIB (9) | VIIIB (10) | IB (11) | IIB (12) | IIIA (13) | IVA (14) | VA (15) | VIA (16) | VIIA (17) | VIIIA (18) |

Period 1
| H — Hydrogen, 1, 1.008 | | | | | | | | | | | | | | | | | He — Helium, 2, 4.003 |

Period 2
Lithium 3 Li 6.941 | Beryllium 4 Be 9.012 | Boron 5 B 10.81 | Carbon 6 C 12.01 | Nitrogen 7 N 14.01 | Oxygen 8 O 16.00 | Fluorine 9 F 19.00 | Neon 10 Ne 20.18

Period 3
Sodium 11 Na 22.99 | Magnesium 12 Mg 24.31 | Aluminum 13 Al 26.98 | Silicon 14 Si 28.09 | Phosphorus 15 P 30.97 | Sulfur 16 S 32.07 | Chlorine 17 Cl 35.45 | Argon 18 Ar 39.95

Period 4
Potassium 19 K 39.10 | Calcium 20 Ca 40.08 | Scanium 21 Sc 44.96 | Titanium 22 Ti 47.88 | Vanadium 23 V 50.94 | Chromium 24 Cr 52.00 | Maganese 25 Mn 54.94 | Iron 26 Fe 55.85 | Cobalt 27 Co 58.93 | Nickel 28 Ni 58.69 | Copper 29 Cu 63.55 | Zinc 30 Zn 65.39 | Gallium 31 Ga 69.72 | Germanium 32 Ge 72.61 | Arsenic 33 As 74.92 | Selenium 34 Se 78.96 | Bromine 35 Br 79.90 | Krypton 36 Kr 83.80

Period 5
Rubidium 37 Rb 85.47 | Strontium 38 Sr 87.62 | Yttrium 39 Y 88.91 | Zirconium 40 Zr 91.22 | Niobium 41 Nb 92.91 | Molybdenum 42 Mo 95.94 | Technetium 43 Tc (98) | Ruthenium 44 Ru 101.1 | Rhodium 45 Rh 102.9 | Palladium 46 Pd 106.4 | Silver 47 Ag 107.9 | Cadmium 48 Cd 112.4 | Indium 49 In 114.8 | Tin 50 Sn 118.7 | Antimony 51 Sb 121.8 | Tellurium 52 Te 127.6 | Iodine 53 I 126.9 | Xenon 54 Xe 131.3

Period 6
Cesium 55 Cs 132.9 | Barium 56 Ba 137.3 | Lanthanum 57 La 138.9 | Hafnium 72 Hf 178.5 | Tantalum 73 Ta 180.9 | Tungsten 74 W 183.8 | Rhenium 75 Re 186.2 | Osmium 76 Os 190.2 | Iridium 77 Ir 192.2 | Platinum 78 Pt 195.1 | Gold 79 Au 197.0 | Mercury 80 Hg 200.6 | Thallium 81 Tl 204.4 | Lead 82 Pb 207.2 | Bismuth 83 Bi 209.0 | Polonium 84 Po (209) | Astatine 85 At (210) | Radon 86 Rn (222)

Period 7
Francium 87 Fr (223) | Radium 88 Ra (226) | Actinium 89 Ac (227) | Rutherfordium 104 Rf (261) | Dubnium 105 Db (262) | Seaborgium 106 Sg (266) | Bohrium 107 Bh (264) | Hassium 108 Hs (277) | Meitnerium 109 Mt (268) | Darmstadtium 110 Ds (281) | Unununium 111 Uuu (272) | Ununbium 112 Uub (285) | | Ununquadium 114 Uuq (289)

Lanthanides (6)
Cerium 58 Ce 140.1 | Praseodymium 59 Pr 140.9 | Neodymium 60 Nd 144.2 | Promethium 61 Pm (145) | Samarium 62 Sm 150.4 | Europium 63 Eu 152.0 | Gadolinium 64 Gd 157.3 | Terbium 65 Tb 158.9 | Dysprosium 66 Dy 162.5 | Holmium 67 Ho 164.9 | Erbium 68 Er 167.3 | Thulium 69 Tm 168.9 | Ytterbium 70 Yb 173.0 | Lutetium 71 Lu 175.0

Actinides (7)
Thorium 90 Th 232.0 | Protactinium 91 Pa 231.0 | Uranium 92 U 238.0 | Neptunium 93 Np (237) | Plutonium 94 Pu (244) | Americium 95 Am (243) | Curium 96 Cm (247) | Berkelium 97 Bk (247) | Californium 98 Cf (251) | Einsteinium 99 Es (252) | Fermium 100 Fm (257) | Mendelevium 101 Md (258) | Nobelium 102 No (259) | Lawrencium 103 Lr (262)

Legend
- Metals
- Semiconductors
- Nonmetals

Transition Elements
Inner Transition Elements

Values in parentheses are the mass numbers of the most stable or best-known isotopes.

Names and symbols for elements 111–114 are under review.

Key:
element name — Hydrogen
symbol of element — H — atomic number 1
1.008 — atomic weight

Physical Science
CHEMISTRY

Sixth Edition

Physical Science
CHEMISTRY

Bill W. Tillery
Arizona State University

 **Custom Publishing**

Boston   Burr Ridge, IL   Dubuque, IA   Madison, WI   New York   San Francisco   St. Louis
Bangkok   Bogotá   Caracas   Kuala Lumpur   Lisbon   London   Madrid   Mexico City
Milan   Montreal   New Delhi   Santiago   Seoul   Singapore   Sydney   Taipei   Toronto

CHEMISTRY
Physical Science

Copyright © 2005 by The McGraw-Hill Companies, Inc. All rights reserved. Printed in the
United States of America. Except as permitted under the United States Copyright Act of
1976, no part of this publication may be reproduced or distributed in any form or by any
means, or stored in a database or retrieval system, without the prior written consent of the
publisher.

This book contains selected material from *Physical Science, Sixth Edition,* by Bill W. Tillery.
Copyright © 2005 by The McGraw-Hill Companies, Inc. Reprinted with permission of the
publisher.

This book is a McGraw-Hill Custom textbook. Many custom published texts are modified
versions or adaptations of our best-selling textbooks. Some adaptations are printed in black
and white to keep prices at a minimum, while others are in color.

1 2 3 4 5 6 7 8 9 0 QSR QSR 0 1

ISBN 0–07–296541–2

Printer/Binder: Quebecor World

Brief Contents

Contents

Preface

Physical Science is a straightforward, easy-to-read, but substantial introduction to the fundamental behavior of matter and energy. It is intended to serve the needs of nonscience majors who are required to complete one or more physical science courses. It introduces basic concepts and key ideas while providing opportunities for students to learn reasoning skills and a new way of thinking about their environment. No prior work in science is assumed. The language, as well as the mathematics, is as simple as can be practical for a college-level science course.

Organization

The *Physical Science* sequence of chapters is flexible, and the instructor can determine topic sequence and depth of coverage as needed. The materials are also designed to support a conceptual approach, or a combined conceptual and problem-solving approach. With laboratory studies, the text contains enough material for the instructor to select a sequence for a two-semester course. It can also serve as a text in a one-semester astronomy and earth science course, or in other combinations.

> *"The text is excellent. I do not think I could have taught the course using any other textbook. I think one reason I really enjoy teaching this course is because of the text. I could say for sure that this is one of the best textbooks I have seen in my career. . . . I love this textbook for the following reasons: (1) it is comprehensive, (2) it is very well written, (3) it is easily readable and comprehendible, (4) it has good graphics."*
> —Ezat Heydari, Jackson State University

Meeting Student Needs

Physical Science is based on two fundamental assumptions arrived at as the result of years of experience and observation from teaching the course: (a) that students taking the course often have very limited background and/or aptitude in the natural sciences; and (b) that this type of student will better grasp the ideas and principles of physical science if they are discussed with minimal use of technical terminology and detail. In addition, it is critical for the student to see relevant applications of the material to everyday life. Most of these everyday-life applications, such as environmental concerns, are not isolated in an arbitrary chapter; they are discussed where they occur naturally throughout the text.

Each chapter presents historical background where appropriate, uses everyday examples in developing concepts, and follows a logical flow of presentation. The historical chronology, of special interest to the humanistically inclined nonscience major, serves to humanize the science being presented. The use of everyday examples appeals to the nonscience major, typically accustomed to reading narration, not scientific technical writing, and also tends to bring relevancy to the material being presented. The logical flow of presentation is helpful to students not accustomed to thinking about relationships between what is being read and previous knowledge learned, a useful skill in understanding the physical sciences. Worked examples help students to integrate concepts and understand the use of relationships called equations. They also serve as a model for problem solving; consequently, special attention is given to *complete* unit work and to the clear, fully expressed use of mathematics. Where appropriate, chapters contain one or more activities, called Concepts Applied, that use everyday materials rather than specialized laboratory equipment. These activities are intended to bring the science concepts closer to the world of the student. The activities are supplemental and can be done as optional student activities or as demonstrations.

> *"It is more readable than any text I've encountered. This has been my first experience teaching university physical science; I picked up the book and found it very user-friendly. The level of detail is one of this text's greatest strengths. It is well suited for a university course."*
> —Richard M. Woolheater, Southeastern Oklahoma State University

"The author's goals and practical approach to the subject matter is exactly what we are looking for in a textbook. . . . The practical approach to problem solving is very appropriate for this level of student."
—Martha K. Newchurch, Nicholls State University

". . . the book engages minimal use of technical language and scientific detail in presenting ideas. It also uses everyday examples to illustrate a point. This approach bonds with the mindset of the nonscience major who is used to reading prose in relation to daily living."
—Ignatius Okafor, Jarvis Christian College

"I was pleasantly surprised to see that the author has written a textbook that seems well suited to introductory physical science at this level. . . . Physical Science seems to strike a nice balance between the two—avoiding unnecessary complications while still maintaining a rigorous viewpoint. I prefer a textbook that goes beyond what I am able to cover in class, but not too much. Tillery seems to have done a good job here."
—T. G. Heil, University of Georgia

New to This Edition

In general, there has been a concerted effort to make the text even more user-friendly and relevant for students:

- A new "Concepts Applied" feature was added throughout the text, adding applications of relevance for students.
- Where needed, Parallel Exercises were reorganized to make Group A and B exercises more physically, as well as conceptually, congruent.
- Then the Parallel Exercises were selectively "tuned" for the intended audience of nonscience majors by revising and replacing some exercises with new, more conceptual exercises.
- Text materials were made more conceptually oriented and student-friendly throughout.
- The overall size of the text was reduced by two chapters through reorganizing and condensing some of the historical background material.
- Old chapter 2, "Motion," and old chapter 3, "Patterns of Motion," were merged into one new chapter ("Motion") for a more intuitive presentation.
- Old chapter 9, "Atomic Structure," was substantially rewritten and merged with old chapter 10, "Elements and the Periodic Table," into one new chapter ("Atoms and Periodic Properties") with a more student-friendly approach.
- Old chapter 13, "Water and Solutions," (new chapter 11) was substantially rewritten to be more conceptual and relevant to students.

- The astronomy chapters were substantially rewritten to be more intuitive, contain less history, and update factual materials.
- To satisfy requests from current users of the text, new "Closer Look" features were added, for example: Freefall, Simple Machines, The Measurement Process, Doppler Radar, Lasers, Radiation and Food Preservation, Three Mile Island and Chernobyl, Dark Energy, Seismic Tomography, Estuary Pollution, and the Health of the Chesapeake Bay.
- Also to satisfy requests from current users of the text, additional "People Behind the Science" features were added, including biographies on Isaac Newton, Michael Faraday, Erwin Schrödinger, Robert Bunsen, Shirley Ann Jackson, Stephen Hawking, Jocelyn (Susan) Bell Burnell, and Carl Sagan.

The Learning System

Physical Science has an effective combination of innovative learning aids intended to make the student's study of science more effective and enjoyable. This variety of aids is included to help students clearly understand the concepts and principles that serve as the foundation of the physical sciences.

Overview

Chapter 1 provides an *overview* or orientation to what the study of physical science in general, and this text in particular, are all about. It discusses the fundamental methods and techniques used by scientists to study and understand the world around us. It also explains the problem-solving approach used throughout the text so that students can more effectively apply what they have learned.

Chapter Opening Tools

Chapter Outline
The chapter outline includes all the major topic headings and subheadings within the body of the chapter. It gives you a quick glimpse of the chapter's contents and helps you locate sections dealing with particular topics.

Chapter Overview
Each chapter begins with an introductory overview. The overview previews the chapter's contents and what you can expect to learn from reading the chapter. It adds to the general outline of the chapter by introducing you to the concepts to be covered, facilitating in the integration of topics, and helping you to stay focused and organized while reading the chapter for the first time. After reading the introduction, browse through the chapter, paying particular attention to the topic headings and illustrations so that you get a feel for the kinds of ideas included within the chapter.

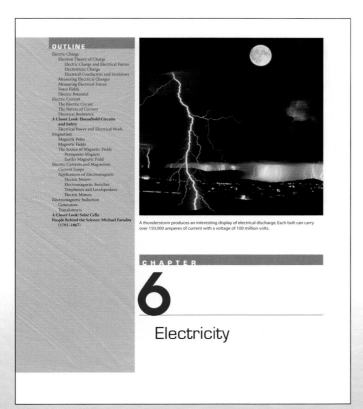

CHAPTER

6

Electricity

A thunderstorm produces an interesting display of electrical discharge. Each bolt can carry over 150,000 amperes of current with a voltage of 100 million volts.

The previous chapters have been concerned with *mechanical* concepts, explanations of the motion of objects that exert forces on one another. These concepts were used to explain straight-line motion, the motion of free fall, and the circular motion of objects on the earth as well as the circular motion of planets and satellites. The mechanical concepts were based on Newton's laws of motion and are sometimes referred to as Newtonian physics. The mechanical explanations were then extended into the submicroscopic world of matter through the kinetic molecular theory. The objects of motion were now particles, molecules that exert force on one another, and concepts associated with heat were interpreted as the motion of these particles. In a further extension of Newtonian concepts, mechanical explanations were given for concepts associated with sound, a mechanical disturbance that follows the laws of motion as it moves through the molecules of matter.

You might wonder, as did the scientists of the 1800s, if mechanical interpretations would also explain other natural phenomena such as electricity, chemical reactions, and light. A mechanical model would be very attractive because it already explained so many other facts of nature, and scientists have always looked for basic, unifying theories. Mechanical interpretations were tried, as electricity was considered a moving fluid, and light was considered a mechanical wave moving through a material fluid. There were many unsolved puzzles with such a model, and gradually it was recognized that electricity, light, and chemical reactions could not be explained by mechanical interpretations. Gradually, the point of view changed from a study of particles to a study of the properties of the space around the particles. In this chapter you will learn about electric charge in terms of the space around particles. This model of electric charge, called the *field model*, will be used to develop concepts about electric current, the electric circuit, and electrical work and power. A relationship between electricity and the fascinating topic of magnetism is discussed next, including what magnetism is and how it is produced. The relationship is then used to explain the mechanical production of electricity (Figure 6.1), how electricity is measured, and how electricity is used in everyday technological applications.

Electric Charge

You are familiar with the use of electricity in many electrical devices such as lights, toasters, radios, and calculators. You are also aware that electricity is used for transportation and for heating and cooling places where you work and live. Many people accept electrical devices as part of their surroundings, with only a hazy notion of how they work. To many people electricity seems to be magical. Electricity is not magical, and it can be understood, just as we understand any other natural phenomenon. There are theories that explain observations, quantities that can be measured, and relationships between these quantities, or laws, that lead to understanding. All of the observations, measurements, and laws begin with an understanding of *electric charge*.

Electron Theory of Charge

It was a big mystery for thousands of years. No one could figure out why a rubbed piece of amber, which is fossilized tree resin, would attract small pieces of paper, thread, and hair. This unexplained attraction was called the "amber effect." Then about one hundred years ago, Joseph J. Thomson found the answer while experimenting with electric currents. From these experiments, Thomson was able to conclude that negatively charged particles were present in all matter, and in fact might be the stuff of which matter is made.

The amber effect was traced to the movement of these particles, so they were called *electrons* after the Greek word for amber. The word *electricity* is also based on the Greek word for amber.

Today, we understand that the basic unit of matter is the *atom*, which is made up of electrons and other particles such as *protons* and *neutrons*. The atom is considered to have a dense center part called a *nucleus* that contains the closely situated protons and neutrons. The electrons move around the nucleus at some relatively greater distance (Figure 6.2). Details on the nature of protons, neutrons, electrons, and models of how the atom is constructed will be considered in chapter 8. For understanding electricity, you need only consider the protons in the nucleus, the electrons that move around the nucleus, and the fact that electrons can be moved from an atom and caused to move to or from one object to another. Basically, the electrical, light, and chemical phenomena involve the *electrons* and not the more massive nucleus. The massive nuclei remain in a relatively fixed position in a solid, but some of the electrons can move about from atom to atom.

Electric Charge and Electrical Forces

Electrons and protons have a property called electric charge. Electrons have a *negative electric charge* and protons have a *positive electric charge*. The negative or positive description simply means that these two properties are opposite; it does not mean that one is better than the other. Charge is as fundamental to

---

Examples

Each topic discussed within the chapter contains one or more concrete, worked *Examples* of a problem and its solution as it applies to the topic at hand. Through careful study of these examples, students can better appreciate the many uses of problem solving in the physical sciences.

> *"I feel this book is written well for our average student. The images correlate well with the text, and the math problems make excellent use of the dimensional analysis method. While it was a toss-up between this book and another one, now that we've taught from the book for the last year, we are extremely happy with it."*
> —Alan Earhart, Three Rivers Community College

A Closer Look	Above It All

The super-speed magnetic levitation (maglev) train is a completely new technology based on magnetically suspending a train 3 to 10 cm (about 1 to 4 in) above a monorail, then moving it along with a magnetic field that travels along the monorail guides. The maglev train does not have friction between wheels and the rails since it does not have wheels. This lack of resistance at the easily manipulated magnetic fields makes very short acceleration distances possible. For example, a German maglev train can accelerate from 0 to 300 km/h (about 185 mi/h) over a distance of just 5 km (about 3 mi). A conventional train with wheels requires about 30 km (about 19 mi) to reach the same speed from a standing start. The maglev is attractive for short runs because of its superior acceleration and braking abilities. It is also attractive for longer runs because of its high top speed—up to about 500 km/h (about 310 mi/h). Today, only an aircraft can match such a speed.

EXAMPLE 2.3

A bicycle moves from rest to 5 m/s in 5 s. What was the acceleration?

Solution

$$v_i = 0 \text{ m/s}$$
$$v_f = 5 \text{ m/s}$$
$$t = 5 \text{ s}$$
$$a = ?$$

$$a = \frac{v_f - v_i}{t}$$
$$= \frac{5 \text{ m/s} - 0 \text{ m/s}}{5 \text{ s}}$$
$$= \frac{5}{5} \frac{\text{m/s}}{\text{s}}$$
$$= 1 \frac{\text{m}}{\text{s}} \cdot \frac{1}{\text{s}}$$
$$= \boxed{1 \frac{\text{m}}{\text{s}^2}}$$

EXAMPLE 2.4

An automobile uniformly accelerates from rest at 15 ft/s^2 for 6 s. What is the final velocity in ft/s? (Answer: 90 ft/s)

So far, you have learned only about straight-line, uniform acceleration that results in an increased velocity. There are also other changes in the motion of an object that are associated with acceleration. One of the more obvious is a change that results in a decreased velocity. Your car's brakes, for example, can slow your car or bring it to a complete stop. This is *negative acceleration*, which is sometimes called *deceleration*. Another change in the motion of an object is a change of direction. Velocity encompasses both the rate of motion and direction, so

a change of direction is an acceleration. The satellite moving with a constant speed in a circular orbit around the earth is constantly changing its direction of movement. It is therefore constantly accelerating because of this constant change in its motion. Your automobile has three devices that could change the state of its motion. Your automobile therefore has three accelerators—the gas pedal (which can increase magnitude of velocity), the brakes (which can decrease magnitude of velocity), and the steering wheel (which can change direction of velocity). (See Figure 2.6.) The important thing to remember is that acceleration results from any *change* in the motion of an object.

The final velocity (v_f) and the initial velocity (v_i) are different variables than the average velocity ($\bar{v}$). You cannot use an initial or final velocity for an average velocity. You may, however, calculate an average velocity ($\bar{v}$) from the other two variables as long as the acceleration taking place between the initial and final velocities is uniform. An example of such a uniform change would be an automobile during a constant, straight-line acceleration. To find an average velocity *during* a uniform

A Constant direction increase speed

B Constant direction decrease speed

C Change direction constant speed

D Change direction change speed

FIGURE 2.6
Four different ways (A–D) to accelerate a car.

Applying Science to the Real World

Concepts Applied

Each chapter also includes one or more *Concepts Applied* boxes. These activities are simple investigative exercises that students can perform at home or in the classroom to demonstrate important concepts and reinforce understanding of them. This feature also describes the application of those concepts to everyday life.

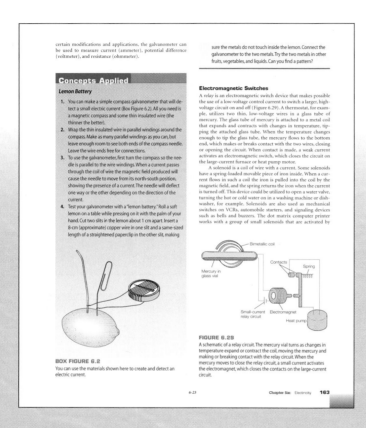

Closer Look

One or more boxed *Closer Look* features can be found in each chapter of *Physical Science*. These readings present topics of special human or environmental concern (the use of seat belts, acid rain, and air pollution, for example). In addition to environmental concerns, topics are presented on interesting technological applications (passive solar homes, solar cells, catalytic converters, etc.), or topics on the cutting edge of scientific research (for example, El Niño and Dark Energy). All boxed features are informative materials that are supplementary in nature. The *Closer Look* readings serve to underscore the relevance of physical science in confronting the many issues we face daily.

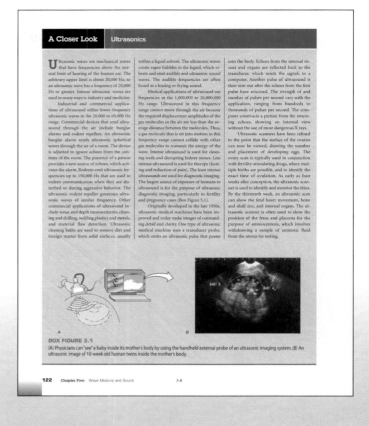

People Behind the Science

Many chapters also have one or two fascinating biographies that spotlight well-known scientists, past or present. From these *People Behind the Science* biographies, students learn about the human side of the science: physical science is indeed relevant, and real people do the research and make the discoveries. These readings present physical science in real-life terms that students can identify with and understand.

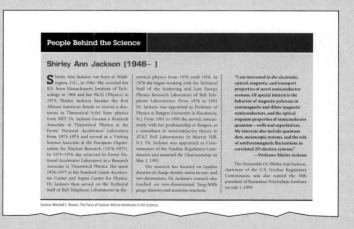

"The People Behind the Science features help relate the history of science and the contributions of the various individuals."
—Richard M. Woolheater, Southeastern Oklahoma State University

End-of-Chapter Features

At the end of each chapter, students will find the following materials:

- *Summary:* highlights the key elements of the chapter.
- *Summary of Equations* (chapters 1–13): reinforces retention of the equations presented.
- *Key Terms:* gives page references for finding the terms defined within the context of the chapter reading.
- *Applying the Concepts:* tests comprehension of the material covered with a multiple-choice quiz.
- *Questions for Thought:* challenges students to demonstrate their understanding of the topics.
- *Parallel Exercises* (chapters 1–13): reinforces problem-solving skills. There are two groups of parallel exercises, Group A and Group B. The Group A parallel exercises have complete solutions worked out, along with useful comments, in appendix D. The Group B parallel exercises are similar to those in Group A but do not contain answers in the text. By working through the Group A parallel exercises and checking the solutions in appendix D, students will gain confidence in tackling the parallel exercises in Group B, and thus reinforce their problem-solving skills.

"I like this [Summary of Equations] feature. It collects the equations together for easy reference. . . . I also like this [Key Terms] feature. It is well organized, thorough and gives the student a tool for review. The instructor can also use it for a checklist of topics. . . . The end-of-chapter features of Summary of Equations and Key Terms make the chapters very user-friendly."
—Richard M. Woolheater, Southeastern Oklahoma State University

"The Parallel Exercises and the explanation in the appendix, the readability of the material, and the depth of coverage are the strongest features of this text."
—Martha K. Newchurch, Nicholls State University

"The provision of solutions to a set of problems as a guide for solving identical problems on an adjacent set is an ingenious learning tool."
—Ignatius Okafor, Jarvis Christian College

End-of-Text Materials

Appendices providing math review, additional background detail, solubility and humidity charts, and solutions for the Group A Parallel Exercises can be found at the back of the text. There is also a glossary of all key terms, an index, and special tables printed on the inside covers for reference use.

Summary

Electromagnetic radiation is emitted from all matter with a temperature above absolute zero, and as the temperature increases, more radiation and shorter wavelengths are emitted. Visible light is emitted from matter hotter than about 700°C, and this matter is said to be *incandescent*. The sun, a fire, and the ordinary lightbulb are incandescent sources of light.

The behavior of light is shown by a light ray model that uses straight lines to show the straight-line path of light. Light that interacts with matter is *reflected* with parallel rays, moves in random directions by *diffuse reflection* from points, or is *absorbed*, resulting in a temperature increase. Matter is *opaque*, reflecting light, or *transparent*, transmitting light.

In reflection, the incoming light, or *incident ray*, has the same angle as the *reflected ray* when measured from a perpendicular from the point of reflection, called the *normal*. That the two angles are equal is called the *law of reflection*. The law of reflection explains how a flat mirror forms a *virtual image*, one from which light rays do not originate. Light rays do originate from the other kind of image, a *real image*.

Light rays are bent, or *refracted*, at the boundary when passing from one transparent medium to another. The amount of refraction depends on the *incident angle* and the *index of refraction*, a ratio of the speed of light in a vacuum to the speed of light in the medium. When the refracted angle is 90°, *total internal reflection* takes place. This limit to the angle of incidence is called the *critical angle*, and all light rays with an incident angle at or beyond this angle are reflected internally.

Each color of light has a range of wavelengths that forms the *spectrum* from red to violet. A glass prism has the property of *dispersion*, separating a beam of white light into a spectrum. Dispersion occurs because the index of refraction is different for each range of colors, with short wavelengths refracted more than larger ones.

A wave model of light can be used to explain interference and polarization. *Interference* occurs when light passes through two small slits or holes and produces an *interference pattern* of bright lines and dark zones. *Polarized* light vibrates in one direction only, in a plane. Light can be polarized by certain materials, by reflection, or by scattering. Polarization can only be explained by a transverse wave model.

A wave model fails to explain observations of light behaviors in the *photoelectric effect* and *blackbody radiation*. Max Planck found that he could modify the wave theory to explain blackbody radiation by assuming that vibrating molecules could only have discrete amounts, or *quanta*, of energy and found that the quantized energy is related to the frequency and a constant known today as *Planck's constant*. Albert Einstein applied Planck's quantum concept to the photoelectric effect and described a light wave in terms of quanta of energy called *photons*. Each photon has an energy that is related to the frequency and Planck's constant.

Today, the properties of light are explained by a model that incorporates both the wave and the particle nature of light. Light is considered to have both wave and particle properties and is not describable in terms of anything known in the everyday-sized world.

Summary of Equations

7.1
$$\text{angle of incidence} = \text{angle of reflection}$$
$$\theta_i = \theta_r$$

7.2
$$\text{index of refraction} = \frac{\text{speed of light in vacuum}}{\text{speed of light in material}}$$
$$n = \frac{c}{v}$$

7.3
$$\text{speed of light in vacuum} = (\text{wavelength})(\text{frequency})$$
$$c = \lambda f$$

7.4
$$\frac{\text{energy of}}{\text{photon}} = \frac{\text{Planck's}}{\text{constant}}(\text{frequency})$$
$$E = hf$$

KEY TERMS

blackbody radiation (p. 180)	polarized (p. 194)
incandescent (p. 180)	quanta (p. 196)
index of refraction (p. 187)	real image (p. 201)
interference (p. 191)	refraction (p. 201)
light ray model (p. 182)	total internal reflection (p. 186)
luminous (p. 180)	unpolarized light (p. 194)
photoelectric effect (p. 195)	virtual image (p. 185)
photons (p. 196)	

APPLYING THE CONCEPTS

1. A luminous object is an object that
   a. reflects a dim blue-green light in the dark.
   b. produces light of its own by any method.
   c. shines by reflected light only, such as the moon.
   d. an object that glows only in the absence of light.
2. An object is hot enough to emit a dull red glow. When this object is heated even more, it will
   a. emit shorter-wavelength, higher-frequency radiation.
   b. emit longer-wavelength, lower-frequency radiation.
   c. emit the same wavelengths as before, but with more energy.
   d. emit more of the same wavelengths with more energy.
3. The difference in the light emitted from a candle, an incandescent lightbulb, and the sun is basically from differences in
   a. energy sources.
   b. materials.
   c. temperatures.
   d. phases of matter.

Appendix A
Mathematical Review

Working with Equations

Many of the problems of science involve an equation, a shorthand way of describing patterns and relationships that are observed in nature. Equations are also used to identify properties and to define certain concepts, but all uses have well-established meanings, symbols that are used by convention, and allowed mathematical operations. This appendix will assist you in better understanding equations and the reasoning that goes with the manipulation of equations in problem-solving activities.

Background

In addition to a knowledge of rules for carrying out mathematical operations, an understanding of certain quantitative ideas and concepts can be very helpful when working with equations. Among these helpful concepts are (1) the meaning of inverse and reciprocal, (2) the concept of a ratio, and (3) fractions.

The term *inverse* means the opposite, or reverse, of something. For example, addition is the opposite, or inverse, of subtraction, and division is the inverse of multiplication. A *reciprocal* is defined as an inverse multiplication relationship between two numbers. For example, if the symbol n represents any number (except zero), then the reciprocal of n is $1/n$. The reciprocal of a number ($1/n$) multiplied by that number (n) always gives a product of 1. Thus, the number multiplied by 5 to give 1 is $1/5$ ($5 \times 1/5 = 5/5 = 1$). So $1/5$ is the reciprocal of 5, and 5 is the reciprocal of $1/5$. Each number is the *inverse* of the other.

The fraction $1/5$ means 1 divided by 5, and if you carry out the division it gives the decimal 0.2. Calculators that have a $1/x$ key will do the operation automatically. If you enter 5, then press the $1/x$ key, the answer of 0.2 is given. If you press the $1/x$ key again, the answer of 5 is given. Each of these numbers is a reciprocal of the other.

A *ratio* is a comparison between two numbers. If the symbols m and n are used to represent any two numbers, then the ratio of the number m to the number n is the fraction m/n. This expression means to divide m by n. For example, if m is 10 and n is 5, the ratio of 10 to 5 is 10/5, or 2:1.

Working with *fractions* is sometimes necessary in problem-solving exercises, and an understanding of these operations is needed to carry out unit calculations. It is helpful in many of

these operations to remember that a number (or a unit) divided by itself is equal to 1, for example,

$$\frac{5}{5} = 1 \qquad \frac{\text{inch}}{\text{inch}} = 1 \qquad \frac{5\text{ inches}}{5\text{ inches}} = 1$$

When one fraction is divided by another fraction, the operation commonly applied is to "invert the denominator and multiply." For example, 2/5 divided by 1/2 is

$$\frac{\frac{2}{5}}{\frac{1}{2}} = \frac{2}{5} \times \frac{2}{1} = \frac{4}{5}$$

What you are really doing when you invert the denominator of the larger fraction and multiply is making the denominator (1/2) equal to 1. Both the numerator (2/5) and the denominator (1/2) are multiplied by 2/1, which does not change the value of the overall expression. The complete operation is

$$\frac{\frac{2}{5}}{\frac{1}{2}} = \frac{\frac{2}{5} \times \frac{2}{1}}{\frac{1}{2} \times \frac{2}{1}} = \frac{\frac{4}{5}}{1} = \frac{4}{5}$$

Symbols and Operations

The use of symbols seems to cause confusion for some students because it seems different from their ordinary experiences with arithmetic. The rules are the same for symbols as they are for numbers, but you cannot do the operations with the symbols until you know what values they represent. The operation signs, such as $+$, $\div$, $\times$, and $-$ are used with symbols to indicate the operation that you would do if you knew the values. Some of the mathematical operations are indicated several ways. For example, $a \times b$, $a \cdot b$, and ab all indicate the same thing, that a is to be multiplied by b. Likewise, $a \div b$, a/b, and $a \times 1/b$ all indicate that a is to be divided by b. Since it is not possible to carry out the operations on symbols alone, they are called *indicated operations*.

Operations in Equations

An equation is a shorthand way of expressing a simple sentence with symbols. The equation has three parts: (1) a left side, (2) an equal sign ($=$), which indicates the equivalence of

Supplements

Physical Science is accompanied by a variety of multimedia supplementary materials, including an interactive website; an Instructor's Testing and Resource CD-ROM, with testing software containing multiple-choice test items for the text and other teacher resources; and a Digital Content Manager CD-ROM, with digital images from the text. The supplement package also contains more traditional supplements: a laboratory manual and overhead transparencies.

Multimedia Supplementary Materials

Online Learning Center

A text-specific website, our *Physical Science* Online Learning Center, offering unlimited resources for both the student and instructor, can be found at: www.mhhe.com/tillery/. By way of this website, students and instructors will be better able to quickly incorporate the Internet into their classrooms. This interactive resource is packaged free with any new textbook.

Student Edition of the Online Learning Center. The *Physical Science, Sixth Edition* Online Learning Center has book-specific study aids organized by chapter. Each chapter includes animations modeling key concepts discussed in the book; interactive questions and problems, such as self-test quizzes and crossword puzzles, flashcards, and matching exercises using key terms and glossary definitions; chapter resources; and web-linked resources. Also included are Exploring Physical Science articles, which expose students to a different viewpoint on a topic or a new research project, as well as links to McGraw-Hill's Access Science and PowerWeb sites, which provide additional research resources.

Instructor's Edition of the Online Learning Center. For instructors, there is an image bank containing the images from the text, PowerPoint lectures, a bank of personal response system questions, the *instructor's manual,* the *instructor's edition of the laboratory manual,* clip art, a database of equations, and much more. From the student edition, instructors can access questions and problems from the text and additional Closer Look questions with e-mail boxes for gradable responses from students.

The *instructor's manual,* also written by the text author, is housed on the Online Learning Center and provides a chapter outline, an introduction/summary of each chapter, suggestions for discussion and demonstrations, multiple-choice questions (with answers) that can be used as resources for cooperative teaching, and answers and solutions to all end-of-chapter questions and exercises not provided in the text.

Instructor's Testing and Resource CD-ROM

The **Instructor's Testing and Resource CD-ROM** contains the *Physical Science* test bank (test questions in a combination of true/false and multiple-choice formats) within the Brownstone DIPLOMA© test generator. The Brownstone software includes a test generator, an online testing program, Internet testing, and a grade management system. This user-friendly software's testing capability is consistently ranked number one in evaluations over other products. Also located on the Instructor's Testing and Resources CD-ROM are Word and PDF files of the test bank, the instructor's manual, instructor's edition of the laboratory manual, the bank of personal response system questions, and the quizzes from the Online Learning Center. Any of these Word files can be used in combination with the Brownstone software or independently.

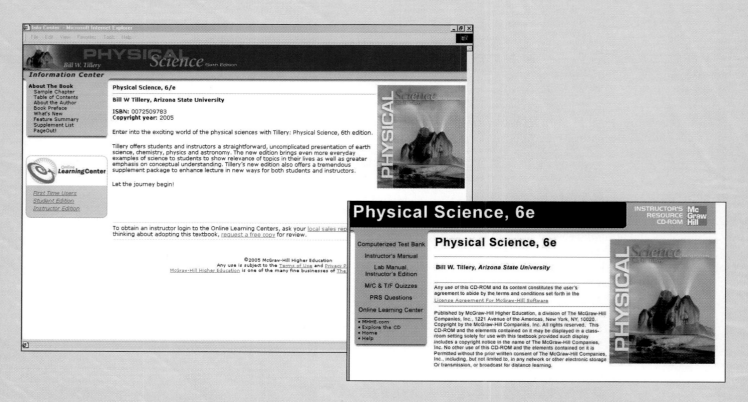

Digital Content Manager CD-ROM

The **Digital Content Manager** contains JPEG files of the four-color illustrations, photos, and tables from the text as well as a collection of animations and video clips. The CD also contains a PowerPoint presentation of the text images and another lecture PowerPoint presentation. These digital assets are contained on a cross-platform CD-ROM and are grouped by chapter within a user-friendly interface. With the help of these valuable resources, instructors can create customized classroom presentations, visually based tests and quizzes, dynamic course website content, and attractive printed support materials.

> "I find Physical Science to be superior to either of the texts that I have used to date. . . . The animations and illustrations are better than those of other textbooks that I have seen, more realistic and less trivial."
> —T. G. Heil, University of Georgia

Printed Supplementary Material

Laboratory Manual

The *laboratory manual,* written and classroom tested by the author, presents a selection of laboratory exercises specifically written for the interests and abilities of nonscience majors. There are laboratory exercises that require measurement, data analysis, and thinking in a more structured learning environment. Alternative exercises that are open-ended "Invitations to Inquiry" are provided for instructors who would like a less structured approach. When the laboratory manual is used with *Physical Science,* students will have an opportunity to master basic scientific principles and concepts, learn new problem-solving and thinking skills, and understand the nature of scientific inquiry from the perspective of hands-on experiences. The *instructor's edition of the laboratory manual* can be found on the *Physical Science* Online Learning Center.

Overhead Transparencies

A set of over 100 full-color transparencies features images from the text. The images have been modified to ensure maximum readability in both small and large classroom settings.

Acknowledgments

We are indebted to the reviewers of the sixth edition for their constructive suggestions, new ideas, and invaluable advice. Special thanks and appreciation goes out to the sixth edition reviewers:

Brian Augustine   *James Madison University*
Charles Blatchley   *Pittsburg State University*
Alan D. Earhart   *Three Rivers Community College*
Carl Frederickson   *University of Central Arkansas*
T. G. Heil   *University of Georgia*
Ezat Heydari   *Jackson State University*
Martha K. Newchurch   *Nicholls State University*
Ignatius Okafor   *Jarvis Christian College*
Karen Savage   *California State University at Northridge*
Ling Jun Wang   *University of Tennessee at Chattanooga*
Richard M. Woolheater   *Southeastern Oklahoma State University*
Heather Woolverton   *University of Central Arkansas*
Michael Young   *Mississippi Delta Community College*

This revision of *Physical Science* has also been made possible by the many users and reviewers of its previous editions. The author and publisher are grateful to the following reviewers of previous editions for their critical reviews, comments, and suggestions:

Lawrence H. Adams   *Polk Community College*
John Akutagawa   *Hawaii Pacific University*
Arthur L. Alt   *University of Great Falls*
Richard Bady   *Marshall University*
David Benin   *Arizona State University*
Charles L. Bissell   *Northwestern State University of Louisiana*
W. H. Breazeale, Jr.   *Francis Marion College*
William Brown   *Montgomery College*
Steven Carey   *Mobile College*
Darry S. Carlston   *University of Central Oklahoma*
Stan Celestian   *Glendale Community College*
Randel Cox   *Arkansas State University*
Paul J. Croft   *Jackson State University*
Keith B. Daniels   *University of Wisconsin–Eau Claire*
Valentina David   *Bethune-Cookman College*
Carl G. Davis   *Danville Area Community College*
Joe D. DeLay   *Freed-Hardeman University*
Renee D. Diehl   *Pennsylvania State University*
Laurencin Dunbar   *Livingstone College*
Dennis Englin   *The Master's College*
Steven S. Funck   *Harrisburg Area Community College*
Lucille B. Garmon   *State University of West Georgia*
Peter K. Glanz   *Rhode Island College*
Nova Goosby   *Philander Smith College*
D. W. Gosbin   *Cumberland County College*
Floretta Haggard   *Rogers State College*
Robert G. Hamerly   *University of Northern Colorado*
Eric Harms   *Brevard Community College*
J. Dennis Hawk   *Navarro College*
L. D. Hendrick   *Francis Marion College*
Christopher Hunt   *Prince George's Community College*
Abe Korn   *New York City Tech College*

Lauree G. Lane   *Tennessee State University*
Robert Larson   *St. Louis Community College*
William Luebke   *Modesto Junior College*
Douglas L. Magnus   *St. Cloud State University*
Stephen Majoros   *Lorain County Community College*
L. Whit Marks   *Central State University*
Richard S. Mitchell   *Arkansas State University*
Jesse C. Moore   *Kansas Newman College*
Michael D. Murphy   *Northwest Alabama Community College*
Oladayo Oyelola   *Lane College*
Harold Pray   *University of Central Arkansas*
Virginia Rawlins   *University of North Texas*
Michael L. Sitko   *University of Cincinnati*
K. W. Trantham   *Arkansas Tech University*
R. Steven Turley   *Brigham Young University*
David L. Vosburg   *Arkansas State University*
Donald A. Whitney   *Hampton University*
Linda Wilson   *Middle Tennessee State University*
David Wingert   *Georgia State University*

We would also like to thank the following contributors to the sixth edition:

Judith Iriarte Gross, Middle Tennessee State University, for her vast knowledge of student conceptual understandings, used in developing and revising the personal response system questions to accompany *Physical Science.*

T. G. Heil, University of Georgia, for his creativity in developing the multimedia lecture PowerPoint presentations on the *Physical Science* Online Learning Center.

Ezat Heydari, Jackson State University, for his thorough review in developing and revising the multiple-choice, true/false, and tutorial self-tests for the *Physical Science* Online Learning Center.

Last, I wish to acknowledge the very special contributions of my wife, Patricia Northrop Tillery, whose assistance and support throughout the revision were invaluable.

Meet the Author

Bill W. Tillery

Bill W. Tillery is a professor in the Department of Physics and Astronomy at Arizona State University, where he has been a member of the faculty since 1973. He earned a bachelor's degree at Northeastern State University (1960), and master's and doctorate degrees from the University of Northern Colorado (1967). Before moving to Arizona State University, he served as director of the Science and Mathematics Teaching Center at the University of Wyoming (1969–73) and as an assistant professor at Florida State University (1967–69). Bill has served on numerous councils, boards, and committees and was honored as the "Outstanding University Educator" at the University of Wyoming in 1972. He was elected the "Outstanding Teacher" in the Department of Physics and Astronomy at Arizona State University in 1995.

During his time at Arizona State, Bill has taught a variety of courses, including general education courses in science and society, physical science, and introduction to physics. He has received more than 40 grants from the National Science Foundation, the U.S. Office of Education, from private industry (Arizona Public Service), and private foundations (The Flinn Foundation) for science curriculum development and science teacher inservice training. In addition to teaching and grant work, Bill has authored or coauthored more than 60 textbooks and many monographs, and has served as editor of three separate newsletters and journals between 1977 and 1996.

Bill also maintains a website dedicated to providing resources for science teachers. This site is named The Science Education Resource Page (SERP) and is funded by a grant from the Flinn Foundation. The URL is http://serp.la.asu.edu.

Bill has attempted to present an interesting, helpful program that will be useful to both students and instructors. Comments and suggestions about how to do a better job of reaching this goal are welcome. Any comments about the text or other parts of the program should be addressed to:

Bill W. Tillery
Department of Physics and Astronomy
Arizona State University
PO Box 871504
Tempe, AZ 85287-1504 USA
Or (preferred) e-mail: bill.tillery@asu.edu

Physical science is concerned with your physical surroundings and your concepts and understanding of these surroundings.

CHAPTER

1

What Is Science?

Have you ever thought about your thinking and what you know? On a very simplified level, you could say that everything you know came to you through your senses. You see, hear, and touch things of your choosing and you can also smell and taste things in your surroundings. Information is gathered and sent to your brain by your sense organs. Somehow, your brain processes all this information in an attempt to find order and make sense of it all. Finding order helps you understand the world and what may be happening at a particular place and time. Finding order also helps you predict what may happen next, which can be very important in a lot of situations.

This is a book on thinking about and understanding your physical surroundings. These surroundings range from the obvious, such as the landscape and the day-to-day weather, to the not so obvious, such as how atoms are put together. Your physical surroundings include natural things as well as things that people have made and used (Figure 1.1). You will learn how to think about your surroundings, whatever your previous experience with thought-demanding situations. This first chapter is about "tools and rules" that you will use in the thinking process.

Objects and Properties

Physical science is concerned with making sense out of the physical environment. The early stages of this "search for sense" usually involve *objects* in the environment, things that can be seen or touched. These could be objects you see every day, such as a glass of water, a moving automobile, or a blowing flag. They could be quite large, such as the sun, the moon, or even the solar system, or invisible to the unaided human eye. Objects can be any size, but people are usually concerned with objects that are larger than a pinhead and smaller than a house. Outside these limits, the actual size of an object is difficult for most people to comprehend.

As you were growing up, you learned to form a generalized mental image of objects called a *concept*. Your concept of an object is an idea of what it is, in general, or what it should be according to your idea (Figure 1.2). You usually have a word stored away in your mind that represents a concept. The word "chair," for example, probably evokes an idea of "something to sit on." Your generalized mental image for the concept that goes with the word "chair" probably includes a four-legged object with a backrest. Upon close inspection, most of your (and everyone else's) concepts are found to be somewhat vague. For example, if the word "chair" brings forth a mental image of something with four legs and a backrest (the concept), what is the difference between a "high chair" and a "bar stool"? When is a chair a chair and not a stool? These kinds of questions can be troublesome for many people.

Not all of your concepts are about material objects. You also have concepts about intangibles such as time, motion, and relationships between events. As was the case with concepts of material objects, words represent the existence of intangible concepts. For example, the words "second," "hour," "day," and "month" represent concepts of time. A concept of the pushes and pulls that come with changes of motion during an airplane flight might be represented with such words as "accelerate" and

FIGURE 1.1

Your physical surroundings include naturally occurring and manufactured objects such as sidewalks and buildings.

FIGURE 1.2
What is your concept of a chair? Are all of these pieces of furniture chairs? Most people have concepts, or ideas of what things in general should be, that are loosely defined. The concept of a chair is one example of a loosely defined concept.

"falling." Intangible concepts might seem to be more abstract since they do not represent material objects.

By the time you reach adulthood you have literally thousands of words to represent thousands of concepts. But most, you would find on inspection, are somewhat ambiguous and not at all clear-cut. That is why you find it necessary to talk about certain concepts for a minute or two to see if the other person has the same "concept" for words as you do. That is why when one person says, "Boy, was it hot!" the other person may respond, "How hot was it?" The meaning of "hot" can be quite different for two people, especially if one is from Arizona and the other from Alaska!

The problem with words, concepts, and mental images can be illustrated by imagining a situation involving you and another person. Suppose that you have found a rock that you believe would make a great bookend. Suppose further that you are talking to the other person on the telephone, and you want to discuss the suitability of the rock as a bookend, but you do not know the name of the rock. If you knew the name, you would simply state that you found a "_____." Then you would probably discuss the rock for a minute or so to see if the other person really understood what you were talking about. But not knowing the name of the rock, and wanting to communicate about the suitability of the object as a bookend, what would you do? You would probably describe the characteristics, or **properties,** of the rock. Properties are the qualities or attributes that, taken together, are usually peculiar to an object. Since you commonly determine properties with your senses (smell, sight, hearing, touch, and taste), you could say that the properties of an object are the effect the object has on your senses. For example, you might say that the rock is a "big, yellow, smooth rock with shiny

gold cubes on one side." But consider the mental image that the other person on the telephone forms when you describe these properties. It is entirely possible that the other person is thinking of something very different from what you are describing (Figure 1.3)!

As you can see, the example of describing a proposed bookend by listing its properties in everyday language leaves much to be desired. The description does not really help the other person form an accurate mental image of the rock. One problem with the attempted communication is that the description of any property implies some kind of *referent.* The word **referent** means that you *refer to,* or think of, a given property in terms of another, more familiar object. Colors, for example, are sometimes stated with a referent. Examples are "sky blue," "grass green," or "lemon yellow." The referents for the colors blue, green, and yellow are, respectively, the sky, living grass, and a ripe lemon.

Referents for properties are not always as explicit as they are with colors, but a comparison is always implied. Since the comparison is implied, it often goes unspoken and leads to assumptions in communications. For example, when you stated that the rock was "big," you assumed that the other person knew that you did not mean as big as a house or even as big as a bicycle. You assumed that the other person knew that you meant that the rock was about as large as a book, perhaps a bit larger.

Another problem with the listed properties of the rock is the use of the word "smooth." The other person would not know if you meant that the rock *looked* smooth or *felt* smooth. After all, some objects can look smooth and feel rough. Other objects can look rough and feel smooth. Thus, here is another assumption, and probably all of the properties lead to implied

FIGURE 1.3
Could you describe this rock to another person over the telephone so that the other person would know *exactly* what you see? This is not likely with everyday language, which is full of implied comparisons, assumptions, and inaccurate descriptions.

As an example of how the measurement process works, consider the property of *length*. Most people are familiar with the concept of the length of something (long or short), the use of length to describe distances (close or far), and the use of length to describe heights (tall or short). The referent units used for measuring length are the familiar inch, foot, and mile from the English system and the centimeter, meter, and kilometer of the metric system. These systems and specific units will be discussed later. For now, imagine that these units do not exist but that you need to measure the length and width of this book. This imaginary exercise will illustrate how the measurement process eliminates vagueness and assumption in communication.

The first requirement in the measurement process is to choose some referent unit of length. You could arbitrarily choose something that is handy, such as the length of a standard paper clip, and you could call this length a "clip." Now you must decide on a procedure to specify how you will use the clip unit. You could define some specific procedures. For example:

1. Place a clip parallel to and on the long edge, or length, of the book so the end of the referent clip is lined up with the bottom edge of the book. Make a small pencil mark at the other end of the clip, as shown in Box Figure 1.1.

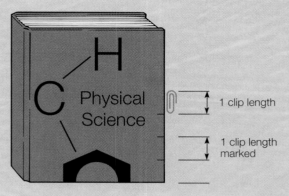

BOX FIGURE 1.1

As an example of the measurement process, a standard paper-clip length is selected as a referent unit. The unit is compared to the property that is being described. In this example, the property of the book length is measured by counting how many clip lengths describe the length.

2. Move the outside end of the clip to the mark and make a second mark at the other end. Continue doing this until you reach the top edge of the book.
3. Compare how many clip replications are in the book length by counting.
4. Record the length measurements by writing (a) how many clip replications were made and (b) the name of the clip length.

If the book length did not measure to a whole number of clips, you might need to divide the clip length into smaller subunits to be more precise. You could develop a *scale* of the basic clip unit and subunits. In fact, you could use multiples of the basic clip unit for an extended scale, using the scale for measurement rather than moving an individual clip unit. You could call the scale a "clipstick" (as in yardstick or meterstick).

comparisons, assumptions, and a not-very-accurate communication. This is the nature of your everyday language and the nature of most attempts at communication.

Quantifying Properties

Typical day-to-day communications are often vague and leave much to be assumed. A communication between two people, for example, could involve one person describing some person, object, or event to a second person. The description is made by using referents and comparisons that the second person may or may not have in mind. Thus, such attributes as "long" fingernails or "short" hair may have entirely different meanings to different people involved in a conversation. Assumptions and

vagueness can be avoided by using **measurement** in a description. Measurement is a process of comparing a property to a well-defined and agreed-upon referent. The well-defined and agreed-upon referent is used as a standard called a **unit.** The measurement process involves three steps: (1) *comparing* the referent unit to the property being described, (2) following a *procedure,* or operation, which specifies how the comparison is made, and (3) *counting* how many standard units describe the property being considered.

The measurement process uses a defined referent unit, which is compared to a property being measured. The *value* of the property is determined by counting the number of referent units. The name of the unit implies the procedure that results in the number. A measurement statement always contains a *number* and *name* for the referent unit. The number answers the question "How much?" and the name answers the question "Of what?" Thus a measurement always tells you "how much of what." You will find that using measurements will sharpen your communications. You will also find that using measurements is one of the first steps in understanding your physical environment.

Measurement Systems

Measurement is a process that brings precision to a description by specifying the "how much" and "of what" of a property in a particular situation. A number expresses the value of the property, and the name of a unit tells you what the referent is as well as implying the procedure for obtaining the number. Referent units must be defined and established, however, if others are to understand and reproduce a measurement. When standards are established the referent unit is called a **standard unit** (Figure 1.4). The use of standard units makes it possible to communicate and duplicate measurements. Standard units are usually defined and established by governments and their agencies that are created for that purpose. In the United States, the agency concerned with measurement standards is named the National Institute of Standards and Technology. In Canada, the Standards Council of Canada oversees the National Standard System.

There are two major *systems* of standard units in use today, the *English system* and the *metric system*. The metric system is used throughout the world except in the United States, where both systems are in use. The continued use of the English system in the United States presents problems in international trade, so there is pressure for a complete conversion to the metric system. More and more metric units are being used in everyday measurements, but a complete conversion will involve an enormous cost. Appendix A contains a method for converting from one system to the other easily. Consult this section if you need to convert from one metric unit to another metric unit or to convert from English to metric units or vice versa. Conversion factors are listed inside the front cover.

People have used referents to communicate about properties of things throughout human history. The ancient Greek

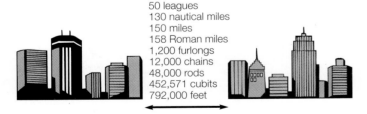

50 leagues
130 nautical miles
150 miles
158 Roman miles
1,200 furlongs
12,000 chains
48,000 rods
452,571 cubits
792,000 feet

FIGURE 1.4

Any of these units and values could have been used at some time or another to describe the same distance between these hypothetical towns. Any unit could be used for this purpose, but when one particular unit is officially adopted, it becomes known as the *standard unit*.

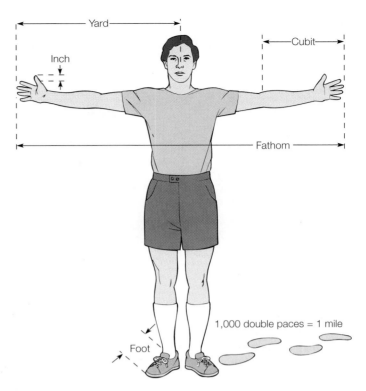

FIGURE 1.5

Many early units for measurement were originally based on the human body. Some of the units were later standardized by governments to become the basis of the English system of measurement.

civilization, for example, used units of *stadia* to communicate about distances and elevations. The "stadium" was a unit of length of the racetrack at the local stadium ("stadia" is the plural of stadium), based on a length of 125 paces. Later civilizations, such as the ancient Romans, adopted the stadia and other referent units from the ancient Greeks. Some of these same referent units were later adopted by the early English civilization, which eventually led to the **English system** of measurement. Some adopted units of the English system were originally based on parts of the human body, presumably because you always had these referents with you (Figure 1.5). The inch, for example, used the end joint of the thumb for a referent. A foot, naturally, was the length of a foot, and a yard was the distance from the tip of the nose to the end of the fingers on an arm held straight out. A cubit was the distance from the end of an elbow to the fingertip, and a fathom was the distance between the fingertips of two arms held straight out.

What body part could be used as a referent for volume? As shown in Table 1.1, all common volume units were based on a "mouthful." Each of the larger volume units was defined as two of the smaller units, making it easier to remember. Some of the units—such as the jack, jill, and pottle—have dropped from use today, leaving us with puzzles such as why there are two pints in a quart, but four quarts in a gallon. Understanding that

TABLE 1.1
English units of volume of 200 years ago

Two Quantities	Equivalent Quantity
2 mouthfuls	= 1 jigger
2 jiggers	= 1 jack
2 jacks	= 1 jill
2 jills	= 1 cup
2 cups	= 1 pint
2 pints	= 1 quart
2 quarts	= 1 pottle
2 pottles	= 1 gallon
2 gallons	= 1 pail
2 pails	= 1 peck
2 pecks	= 1 bushel

at one time there were two quarts in a pottle and two pottles make a gallon seems to make more sense if you understand the old scheme that two of something smaller makes one of the larger units.

As you can imagine, there were problems with these early units because everyone had different-sized arms, legs, and mouths. Beginning in the 1300s, the sizes of the various units were gradually standardized by English kings.

The **metric system** was established by the French Academy of Sciences in 1791. The academy created a measurement system that was based on invariable referents in nature, not human body parts. These referents have been redefined over time to make the standard units more reproducible. The *International System of Units,* abbreviated *SI,* is a modernized version of the metric system. Today, the SI system has seven base units that define standards for the properties of length, mass, time, electric current, temperature, amount of substance, and light intensity (Table 1.2). All units other than the seven basic ones are *derived* units. Area, volume, and speed, for example, are all expressed with derived units. Units for the properties of length, mass, and time are introduced in this chapter. The remaining units will be introduced in later chapters as the properties they measure are discussed.

TABLE 1.2
The SI base units

Property	Unit	Symbol
Length	meter	m
Mass	kilogram	kg
Time	second	s
Electric current	ampere	A
Temperature	kelvin	K
Amount of substance	mole	mol
Luminous intensity	candela	cd

Standard Units for the Metric System

If you consider all the properties of all the objects and events in your surroundings, the number seems overwhelming. Yet, close inspection of how properties are measured reveals that some properties are combinations of other properties (Figure 1.6). Volume, for example, is described by the three length measurements of length, width, and height. Area, on the other hand, is described by just the two length measurements of length and width. Length, however, cannot be defined in simpler terms of any other property. There are four properties that cannot be described in simpler terms, and all other properties are combinations of these four. For this reason they are called the **fundamental properties.** A fundamental property cannot be defined in simpler terms other than to describe how it is measured. These four fundamental properties are (1) *length,* (2) *mass,* (3) *time,* and (4) *charge.* Used individually or in combinations, these four properties will describe or measure what you observe in nature. Metric units for measuring the fundamental properties of length, mass, and time will be described next. The fourth fundamental property, charge, is associated with electricity, and a unit for this property will be discussed in a future chapter.

Length

The standard unit for length in the metric system is the **meter** (the symbol or abbreviation is m). The meter is defined as the distance that light travels in a vacuum during a certain time period, 1/299,792,458 second. The important thing to remember, however, is that the meter is the metric *standard unit* for length. A meter is slightly longer than a yard, 39.3 inches. It is approximately the distance from your left shoulder to the tip of your right hand when your arm is held straight out. Many doorknobs are about one meter above the floor. Think about these distances when you are trying to visualize a meter length.

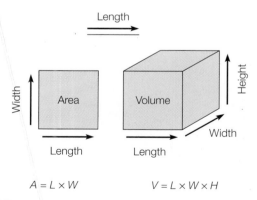

$$A = L \times W \qquad V = L \times W \times H$$

FIGURE 1.6

Area, or the extent of a surface, can be described by two length measurements. Volume, or the space that an object occupies, can be described by three length measurements. Length, however, can be described only in terms of how it is measured, so it is called a *fundamental property.*

Most people have heard of a leap year, but not a leap second. A *leap year* is needed because the earth does not complete an exact number of turns on its axis while completing one trip around the Sun. Our calendar system was designed to stay in step with the seasons with 365-day years and a 366-day year (leap year) every fourth year.

Likewise, our clocks are occasionally adjusted by a one-second increment known as a *leap second*. The leap second is needed because the earth does not have a constant spin. Coordinated Universal Time is the worldwide scientific standard of time keeping. It is based upon Earth's rotation and is kept accurate to within microseconds with carefully maintained atomic clocks. A leap second is a second added to Coordinated Universal Time to make it agree with astronomical time to within 0.9 second.

In 1955, astronomers at the U.S. Naval Observatory and the National Physical Laboratory in England measured the relationship between the frequency of the cesium atom (the standard of time) and the rotation of the earth at a particular period of time. The standard atomic clock second was defined to be equivalent to the fraction 1/31,556,925.9747 of the year 1900—or, an average second for that year. This turned out to be the time required for 9,192,631,770 vibrations of the cesium 133 atom. The second was defined in 1967 in terms of the length of time required for 9,192,631,770 vibrations of the cesium 133 atom. So, the atomic second was set equal to an average second of Earth rotation time near the turn of the twentieth century, but defined in terms of the frequency of a cesium atom.

The earth is constantly slowing from the frictional effects of the tides. Evidence of this slowing can be found in records of ancient observations of eclipses. From these records it is possible to determine the slowing of the earth. One day was only 18 hours long 900 million years ago.

It has been nearly a century since the referent year used for the definition of a second, and the difference is now roughly 2 milliseconds per day. Other factors also affect the earth's spin, such as the wind from hurricanes, so that it is necessary to monitor the earth's rotation continuously and add or subtract leap seconds when needed.

Mass

The standard unit for mass in the metric system is the **kilogram** (kg). The kilogram is defined as the mass of a certain metal cylinder kept by the International Bureau of Weights and Measures in France. This is the only standard unit that is still defined in terms of an object. The property of mass is sometimes confused with the property of weight since they are directly proportional to each other at a given location on the surface of the earth. They are, however, two completely different properties and are measured with different units. All objects tend to maintain their state of rest or straight-line motion, and this property is called "inertia." The *mass* of an object is a measure of the inertia of an object. The *weight* of the object is a measure of the force of gravity on it. This distinction between weight and mass will be discussed in detail in chapter 2. For now, remember that weight and mass are not the same property.

Time

The standard unit for time is the **second** (s). The second was originally defined as 1/86,400 of a solar day ($1/60 \times 1/60 \times 1/24$). The earth's spin was found not to be as constant as thought, so the second was redefined to be the duration required for a certain number of vibrations of a certain cesium atom. A special spectrometer called an "atomic clock" measures these vibrations and keeps time with an accuracy of several millionths of a second per year.

Metric Prefixes

The metric system uses prefixes to represent larger or smaller amounts by factors of 10. Some of the more commonly used prefixes, their abbreviations, and their meanings are listed in Table 1.3. Suppose you wish to measure something smaller than the standard unit of length, the meter. The meter is subdivided into ten equal-sized subunits called *decimeters*. The prefix *deci-* has a meaning of "one-tenth of," and it takes 10 decimeters to equal the length of 1 meter. For even smaller measurements,

TABLE 1.3
Some metric prefixes

Prefix	Symbol	Meaning	Unit Multiplier
exa-	E	quintillion	10^{18}
peta-	P	quadrillion	10^{15}
tera-	T	trillion	10^{12}
giga-	G	billion	10^{9}
mega-	M	million	10^{6}
kilo-	k	thousand	10^{3}
hecto-	h	hundred	10^{2}
deka-	da	ten	10^{1}
unit			
deci-	d	one-tenth	10^{-1}
centi-	c	one-hundredth	10^{-2}
milli-	m	one-thousandth	10^{-3}
micro-	μ	one-millionth	10^{-6}
nano-	n	one-billionth	10^{-9}
pico-	p	one-trillionth	10^{-12}
femto-	f	one-quadrillionth	10^{-15}
atto-	a	one-quintillionth	10^{-18}

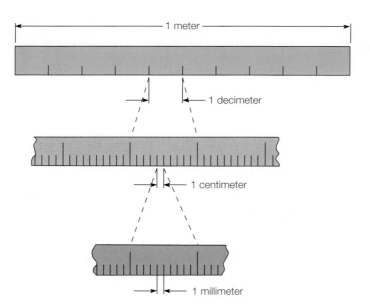

FIGURE 1.7

Compare the units shown here. How many millimeters fit into the space occupied by one centimeter? How many millimeters fit into the space of one decimeter? How many millimeters fit into the space of one meter? Can you express all of this as multiples of ten?

each decimeter is divided into ten equal-sized subunits called *centimeters*. It takes 10 centimeters to equal 1 decimeter and 100 to equal 1 meter. In a similar fashion, each prefix up or down the metric ladder represents a simple increase or decrease by a factor of 10 (Figure 1.7).

When the metric system was established in 1791, the standard unit of mass was defined in terms of the mass of a certain volume of water. A cubic decimeter (dm^3) of pure water at 4°C

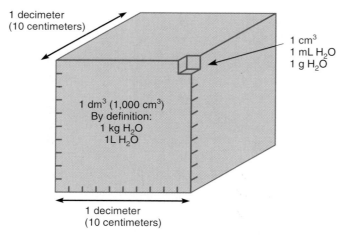

FIGURE 1.8

A cubic decimeter of water (1,000 cm^3) has a liquid volume of 1 L (1,000 mL) and a mass of 1 kg (1,000 g). Therefore, 1 cm^3 of water has a liquid volume of 1 mL and a mass of 1 g.

was *defined* to have a mass of 1 kilogram (kg). This definition was convenient because it created a relationship between length, mass, and volume. As illustrated in Figure 1.8, a cubic decimeter is 10 cm on each side. The volume of this cube is therefore 10 cm $\times$ 10 cm $\times$ 10 cm, or 1,000 cubic centimeters (abbreviated as cc or cm^3). Thus, a volume of 1,000 cm^3 of water has a mass of 1 kg. Since 1 kg is 1,000 g, 1 cm^3 of water has a mass of 1 g.

The volume of 1,000 cm^3 also defines a metric unit that is commonly used to measure liquid volume, the **liter** (L). For smaller amounts of liquid volume, the milliliter (mL) is used. The relationship between liquid volume, volume, and mass of water is therefore

$$1.0 \text{ L} \Rightarrow 1.0 \text{ dm}^3 \text{ and has a mass of } 1.0 \text{ kg}$$

or, for smaller amounts,

$$1.0 \text{ mL} \Rightarrow 1.0 \text{ cm}^3 \text{ and has a mass of } 1.0 \text{ g}$$

Understandings from Measurements

One of the more basic uses of measurement is to *describe* something in an exact way that everyone can understand. For example, if a friend in another city tells you that the weather has been "warm," you might not understand what temperature is being described. A statement that the air temperature is 70°F carries more exact information than a statement about "warm weather." The statement that the air temperature is 70°F contains two important concepts: (1) the numerical value of 70 and (2) the referent unit of degrees Fahrenheit. Note that both a numerical value and a unit are necessary to communicate a measurement correctly. Thus, weather reports describe weather conditions with numerically specified units; for example, 70° Fahrenheit for air temperature, 5 miles per hour for wind speed, and 0.5 inches for rainfall (Figure 1.9). When such numerically specified units are used in a description, or a weather report, everyone understands *exactly* the condition being described.

Weather Report

Friday (24 hours ended at 5 P.M.)
Highs—airport 73°F, downtown 76°F
Lows—airport 68°F, downtown 70°F
Rainfall .... 0.26 in
Average wind speed .... 5.2 mph
Relative humidity .... High 85%
                                      Low 75%
Rainfall ± normal to date.....+0.94 in

FIGURE 1.9

A weather report gives exact information, data that describes the weather by reporting numerically specified units for each condition being described.

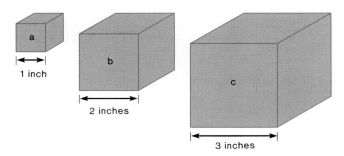

FIGURE 1.10
Cube *a* is 1 inch on each side, cube *b* is 2 inches on each side, and cube *c* is 3 inches on each side. These three cubes can be described and compared with data, or measurement information, but some form of analysis is needed to find patterns or meaning in the data.

Data

Measurement information used to describe something is called **data.** Data can be used to describe objects, conditions, events, or changes that might be occurring. You really do not know if the weather is changing much from year to year until you compare the yearly weather data. The data will tell you, for example, if the weather is becoming hotter or dryer or is staying about the same from year to year.

Let's see how data can be used to describe something and how the data can be analyzed for further understanding. The cubes illustrated in Figure 1.10 will serve as an example. Each cube can be described by measuring the properties of size and surface area.

First, consider the size of each cube. Size can be described by **volume,** which means *how much space something occupies.* The volume of a cube can be obtained by measuring and multiplying the length, width, and height. The data is

volume of cube *a*	1 in^3
volume of cube *b*	8 in^3
volume of cube *c*	27 in^3

Now consider the surface area of each cube. **Area** means *the extent of a surface,* and each cube has six surfaces, or faces (top, bottom, and four sides). The area of any face can be obtained by measuring and multiplying length and width. The data for the three cubes describes them as follows:

	Volume	**Surface Area**
cube *a*	1 in^3	6 in^2
cube *b*	8 in^3	24 in^2
cube *c*	27 in^3	54 in^2

Ratios and Generalizations

Data on the volume and surface area of the three cubes in Figure 1.10 describes the cubes, but whether it says anything about a relationship between the volume and surface area of a cube is difficult to tell. Nature seems to have a tendency to camouflage relationships, making it difficult to extract meaning from raw data. Seeing through the camouflage requires the use of mathematical techniques to expose patterns. Let's see how such techniques can be applied to the data on the three cubes and what the pattern means.

One mathematical technique for reducing data to a more manageable form is to expose patterns through a **ratio.** A ratio is a relationship between two numbers that is obtained when one number is divided by another number. Suppose, for example, that an instructor has 50 sheets of graph paper for a laboratory group of 25 students. The relationship, or ratio, between the number of sheets and the number of students is 50 papers to 25 students, and this can be written as 50 papers/25 students. This ratio is *simplified* by dividing 25 into 50, and the ratio becomes 2 papers/1 student. The 1 is usually understood (not stated), and the ratio is written as simply 2 papers/student. It is read as 2 papers "for each" student, or 2 papers "per" student. The concept of simplifying with a ratio is an important one, and you will see it time and time again throughout science. It is important that you understand the meaning of "per" and "for each" when used with numbers and units.

Applying the ratio concept to the three cubes in Figure 1.10, the ratio of surface area to volume for the smallest cube, cube *a*, is 6 in^2 to 1 in^3, or

$$\frac{6 \text{ in}^2}{1 \text{ in}^3} = 6 \frac{\text{in}^2}{\text{in}^3}$$

meaning there are 6 square inches of area *for each* cubic inch of volume.

The middle-sized cube, cube *b*, had a surface area of 24 in^2 and a volume of 8 in^3. The ratio of surface area to volume for this cube is therefore

$$\frac{24 \text{ in}^2}{8 \text{ in}^3} = 3 \frac{\text{in}^2}{\text{in}^3}$$

meaning there are 3 square inches of area *for each* cubic inch of volume.

The largest cube, cube *c*, had a surface area of 54 in^2 and a volume of 27 in^3. The ratio is

$$\frac{54 \text{ in}^2}{27 \text{ in}^3} = 2 \frac{\text{in}^2}{\text{in}^3}$$

or 2 square inches of area *for each* cubic inch of volume. Summarizing the ratio of surface area to volume for all three cubes, you have

small cube	*a*—6:1
middle cube	*b*—3:1
large cube	*c*—2:1

Now that you have simplified the data through ratios, you are ready to generalize about what the information means. You can generalize that the surface-area-to-volume ratio of a cube *decreases* as the volume of a cube becomes larger. Reasoning from this generalization will provide an

Chapter One: What Is Science? **9**

explanation for a number of related observations. For example, why does crushed ice melt faster than a single large block of ice with the same volume? The explanation is that the crushed ice has a larger surface-area-to-volume ratio than the large block, so more surface is exposed to warm air. If the generalization is found to be true for shapes other than cubes, you could explain why a log chopped into small chunks burns faster than the whole log. Further generalizing might enable you to predict if 10 lb of large potatoes would require more or less peeling than 10 lb of small potatoes. When generalized explanations result in predictions that can be verified by experience, you gain confidence in the explanation. Finding patterns of relationships is a satisfying intellectual adventure that leads to understanding and generalizations that are frequently practical.

The Density Ratio

The power of using a ratio to simplify things, making explanations more accessible, is evident when you compare the simplified ratio 6 to 3 to 2 with the hodgepodge of numbers that you would have to consider without using ratios. The power of using the ratio technique is also evident when considering other properties of matter. Volume is a property that is sometimes confused with mass. Larger objects do not necessarily contain more matter than smaller objects. A large balloon, for example, is much larger than this book but the book is much more massive than the balloon. The simplified way of comparing the mass of a particular volume is to find the ratio of mass to volume. This ratio is called mass **density**, which is defined as *mass per unit volume.* The "per" means "for each" as previously discussed, and "unit" means one, or each. Thus "mass per unit volume" literally means the "mass of one volume" (Figure 1.11). The relationship can be written as

$$\text{mass density} = \frac{\text{mass}}{\text{volume}}$$

or

$$\rho = \frac{m}{V}$$

(ρ is the symbol for the Greek letter rho.)

equation 1.1

As with other ratios, density is obtained by dividing one number and unit by another number and unit. Thus, the density of an object with a volume of 5 cm³ and a mass of 10 g is

$$\text{density} = \frac{10 \text{ g}}{5 \text{ cm}^3} = 2 \frac{\text{g}}{\text{cm}^3}$$

The density in this example is the ratio of 10 g to 5 cm³, or 10 g/5 cm³, or 2 g to 1 cm³. Thus, the density of the example object is the mass of *one* volume (a unit volume), or 2 g *for each* cm³.

Any unit of mass and any unit of volume may be used to express density. The densities of solids, liquids, and gases are

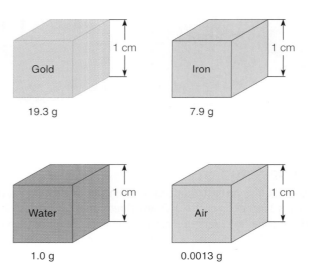

FIGURE 1.11
Equal volumes of different substances do not have the same mass, as these cube units show. Calculate the densities in g/cm³. Do equal volumes of different substances have the same density? Explain.

usually expressed in grams per cubic centimeter (g/cm³), but the densities of liquids are sometimes expressed in grams per milliliter (g/mL). Using SI standard units, densities are expressed as kg/m³. Densities of some common substances are shown in Table 1.4.

If matter is distributed the same throughout a volume, the *ratio* of mass to volume will remain the same no matter what mass and volume are being measured. Thus, a teaspoonful, a cup, and a lake full of freshwater at the same temperature will all have a density of about 1 g/cm³ or 1 kg/L. A given material will have its own unique density; example 1.1 shows how density can be used to identify an unknown substance. For help with significant figures, see appendix A (p. 595).

TABLE 1.4	
Mass densities (ρ) of some common substances	
	g/cm³
Aluminum	2.70
Copper	8.96
Iron	7.87
Lead	11.4
Water	1.00
Seawater	1.03
Mercury	13.6
Gasoline	0.680

Concepts Applied

Density Matters—Fish, Nuclear Fusion, and Cola

Sharks and rays are marine animals that have an internal skeleton made entirely of cartilage. These animals have no swim bladder to adjust their body density in order to maintain their position in the water; therefore, they must constantly swim or they will sink. The bony fish, on the other hand, have a skeleton composed of bone and most also have a swim bladder. These fish can regulate the amount of gas in the bladder to control their density. Thus, the fish can remain at a given level in the water without expending large amounts of energy.

Scientists plan to study nuclear fusion involving extremely dense plasmas, up to six times denser than the center of the sun. The proposed National Ignition Facility (NIF) is expected to be funded and completed by 2010. This research facility will use high-power lasers to achieve self-sustaining fusion reactions, with densities reaching up to 1,000 g/cm^3. Existing research facilities today are capable of creating plasma densities up to about 50 g/cm^3. Compare these densities with what you might find at the center of the sun, about 160 g/cm^3.

Finally, have you ever noticed the different floating characteristics of cans of the normal version of a carbonated cola beverage and a diet version? The surprising result is that the normal version usually sinks and the diet version usually floats. This has nothing to do with the amount of carbon dioxide in the two drinks. It is a result of the increase in density from the sugar added to the normal version, while the diet version has much less of an artificial sweetener that is much sweeter than sugar.

EXAMPLE 1.1

Two blocks are on a table. Block A has a volume of 30.0 cm^3 and a mass of 81.0 g. Block B has a volume of 50.0 cm^3 and a mass of 135 g. Which block has the greater density? If the two blocks have the same density, what material are they? (See Table 1.4.)

Solution

Density is defined as the ratio of the mass or weight of a substance per unit volume. Assuming the mass is distributed equally throughout the volume, you could assume that the ratio of mass to volume is the same no matter what quantity of mass and volume are measured. If you can accept this assumption, you can use equation 1.1 to determine the mass density—choosing this equation because the mass of the substance is given, not the weight:

Block A

mass (m) = 81.0 g
volume (V) = 30.0 cm^3
density = ?

$$\rho = \frac{m}{V}$$
$$= \frac{81.0 \text{ g}}{30.0 \text{ cm}^3}$$
$$= 2.70 \frac{\text{g}}{\text{cm}^3}$$

Block B

mass (m) = 135 g
volume (V) = 50.0 cm^3
density = ?

$$\rho = \frac{m}{V}$$
$$= \frac{135 \text{ g}}{50.0 \text{ cm}^3}$$
$$= 2.70 \frac{\text{g}}{\text{cm}^3}$$

As you can see, both blocks have the same mass density. Inspecting Table 1.4, you can see that aluminum has a mass density of 2.70 g/cm^3, so both blocks must be aluminum.

EXAMPLE 1.2

A rock with a volume of 4.50 cm^3 has a mass of 15.0 g. What is the density of the rock? (Answer: 3.33 g/cm^3)

Concepts Applied

A Dense Textbook?

What is the mass density of this book? Measure the length, width, and height of this book in cm, then multiply to find the volume in cm^3. Use a balance to find the mass of this book in grams. Compute the density of the book by dividing the mass by the volume. Compare the density in g/cm^3 with other substances listed in Table 1.4.

Symbols and Equations

In the previous section, the relationship of density, mass, and volume was written with symbols. Mass density was represented by ρ, the lowercase letter rho in the Greek alphabet, mass was represented by m, and volume by V. The use of such symbols is established and accepted by convention, and these symbols are like the vocabulary of a foreign language. You learn what the symbols mean by use and practice, with the understanding that each symbol stands for a very specific property or concept. The symbols actually represent **quantities,** or *measured properties.* The symbol m thus represents a quantity of mass that is specified by a number and a unit, for example, 16 g. The symbol V represents a quantity of volume that is specified by a number and a unit, such as 17 cm³.

Symbols

Symbols usually provide a clue about which quantity they represent, such as m for mass and V for volume. However, in some cases two quantities start with the same letter, such as volume and velocity, so the uppercase letter is used for one (V for volume) and the lowercase letter is used for the other (v for velocity). There are more quantities than upper- and lowercase letters, however, so letters from the Greek alphabet are also used, for example, ρ for mass density. Sometimes a subscript is used to identify a quantity in a particular situation, such as v_i for initial, or beginning, velocity and v_f for final velocity. Some symbols are also used to carry messages; for example, the Greek letter delta (Δ) is a message that means "the change in" a value. Other message symbols are the symbol $\therefore$, which means "therefore," and the symbol $\propto$, which means "is proportional to."

Equations

Symbols are used in an **equation,** a statement that describes a relationship where *the quantities on one side of the equal sign are identical to the quantities on the other side.* Identical refers to both the numbers and the units. Thus, in the equation describing the property of density, $\rho = m/V$, the numbers on both sides of the equal sign are identical (e.g., $5 = 10/2$). The units on both sides of the equal sign are also identical (e.g., g/cm³ = g/cm³).

Equations are used to (1) *describe a property,* (2) *define a concept,* or (3) *describe how quantities change relative to each other.* Understanding how equations are used in these three classes is basic to successful problem solving and comprehension of physical science. Each class of uses is considered separately in the following discussion.

Describing a property. You have already learned that the compactness of matter is described by the property called density. Density is a ratio of mass to a unit volume, or $\rho = m/V$. The key to understanding this property is to understand the meaning of a ratio and what "per" or "for each" means. Other examples of properties that can be defined by ratios are how fast something is moving (speed) and how rapidly a speed is changing (acceleration).

Defining a concept. A physical science concept is sometimes defined by specifying a measurement procedure. This is called an *operational definition* because a procedure is established that defines a concept as well as telling you how to measure it. Concepts of what is meant by force, mechanical work, and mechanical power and concepts involved in electrical and magnetic interactions can be defined by measurement procedures.

Describing how quantities change relative to each other. The term **variable** refers to a specific quantity of an object or event that can have different values. Your weight, for example, is a variable because it can have a different value on different days. The rate of your heartbeat, the number of times you breathe each minute, and your blood pressure are also variables. Any quantity describing an object or event can be considered a variable, including the conditions that result in such things as your current weight, pulse, breathing rate, or blood pressure.

As an example of relationships between variables, consider that your weight changes in size in response to changes in other variables, such as the amount of food you eat. With all other factors being equal, a change in the amount of food you eat results in a change in your weight, so the variables of amount of food eaten and weight change together in the same ratio.

When two variables increase (or decrease) together in the same ratio, they are said to be in **direct proportion.** When two variables are in direct proportion, *an increase or decrease in one variable results in the same relative increase or decrease in a second variable.* Recall that the symbol $\propto$ means "is proportional to," so the relationship is

$$\text{amount of food consumed} \propto \text{weight gain}$$

Variables do not always increase or decrease together in direct proportion. Sometimes one variable *increases* while a second variable *decreases* in the same ratio. This is an **inverse proportion** relationship. Other common relationships include one variable increasing in proportion to the *square* or to the *inverse square* of a second variable. Here are the forms of these four different types of proportional relationships:

Direct	$a \propto b$
Inverse	$a \propto 1/b$
Square	$a \propto b^2$
Inverse square	$a \propto 1/b^2$

Proportionality Statements

Proportionality statements describe in general how two variables change relative to each other, but a proportionality statement is *not* an equation. For example, consider the last time you filled your fuel tank at a service station (Figure 1.12). You could say that the volume of gasoline in an empty tank you are filling is directly proportional to the amount of time that the fuel pump was running, or

$$\text{volume} \propto \text{time}$$

This is not an equation because the numbers and units are not identical on both sides. Considering the units, for example, it should be clear that minutes do not equal liters; they are two different quantities. To make a statement of proportionality into an equation, you need to apply a **proportionality constant,** which is sometimes given the symbol k. For the fuel pump example the equation is

$$\text{volume} = (\text{time})(\text{constant})$$

or

$$V = tk$$

Concepts Applied

Inverse Square Relationship

An inverse square relationship between energy and distance is found in light, sound, gravitational force, electric fields, nuclear radiation, and any other phenomena that spread equally in all directions from a source.

Box Figure 1.2 could represent any of the phenomena that have an inverse square relationship, but let us assume it is showing a light source and how the light spreads at a certain distance (d), at twice that distance (2d), and at three times that distance (3d). As you can see, light twice as far from the source is spread over four times the area, and will therefore have one-fourth the intensity. This is the same as $\frac{1}{2^2}$, or $\frac{1}{4}$.

Light three times as far from the source is spread over nine times the area and will therefore have one-ninth the intensity. This is the same as $\frac{1}{3^2}$, or $\frac{1}{9}$, again showing an inverse square relationship.

You can measure the inverse square relationship by moving an overhead projector so its light is shining on a wall (see distance d in Box Figure 1.2). Use a light meter or some other way of measuring the intensity of light. Now move the projector to double the distance from the wall. Measure the increased area of the projected light on the wall, and again measure the intensity of the light. What relationship did you find between the light intensity and distance?

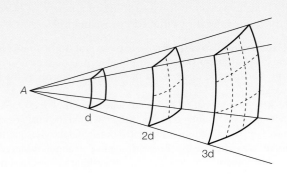

BOX FIGURE 1.2

How much would light moving from point A spread out at twice the distance (2d) and three times the distance (3d)? What would this do to the brightness of the light?

FIGURE 1.12

The volume of fuel you have added to the fuel tank is directly proportional to the amount of time that the fuel pump has been running. This relationship can be described with an equation by using a proportionality constant.

In the example, the constant is the flow of gasoline from the pump in L/min (a ratio). Assume the rate of flow is 40 L/min. In units, you can see why the statement is now an equality.

$$L = (\text{min})\left(\frac{L}{\text{min}}\right)$$

$$L = \frac{\text{min} \times L}{\text{min}}$$

$$L = L$$

A proportionality constant in an equation might be a **numerical constant,** a constant that is without units. Such numerical constants are said to be dimensionless, such as 2 or 3. Some of the more important numerical constants have their own symbols, for example, the ratio of the circumference of a circle to its diameter is known as π (pi). The numerical constant of π does not have units because the units cancel when the ratio is simplified by division (Figure 1.13). The value of π is usually rounded to 3.14, and an example of using this numerical constant in an equation is that the area of a circle equals π times the radius squared ($A = \pi r^2$).

The flow of gasoline from a pump is an example of a constant that has dimensions (40 L/min). Of course the value of this constant will vary with other conditions, such as the particular fuel pump used and how far the handle on the pump hose is depressed, but it can be considered to be a constant under the same conditions for any experiment.

Problem Solving Made Easy

The activity of problem solving is made easier by using certain techniques that help organize your thinking. One such technique is to follow a format, such as the following procedure:

Step 1: Read through the problem and *make a list* of the variables with their symbols on the left side of the page, including the unknown with a question mark.

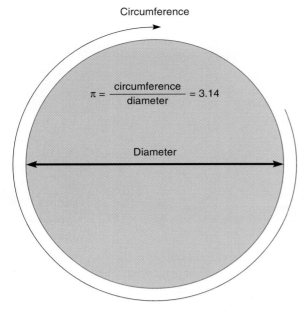

Circumference

$$\pi = \frac{\text{circumference}}{\text{diameter}} = 3.14$$

Diameter

FIGURE 1.13

The ratio of the circumference of *any* circle to the diameter of that circle is always π, a numerical constant that is usually rounded to 3.14. Pi does not have units because they cancel in the ratio.

Step 2: Inspect the list of variables and the unknown and identify the equation that expresses a relationship between these variables. A list of equations discussed in each chapter is found at the end of that chapter. *Write the equation* on the right side of your paper, opposite the list of symbols and quantities.

Step 3: If necessary, *solve the equation* for the variable in question. This step must be done before substituting any numbers or units in the equation. This simplifies things and keeps down confusion that might otherwise result. If you need help solving an equation, see the section on this topic in appendix A.

Step 4: If necessary, *convert unlike units* so they are all the same. For example, if a time is given in seconds and a speed is given in kilometers per hour, you should convert the km/h to m/s. Again, this step should be done at this point in the procedure to avoid confusion or incorrect operations in a later step. If you need help converting units, see the section on this topic in appendix A.

Step 5: Now you are ready to *substitute the number value and unit* for each symbol in the equation (except the unknown). Note that it might sometimes be necessary to perform a "subroutine" to find a missing value and unit for a needed variable.

Step 6: Do the indicated *mathematical operations* on the numbers and on the units. This is easier to follow if you first separate the numbers and units, as shown in the example that follows and in the examples throughout this text. Then perform the indicated operations on the numbers and units as separate steps, showing all work. If you are not sure how to read the indicated operations, see the section on "Symbols and Operations" in appendix A.

Step 7: *Draw a box* around your answer (numbers and units) to communicate that you have found what you were looking for. The box is a signal that you have finished your work on this problem.

For an example problem, use the equation from the previous section describing the variables of a fuel pump, $V = tk$, to predict how long it will take to fill an empty 80-liter tank. Assume $k = 40$ L/min.

Step 1

$V = 80$ L	$V = tk$	**Step 2**
$k = 80$ L min		
$t = ?$	$\dfrac{V}{k} = \dfrac{tk}{k}$	**Step 3**
	$t = \dfrac{V}{k}$	

(no conversion needed for this problem) **Step 4**

$$t = \frac{80 \text{ L}}{40 \dfrac{\text{L}}{\text{min}}}$$ **Step 5**

$$= \frac{80}{40} \; \frac{\text{L}}{1} \times \frac{\text{min}}{\text{L}}$$ **Step 6**

$$= \boxed{2 \text{ min}}$$ **Step 7**

Note that procedure step 4 was not required in this solution.

This formatting procedure will be demonstrated throughout this text in example problems and in the solutions to problems found in appendix D. Note that each of the chapters with problems has parallel exercises. The exercises in groups A and B cover the same concepts. If you cannot work a problem in group B, look for the parallel problem in group A. You will find a solution to this problem, in the previously described format, in appendix D. Use this parallel problem solution as a model to help you solve the problem in group B. If you follow the suggested formatting procedures and seek help from the appendix as needed, you will find that problem solving is a simple, fun activity that helps you to learn to think in a new way. Here are some more considerations that will prove helpful.

1. Read the problem carefully, perhaps several times, to understand the problem situation. If possible, make a sketch to help you visualize and understand the problem in terms of the real world.
2. Be alert for information that is not stated directly. For example, if a moving object "comes to a stop," you know that the final velocity is zero, even though this was not stated outright. Likewise, questions about "how far?" are usually asking a question about distance, and questions about "how long?" are usually asking a question about time. Such information can be very important in procedure step 1, the listing of quantities and their symbols. Overlooked or missing quantities and symbols can make it difficult to identify the appropriate equation.
3. Understand the meaning and concepts that an equation represents. An equation represents a *relationship* that exists between variables. Understanding the relationship helps you to identify the appropriate equation or equations by inspection of the list of known and unknown quantities (procedure step 2). You will find a list of the equations being considered at the end of each chapter. Information about the meaning and the concepts that an equation represents is found within each chapter.

4. Solve the equation before substituting numbers and units for symbols (procedure step 3). A helpful discussion of the mathematical procedures required, with examples, is in appendix A.

5. Note whether the quantities are in the same units. A mathematical operation requires the units to be the same; for example, you cannot add nickels, dimes, and quarters until you first convert them all to the same unit of money. Likewise, you cannot correctly solve a problem if one time quantity is in seconds and another time quantity is in hours. The quantities must be converted to the same units before anything else is done (procedure step 4). There is a helpful section on how to use conversion ratios in appendix A.

6. Perform the required mathematical operations on the numbers and the units as if they were two separate problems (procedure step 6). You will find that following this step will facilitate problem-solving activities because the units you obtain will tell you if you have worked the problem correctly. If you just write the units that you think should appear in the answer, you have missed this valuable self-check.

7. Be aware that not all learning takes place in a given time frame and that solutions to problems are not necessarily arrived at "by the clock." If you have spent a half an hour or so unsuccessfully trying to solve a particular problem, move on to another problem or do something entirely different for a while. Problem solving often requires time for something to happen in your brain. If you move on to some other activity, you might find that the answer to a problem that you have been stuck on will come to you "out of the blue" when you are not even thinking about the problem. This unexpected revelation of solutions is common to many real-world professions and activities that involve thinking.

Example Problem

Mercury is a liquid metal with a mass density of 13.6 g/cm^3. What is the mass of 10.0 cm^3 of mercury?

Solution

The problem gives two known quantities, the mass density (ρ) of mercury and a known volume (V), and identifies an unknown quantity, the mass (m) of that volume. Make a list of these quantities:

$\rho = 13.6 \text{ g/cm}^3$
$V = 10.0 \text{ cm}^3$
$m = ?$

The appropriate equation for this problem is the relationship between mass density (ρ), mass (m), and volume (V):

$$\rho = \frac{m}{V}$$

The unknown in this case is the mass, m. Solving the equation for m, by multiplying both sides by V, gives:

$$V\rho = \frac{mV}{V}$$

$$V\rho = m, \text{ or}$$

$$m = V\rho$$

Now you are ready to substitute the known quantities in the equation:

$$m = \left(13.6 \, \frac{\text{g}}{\text{cm}^3}\right)(10.0 \text{ cm}^3)$$

And perform the mathematical operations on the numbers and on the units:

$$m = (13.6)(10.0) \quad \left(\frac{\text{g}}{\text{cm}^3}\right)(\text{cm}^3)$$

$$= 136 \, \frac{\text{g·cm}^3}{\text{cm}^3}$$

$$= \boxed{136 \text{ g}}$$

The Nature of Science

Most humans are curious, at least when they are young, and are motivated to understand their surroundings. These traits have existed since antiquity and have proven to be a powerful motivation. In recent times the need to find out has motivated the launching of space probes to learn what is "out there," and humans have visited the moon to satisfy their curiosity. Curiosity and the motivation to understand nature were no less powerful in the past than today. Over two thousand years ago, the Greeks lacked the tools and technology of today and could only make conjectures about the workings of nature. These early seekers of understanding are known as *natural philosophers,* and they observed, thought about, and wrote about the workings of all of nature. They are called philosophers because their understandings came from reasoning only, without experimental evidence. Nonetheless, some of their ideas were essentially correct and are still in use today. For example, the idea of matter being composed of *atoms* was first reasoned by certain Greeks in the fifth century B.C. The idea of *elements,* basic components that make up matter, was developed much earlier but refined by the ancient Greeks in the fourth century B.C. The concept of what the elements are and the concept of the nature of atoms have changed over time, but the idea first came from ancient natural philosophers.

The Scientific Method

Some historians identify the time of Galileo and Newton, approximately three hundred years ago, as the beginning of modern science. Like the ancient Greeks, Galileo and Newton were interested in studying all of nature. Since the time of Galileo and Newton, the content of physical science has increased in scope and specialization, but the basic means of acquiring understanding, the scientific investigation, has changed little. A *scientific investigation* provides understanding through *experimental evidence* as opposed to the conjectures based on the "thinking only" approach of the ancient natural philosophers. In the next chapter, for example, you will learn how certain ancient Greeks described how objects fall toward the earth with a thought-out, or reasoned, explanation. Galileo, on the other hand, changed how people thought of falling objects by developing explanations from both creative thinking and precise measurement of physical quantities, providing experimental evidence for his explanations. Experimental evidence provides explanations today, much as it did for Galileo, as relationships are found from precise measurements of physical quantities. Thus, scientific knowledge about nature has grown as measurements and investigations have led to understandings that lead to further measurements and investigations.

What is a scientific investigation, and what methods are used to conduct one? Attempts have been made to describe scientific methods in a series of steps (define problem, gather data, make hypothesis, test, make conclusion), but no single description has ever been satisfactory to all concerned. Scientists do similar things in investigations, but there are different approaches and different ways to evaluate what is found. Overall, the similar things might look like this:

1. Observe some aspect of nature.
2. Propose an explanation for something observed.
3. Use the explanation to make predictions.
4. Test predictions by doing an experiment or by making more observations.
5. Modify explanation as needed.
6. Return to step 3.

The exact approach used depends on the individual doing the investigation and on the field of science being studied.

Another way to describe what goes on during a scientific investigation is to consider what can be generalized. There are at least three separate activities that seem to be common to scientists in different fields as they conduct scientific investigations, and these generalizations look like this:

• Collecting observations
• Developing explanations
• Testing explanations

No particular order or routine can be generalized about these common elements. In fact, individual scientists might not even be involved in all three activities. Some, for example, might spend all of their time out in nature, "in the field" collecting data and generalizing about their findings. This is an acceptable means of investigation in some fields of science. Other scientists might spend all of their time indoors at computer terminals developing theoretical equations to explain the generalizations made by others. Again, the work at a computer terminal is an acceptable means of scientific investigation. Thus, many of today's specialized scientists never engage in a five-step process. This is one reason why many philosophers of science argue that there is no such thing as *the* scientific method. There are common activities of observing, explaining, and testing in scientific investigations in different fields, and these activities will be discussed next.

Explanations and Investigations

Explanations in the natural sciences are concerned with things or events observed, and there can be several different ways to develop or create explanations. In general, explanations can come from the results of experiments, from an educated guess, or just from imaginative thinking. In fact, there are even several examples in the history of science of valid explanations being developed from dreams. Explanations go by various names, each depending on intended use or stage of development. For example, an explanation in an early stage of development is sometimes called a *hypothesis*. A **hypothesis** is a tentative thought- or experiment-derived explanation. It must be compatible with all observations and provide understanding of some aspect of nature, but the key word here is *tentative*. A hypothesis is tested by experiment and is rejected, or modified, if a single observation or test does not fit. The successful testing of a hypothesis may lead to the design of experiments, or it could lead to the development of another hypothesis, which could, in turn, lead to the design of yet more experiments, which could lead to . . . As you can see, this is a branching, ongoing process that is very difficult to describe in specific terms. In addition, it can be difficult to identify an endpoint in the process that you could call a conclusion. The search for new concepts to explain experimental evidence may lead from hypothesis to a new theory, which results in more new hypotheses. This is why one of the best ways to understand scientific methods is to study the history of science. Or do the activity of science yourself by planning, then conducting experiments.

Testing a Hypothesis

In some cases a hypothesis may be tested by simply making some simple observations. For example, suppose you hypothesized that the height of a bounced ball depends only on the height from which the ball is dropped. You could test this by observing different balls being dropped from several different heights and recording how high each bounced.

Another common method for testing a hypothesis involves devising an experiment. An **experiment** is a re-creation of an event or occurrence in a way that enables a scientist to support or disprove a hypothesis. This can be difficult, since an event can be influenced by a great many different things. For example, suppose someone tells you that soup heats to the boiling point faster than water. Is this true? How can you find the answer to this question? The time required to boil a can of soup might depend on a number of things: the composition of the soup, how much soup is in the pan, what kind of pan is used, the nature of the stove, the size of the burner, how high the temperature is set, environmental factors such as the humidity and temperature, and more factors. It might seem that answering a simple question about the time involved in boiling soup is an impossible task. To help unscramble such situations, scientists use what is known as a *controlled experiment*. A **controlled experiment** compares two situations in which all the influencing factors are identical except one. The situation used as the basis of comparison is called the *control group* and the other is called the *experimental group*. The single influencing factor that is allowed to be different in the experimental group is called the *experimental variable*.

The situation involving the time required to boil soup and water would have to be broken down into a number of simple questions. Each question would provide the basis on which experimentation would occur. Each experiment would provide information about a small part of the total process of heating liquids. For example, in order to test the hypothesis that soup boils faster than water, an experiment could be performed in which soup is brought to a boil (the experimental group), while water is brought to a boil in the control group. Every factor in the control group is *identical* to the factors in the experimental group except the experimental variable—the soup factor. After the experiment, the new data (facts) are gathered and analyzed. If there were no differences between the two groups, you could conclude that the soup variable evidently did not have a cause-and-effect relationship with the time needed to come to a boil (i.e., soup was not responsible for the time to boil). However, if there were a difference, it would be likely that this variable was responsible for the difference between the control and experimental groups. In the case of the time to come to a boil, you would find that soup indeed does boil faster than water alone. If you doubt this, why not do the experiment yourself?

Accept Results?

Scientists are not likely to accept the results of a single experiment, since it is possible that a random event that had nothing to do with the experiment could have affected the results and caused people to think there was a cause-and-effect relationship when none existed. For example, the density of soup is greater than the density of water, and this might be the important factor. A way to overcome this difficulty would be to test a number of different kinds of soup with different densities. When there is only one variable, many replicates (copies) of the same experiment are conducted, and the consistency of the results determines how convincing the experiment is.

Furthermore, scientists often apply statistical tests to the results to help decide in an impartial manner if the results obtained are *valid* (meaningful; fit with other knowledge), *reliable* (give the same results repeatedly), and show cause-and-effect or if they are just the result of random events.

Other Considerations

As you can see from the discussion of the nature of science, a scientific approach to the world requires a certain way of thinking. There is an insistence on ample supporting evidence by numerous studies rather than easy acceptance of strongly stated opinions. Scientists must separate opinions from statements of fact. A scientist is a healthy skeptic.

Careful attention to detail is also important. Since scientists publish their findings and their colleagues examine their work, there is a strong desire to produce careful work that can be easily defended. This does not mean that scientists do not speculate and state opinions. When they do, however, they take great care to clearly distinguish fact from opinion.

There is also a strong ethic of honesty. Scientists are not saints, but the fact that science is conducted out in the open in front of one's peers tends to reduce the incidence of dishonesty. In addition, the scientific community strongly condemns and severely penalizes those who steal the ideas of others, perform shoddy science, or falsify data. Any of these infractions could lead to the loss of one's job and reputation.

Science is also limited by the ability of people to pry understanding from the natural world. People are fallible and do not always come to the right conclusions, because information is lacking or misinterpreted, but science is self-correcting. As new information is gathered, old, incorrect ways of thinking must be changed or discarded. For example, at one time people were sure that the sun went around the earth. They observed that the sun rose in the east and traveled across the sky to set in the west. Since they could not feel the earth moving, it seemed perfectly logical that the sun traveled around the earth. Once they understood that the earth rotated on its axis, people began to understand that the rising and setting of the sun could be explained in other ways. A completely new concept of the relationship between the sun and the earth developed.

Although this kind of study seems rather primitive to us today, this change in thinking about the sun and the earth was a very important step in understanding the universe and how the various parts are related to one another. This background information was built upon by many generations of astronomers and space scientists, and it finally led to space exploration.

People also need to understand that science cannot answer all the problems of our time. Although science is a powerful tool, there are many questions it cannot answer and many problems it cannot solve. The behavior and desires of people generate most of the problems societies face. Famine, drug abuse, and pollution are human-caused and must be resolved by humans. Science may provide some tools for social planners, politicians, and ethical thinkers, but science does not have, nor does it attempt to provide, answers for the problems of the human race. Science is merely one of the tools at our disposal.

Pseudoscience

Pseudoscience (*pseudo-* means false) is not science, but it uses the appearance or language of science to mislead people into thinking that something has scientific validity. Absurd claims that are clearly pseudoscience sometimes appear to gain public acceptance because of promotion in the press. Thus some people continue to believe stories that psychics can really help solve puzzling crimes, that perpetual energy machines exist, or that sources of water can be found by a person with a forked stick. Such claims could be directly tested and disposed of if they fail the test, but this process is generally ignored. In addition to experimentally testing a claim that appears to be pseudoscience, here are some questions that should be considered:

1. What is the academic background and scientific experience of the person promoting the claim?
2. How many articles have been published by the person in peer-reviewed scientific journals?
3. Has the person given invited scientific talks at universities and national professional organization meetings?
4. Has the claim been researched and published by the person in a peer-reviewed scientific journal *and* have other scientists independently validated the claim?

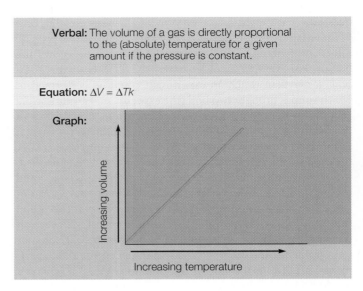

Verbal: The volume of a gas is directly proportional to the (absolute) temperature for a given amount if the pressure is constant.

Equation: $\Delta V = \Delta Tk$

Graph:

(y-axis: Increasing volume; x-axis: Increasing temperature)

FIGURE 1.14

A relationship between variables can be described in at least three different ways: (1) verbally, (2) with an equation, and (3) with a graph. This figure illustrates the three ways of describing the relationship known as Charles' law.

Laws and Principles

Sometimes you can observe a series of relationships that seem to happen over and over again. There is a popular saying, for example, that "if anything can go wrong, it will." This is called Murphy's law. It is called a *law* because it describes a relationship between events that seems to happen time after time. If you drop a slice of buttered bread, for example, it can land two ways, butter side up or butter side down. According to Murphy's law, it will land butter side down. With this example, you know at least one way of testing the validity of Murphy's law.

Another "popular saying" type of relationship seems to exist between the cost of a houseplant and how long it lives. You could call it the "law of houseplant longevity" that the life span of a houseplant is inversely proportional to its purchase price. This "law" predicts that a ten-dollar houseplant will wilt and die within a month, but a fifty-cent houseplant will live for years. The inverse relationship is between the variables of (1) cost and (2) life span, meaning the more you pay for a plant, the shorter the time it will live. This would also mean that inexpensive plants will live for a long time. Since the relationship seems to occur time after time, it is called a "law."

A **scientific law** describes an important relationship that is observed in nature to occur consistently time after time. Basically, scientific laws describe *what* happens in nature. The law is often identified with the name of a person associated with the formulation of the law. For example, with all other factors being equal, an increase in the temperature of the air in a balloon results in an increase in its volume. Likewise, a decrease in the temperature results in a decrease in the total volume of the balloon. The volume of the balloon varies directly with the temperature of the air in the balloon, and this can be observed to occur consistently time after time. This relationship was first discovered in the latter part of the eighteenth century by two French scientists, A.C. Charles and Joseph Gay-Lussac. Today, the relationship is sometimes called *Charles' law* (Figure 1.14). When you read about a scientific *law*, you should remember that a law is a statement that means something about a relationship that you can observe time after time in nature.

Have you ever heard someone state that something behaved a certain way *because* of a scientific principle or law? For example,

a big truck accelerated slowly *because* of Newton's laws of motion. Perhaps this person misunderstands the nature of scientific principles and laws. Scientific principles and laws do not dictate the behavior of objects; they simply describe it. They do not say how things ought to act but rather how things *do* act. A scientific principle or law is *descriptive;* it describes how things act.

A **scientific principle** describes a more specific set of relationships than is usually identified in a law. The difference between a scientific principle and a scientific law is usually one of the extent of the phenomena covered by the explanation, but there is not always a clear distinction between the two. As an example of a scientific principle, consider Archimedes' principle. This principle is concerned with the relationship between an object, a fluid, and buoyancy, which is a specific phenomenon.

Models and Theories

Often the part of nature being considered is too small or too large to be visible to the human eye, and the use of a *model* is needed. A **model** (Figure 1.15) is a description of a theory or idea that accounts for all known properties. The description can come in many different forms, such as a physical model, a computer model, a sketch, an analogy, or an equation. No one has ever seen the whole solar system, for example, and all you can see in the real world is the movement of the sun, moon, and planets against a background of stars. A physical model or sketch of the solar system, however, will give you a pretty good idea of what the solar system might look like. The physical model and the sketch are both models, since they both give you a mental picture of the solar system.

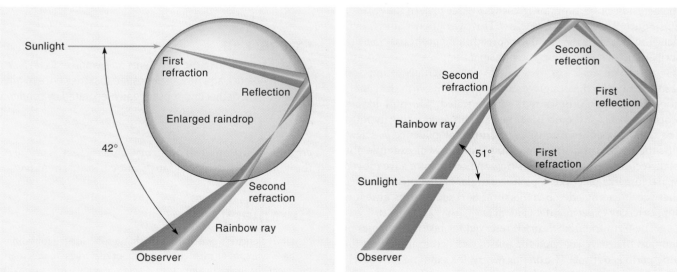

FIGURE 1.15

A model helps you visualize something that cannot be observed. You cannot observe what is making a double rainbow, for example, but models of light entering the upper and lower surfaces of a raindrop help you visualize what is happening. The drawings in B serve as a model that explains how a double rainbow is produced (also see "The Rainbow" in chapter 7).

At the other end of the size scale, models of atoms and molecules are often used to help us understand what is happening in this otherwise invisible world. A container of small, bouncing rubber balls can be used as a model to explain the relationships of Charles' law. This model helps you see what happens to invisible particles of air as the temperature, volume, or pressure of the gas changes. Some models are better than others are, and models constantly change as our understanding evolves. Early twentieth-century models of atoms, for example, were based on a "planetary model," in which electrons moved around the nucleus like planets around the sun. Today, the model has changed as our understanding of the nature of atoms has changed. Electrons are now pictured as vibrating with certain wavelengths, which can make standing waves only at certain distances from the nucleus. Thus the model of the atom changed from one that views electrons as solid particles to one that views them like vibrations on a string.

The most recently developed scientific theory was refined and expanded during the 1970s. This theory concerns the surface of the earth, and it has changed our model of what the earth is like. At first, the basic idea of today's accepted theory was pure and simple conjecture. The term *conjecture* usually means an explanation or idea based on speculation, or one based on trivial grounds without any real evidence. Scientists would look at a map of Africa and South America, for example, and mull over how the two continents look like pieces of a picture puzzle that had moved apart (Figure 1.16). Any talk of moving continents was considered conjecture, because it was not based on anything acceptable as real evidence.

Many years after the early musings about moving continents, evidence was collected from deep-sea drilling rigs that the ocean floor becomes progressively older toward the African and South American continents. This was good enough evidence to establish the "seafloor spreading hypothesis" that described the two continents moving apart.

If a hypothesis survives much experimental testing and leads, in turn, to the design of new experiments with the generation of new hypotheses that can be tested, you now have a working *theory*. A **theory** is defined as a broad working hypothesis that is based on extensive experimental evidence. A scientific theory tells you *why* something happens. For example, the plate tectonic theory describes how the continents have moved apart, just like pieces of a picture puzzle. Is this the same idea that was once considered conjecture? Sort of, but this time it is supported by experimental evidence.

The term *scientific theory* is reserved for historic schemes of thought that have survived the test of detailed examination for long periods of time. The *atomic theory,* for example, was developed in the late 1800s and has been the subject of extensive investigation and experimentation over the last century. The atomic theory and other scientific theories form the framework of scientific thought and experimentation today. Scientific theories point to new ideas about the behavior of nature, and these ideas result in more experiments, more data to collect, and more explanations to develop. All of this may lead to a slight modification of an existing theory, a major modification, or perhaps the

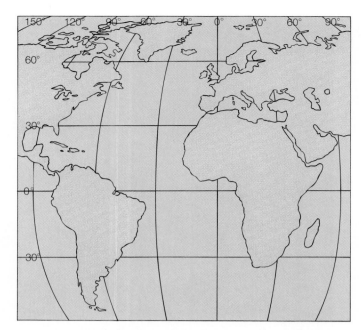

A

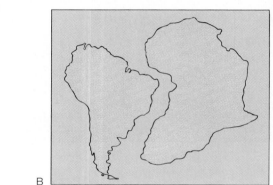

B

FIGURE 1.16

(*A*) Normal position of the continents on a world map. (*B*) A sketch of South America and Africa, suggesting that they once might have been joined together and subsequently separated by continental drift.

creation of an entirely new theory. These activities are all part of the continuing attempt to satisfy our curiosity about nature.

Summary

Physical science is a search for order in our physical surroundings. People have *concepts,* or mental images, about material *objects* and intangible *events* in their surroundings. Concepts are used for thinking and communicating. Concepts are based on *properties,* or attributes that describe a thing or event. Every property implies a *referent* that describes the property. Referents are not always explicit, and most communications require assumptions. Measurement brings precision to descriptions by using numbers and standard units for referents to communicate "exactly how much of exactly what."

Measurement is a process that uses a well-defined and agreed-upon *referent* to describe a *standard unit.* The unit is compared to the property being defined by an *operation* that determines the *value* of the unit by *counting.* Measurements are always reported with a *number,* or value, and a *name* for the unit.

The two major *systems* of standard units are the *English system* and the *metric system.* The English system uses standard units that were originally based on human body parts, and the metric system uses standard units based on referents found in nature. The metric system also uses a system of prefixes to express larger or smaller amounts of units. The metric standard units for length, mass, and time are the *meter, kilogram,* and *second.*

Measurement information used to describe something is called *data.* One way to extract meanings and generalizations from data is to use a *ratio,* a simplified relationship between two numbers. Density is a ratio of mass to volume, or $\rho = m/V$.

Symbols are used to represent *quantities,* or measured properties. Symbols are used in *equations,* which are shorthand statements that describe a relationship where the quantities (both number values and units) are identical on both sides of the equal sign. Equations are used to (1) *describe* a property, (2) *define* a concept, or (3) *describe* how *quantities change* together.

Quantities that can have different values at different times are called *variables.* Variables that increase or decrease together in the same ratio are said to be in *direct proportion.* If one variable increases while the other decreases in the same ratio, the variables are in *inverse proportion.* Proportionality statements are not necessarily equations. A *proportionality constant* can be used to make such a statement into an equation. Proportionality constants might have numerical value only, without units, or they might have both value and units.

Modern science began about three hundred years ago during the time of Galileo and Newton. Since that time, *scientific investigation* has been used to provide *experimental evidence* about nature. *Methods* used to conduct scientific investigations can be generalized as *collecting observations, developing explanations,* and *testing explanations.*

A *hypothesis* is a tentative explanation that is accepted or rejected based on experimental data. Experimental data can come from *observations* or from a *controlled experiment.* The controlled experiment compares two situations that have all the influencing factors identical except one. The single influencing variable being tested is called the *experimental variable,* and the group of variables that form the basis of comparison is called the *control group.*

An accepted hypothesis may result in a *principle,* an explanation concerned with a specific range of phenomena, or a *scientific law,* an explanation concerned with important, wider-ranging phenomena. Laws are sometimes identified with the name of a scientist and can be expressed verbally, with an equation, or with a graph.

A *model* is used to help understand something that cannot be observed directly, explaining the unknown in terms of things already understood. Physical models, mental models, and equations are all examples of models that explain how nature behaves. A *theory* is a broad, detailed explanation that guides development and interpretations of experiments in a field of study.

Summary of Equations

1.1

$$\text{mass density} = \frac{\text{mass}}{\text{volume}}$$

$$\rho = \frac{m}{V}$$

area (p. **9**)

controlled
  experiment (p. **16**)

data (p. **9**)

density (p. **10**)

direct proportion (p. **12**)

English system (p. **5**)

equation (p. **12**)

experiment (p. **16**)

fundamental
  properties (p. **6**)

hypothesis (p. **16**)

inverse proportion (p. **12**)

kilogram (p. **7**)

liter (p. **8**)

measurement (p. **4**)

meter (p. **6**)

metric system (p. **6**)

model (p. **18**)

numerical constant (p. **13**)

properties (p. **3**)

proportionality
  constant (p. **13**)

quantities (p. **12**)

ratio (p. **9**)

referent (p. **3**)

scientific law (p. **18**)

scientific principle (p. **18**)

second (p. **7**)

standard unit (p. **5**)

theory (p. **20**)

unit (p. **4**)

variable (p. **12**)

volume (p. **9**)

1. The process of comparing a property of an object to a well-defined and agreed-upon referent is called
   a. generalizing.
   b. measurement.
   c. graphing.
   d. scientific investigation.

2. The height of an average person is closest to
   a. 1.0 m.
   b. 1.5 m.
   c. 2.5 m.
   d. 3.5 m.

3. Which of the following standard units is defined in terms of an object as opposed to an event?
   a. kilogram
   b. meter
   c. second
   d. None of the above is correct.

4. One-half liter of water has a mass of
   a. 0.5 g.
   b. 5 g.
   c. 50 g.
   d. 500 g.

5. A cubic centimeter (cm^3) of water has a mass of about 1
   a. mL.
   b. kg.
   c. g.
   d. dm.

6. Measurement information that is used to describe something is called
   a. referents.
   b. properties.
   c. data.
   d. a scientific investigation.

7. The property of volume is a measure of
   a. how much matter an object contains.
   b. how much space an object occupies.
   c. the compactness of matter in a certain size.
   d. the area on the outside surface.

8. As the volume of a cube becomes larger and larger, the surface-area-to-volume ratio
   a. increases.
   b. decreases.
   c. remains the same.
   d. sometimes increases and sometimes decreases.

9. If you consider a very small portion of a material that is the same throughout, the density of the small sample will be
   a. much less.
   b. slightly less.
   c. the same.
   d. greater.

10. Symbols that are used in equations represent
    a. a message.
    b. specific properties.
    c. quantities, or measured properties.
    d. all of the above.

11. An equation is composed of symbols in such a way that
    a. the numbers and units on both sides are always equal.
    b. the units are equal, but the numbers are not because one is unknown.
    c. the numbers are equal, but the units are not equal.
    d. neither the numbers nor units are equal because of the unknown.

12. The symbol $\therefore$ has a meaning of
    a. is proportional to.
    b. the change in.
    c. therefore.
    d. however.

13. Quantities, or measured properties, that are capable of changing values are called
    a. data.
    b. variables.
    c. proportionality constants.
    d. dimensionless constants.

14. A proportional relationship that is represented by the symbols $a \propto 1/b$ represents which of the following relationships?
    a. direct proportion
    b. inverse proportion
    c. direct square proportion
    d. inverse square proportion

15. A hypothesis concerned with a specific phenomenon is found to be acceptable through many experiments over a long period of time. This hypothesis usually becomes known as a
    a. scientific law.
    b. scientific principle.
    c. theory.
    d. model.

16. A scientific law can be expressed as
    a. a written concept.
    b. an equation.
    c. a graph.
    d. all of the above.

17. The symbol $\propto$ has a meaning of
    a. almost infinity.
    b. the change in.
    c. is proportional to.
    d. therefore.

18. Which of the following symbols represents a measured property of the compactness of matter?
    a. m
    b. ρ
    c. V
    d. Δ

Answers

1. b **2.** b **3.** a **4.** d **5.** c **6.** c **7.** b **8.** b **9.** c **10.** d **11.** a **12.** c **13.** b **14.** b **15.** b **16.** d **17.** c **18.** b

QUESTIONS FOR THOUGHT

1. What is a concept?
2. What two things does a measurement statement always contain? What do the two things tell you?
3. Other than familiarity, what are the advantages of the English system of measurement?
4. Describe the metric standard units for length, mass, and time.
5. Does the density of a liquid change with the shape of a container? Explain.
6. Does a flattened pancake of clay have the same density as the same clay rolled into a ball? Explain.
7. What is an equation? How are equations used in the physical sciences?
8. Compare and contrast a scientific principle and a scientific law.
9. What is a model? How are models used?
10. Are all theories always completely accepted or completely rejected? Explain.

The exercises in groups A and B cover the same concepts. Solutions to group A exercises are located in appendix D.

Note: *You will need to refer to Table 1.4 to complete some of the following exercises.*

Group A

1. What is your height in meters? In centimeters?
2. What is the mass density of mercury if 20.0 cm^3 has a mass of 272 g?
3. What is the mass of a 10.0 cm^3 cube of lead?
4. What is the volume of a rock with a mass density of 3.00 g/cm^3 and a mass of 600 g?
5. If you have 34.0 g of a 50.0 cm^3 volume of one of the substances listed in Table 1.4, which one is it?
6. What is the mass of water in a 40 L aquarium?

7. A 2.1 kg pile of aluminum cans is melted, then cooled into a solid cube. What is the volume of the cube?
8. A cubic box contains 1,000 g of water. What is the length of one side of the box in meters? Explain your reasoning.
9. A loaf of bread (volume 3,000 cm^3) with a density of 0.2 g/cm^3 is crushed in the bottom of a grocery bag into a volume of 1,500 cm^3. What is the density of the mashed bread?
10. According to Table 1.4, what volume of copper would be needed to balance a 1.00 cm^3 sample of lead on a two-pan laboratory balance?

Group B

1. What is your mass in kilograms? In grams?
2. What is the mass density of iron if 5.0 cm^3 has a mass of 39.5 g?
3. What is the mass of a 10.0 cm^3 cube of copper?
4. If ice has a mass density of 0.92 g/cm^3, what is the volume of 5,000 g of ice?
5. If you have 51.5 g of a 50.0 cm^3 volume of one of the substances listed in Table 1.4, which one is it?
6. What is the mass of gasoline ($\rho = 0.680$ g/cm^3) in a 94.6 L gasoline tank?
7. What is the volume of a 2.00 kg pile of iron cans that are melted, then cooled into a solid cube?
8. A cubic tank holds 1,000.0 kg of water. What are the dimensions of the tank in meters? Explain your reasoning.
9. A hot dog bun (volume 240 cm^3) with a density of 0.15 g/cm^3 is crushed in a picnic cooler into a volume of 195 cm^3. What is the new density of the bun?
10. According to Table 1.4, what volume of iron would be needed to balance a 1.00 cm^3 sample of lead on a two-pan laboratory balance?

This is a picture of pure zinc, one of the 89 naturally occurring elements found on the earth.

CHAPTER

8

Atoms and Periodic Properties

The development of the modern atomic model illustrates how modern scientific understanding comes from many different fields of study. For example, you will learn how studies of electricity led to the discovery that atoms have subatomic parts called *electrons*. The discovery of radioactivity led to the discovery of more parts, a central nucleus that contains protons and neutrons. Information from the absorption and emission of light was used to construct a model of how these parts are put together, a model resembling a miniature solar system with electrons circling the nucleus. The solar system model had initial, but limited, success and was inconsistent with other understandings about matter and energy. Modifications of this model were attempted, but none solved the problems. Then the discovery of wave properties of matter led to an entirely new model of the atom (Figure 8.1).

The atomic model will be put to use in later chapters to explain the countless varieties of matter and the changes that matter undergoes. In addition, you will learn how these changes can be manipulated to make new materials, from drugs to ceramics. In short, you will learn how understanding the atom and all the changes it undergoes not only touches your life directly but shapes and affects all parts of civilization.

Atomic Structure Discovered

Did you ever wonder how scientists could know about something so tiny that you cannot see it, even with the most powerful optical microscope? The atom is a tiny unit of matter, so small that one gram of hydrogen contains about 600,000,000,000,000,000,000,000 (six-hundred-thousand-billion-billion or 6×10^{23}) atoms. Even more unbelievable is that atoms are not individual units, but are made up of even smaller particles. How is it possible that scientists are able to tell you about the parts of something so small that it cannot be seen? The answer is that these things cannot be observed directly, but their existence can be inferred from experimental evidence. The following story describes the evidence and how scientists learned about the parts—electrons, the nucleus, protons, and neutrons—and how they are all arranged in the atom.

The atomic concept is very old, dating back to ancient Greek philosophers some 2,500 years ago. The ancient Greeks also reasoned about the way that pure substances are put together. A glass of water, for example, appears to be the same throughout. Is it the same? Two plausible, but conflicting, ideas were possible as an intellectual exercise. The water could have a continuous structure, that is, it could be completely homogeneous throughout. The other idea was that the water only appears to be continuous but is actually *discontinuous*. This means that if you continue to divide the water into smaller and smaller volumes, you would eventually reach a limit to this dividing, a particle that could not be further subdivided. The Greek philosopher Democritus (460–362 B.C.) developed this model in the fourth century B.C., and he called the indivisible particle an *atom*, from a Greek word meaning "uncuttable." However, neither Plato nor Aristotle accepted the atomic theory of matter, and it was not until about 2,000 years later that the atomic concept of matter was reintroduced. In the early 1800s, the English chemist John Dalton brought back the ancient Greek idea of hard, indivisible atoms to explain chemical reactions. Five

statements will summarize his theory. As you will soon see, today we know that statement 2 is not strictly correct:

1. Indivisible minute particles called atoms make up all matter.
2. All the atoms of an element are exactly alike in shape and mass.
3. The atoms of different elements differ from one another in their masses.
4. Atoms chemically combine in definite whole-number ratios to form chemical compounds.
5. Atoms are neither created nor destroyed in chemical reactions.

During the 1800s, Dalton's concept of hard, indivisible atoms was familiar to most scientists. Yet, the existence of atoms was not generally accepted by all scientists. There was skepticism about something that could not be observed directly. Strangely, full acceptance of the atom came in the early 1900s with the discovery that the atom was not indivisible after all. The atom has parts that give it an internal structure. The first part to be discovered was the *electron*, a part that was discovered through studies of electricity.

Discovery of the Electron

Scientists of the late 1800s were interested in understanding the nature of the recently discovered electric current. To observe a current directly they tried to produce a current by itself, away from wires, by removing the air from a tube and then running a current through the vacuum. When metal plates inside a tube were connected to the negative and positive terminals of a high-voltage source (Figure 8.2), a greenish beam was observed that seemed to move from the cathode (negative terminal) through the empty tube and collect at the anode (positive terminal). Since this mysterious beam seemed to come out of the cathode it was said to be a *cathode ray*.

The English physicist J. J. Thomson figured out what the cathode ray was in 1897. He placed charged metal plates on each side of the beam (Figure 8.3) and found that the beam was

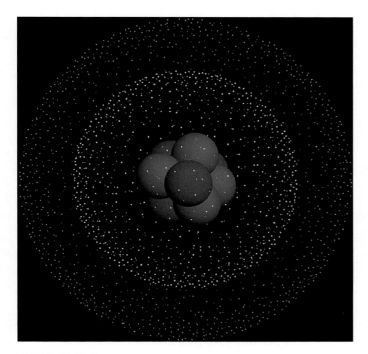

FIGURE 8.1

This is a computer-generated model of a beryllium atom, showing the nucleus and 1s, 2s electron orbitals. This configuration can also be predicted from information on a periodic table.

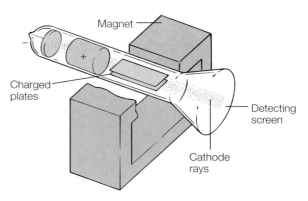

FIGURE 8.3

A cathode ray passed between two charged plates is deflected toward the positively charged plate. The ray is also deflected by a magnetic field. By measuring the deflection by both, J. J. Thomson was able to calculate the ratio of charge to mass. He was able to measure the deflection because the detecting screen was coated with zinc sulfide, a substance that produces a visible light when struck by a charged particle.

deflected away from the negative plate. Since it was known that like charges repel, this meant that the beam was composed of negatively charged particles.

The cathode ray was also deflected when caused to pass between the poles of a magnet. By balancing the deflections made by the magnet with the deflections made by the electric field, Thomson could determine the ratio of the charge to mass for an individual particle. Today, the charge-to-mass ratio is

considered to be 1.7584×10^{11} coulomb/kilogram. A significant part of Thomson's experiments was that he found the charge-to-mass ratio was the same no matter what gas was in the tube and of what materials the electrodes were made. Thomson had discovered the **electron**, a fundamental particle of matter.

A method for measuring the charge and mass of the electron was worked out by an American physicist, Robert A. Millikan, around 1906. Millikan used an apparatus like the one illustrated in Figure 8.4 to measure the charge on tiny droplets of oil.

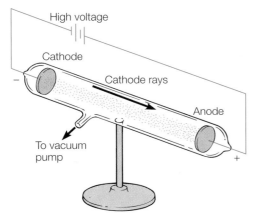

FIGURE 8.2

A vacuum tube with metal plates attached to a high-voltage source produces a greenish beam called *cathode rays*. These rays move from the cathode (negative charge) to the anode (positive charge).

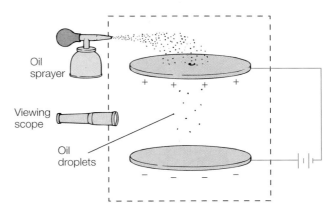

FIGURE 8.4

Millikan measured the charge of an electron by balancing the pull of gravity on oil droplets with an upward electrical force. Knowing the charge-to-mass ratio that Thomson had calculated, Millikan was able to calculate the charge on each droplet. He found that all the droplets had a charge of 1.60×10^{-19} coulomb or multiples of that charge. The conclusion was that this had to be the charge of an electron.

Millikan found that none of the droplets had a charge less than one particular value (1.60×10^{-19} coulomb) and that larger charges on various droplets were always multiples of this unit of charge. Since all of the droplets carried the single unit of charge or multiples of the single unit, the unit of charge was understood to be the charge of a single electron.

Knowing the charge of a single electron and knowing the charge-to-mass ratio that Thomson had measured now made it possible to calculate the mass of a single electron. The mass of an electron was thus determined to be about 9.11×10^{-31} kg, or about 1/1,840 of the mass of the lightest atom, hydrogen.

Thomson had discovered the negatively charged electron, and Millikan had measured the charge and mass of the electron. But atoms themselves are electrically neutral. If an electron is part of an atom, there must be something else that is positively charged, canceling the negative charge of the electron. The next step in the sequence of understanding atomic structure would be to find what is neutralizing the negative charge and to figure out how all the parts are put together.

Thomson had proposed a model for what was known about the atom at the time. He suggested that an atom could be a blob of positively charged matter in which electrons were stuck like "raisins in plum pudding." If the mass of a hydrogen atom is due to the electrons embedded in a positively charged matrix, then 1,840 electrons would be needed together with sufficient positive matter to make the atom electrically neutral.

The Nucleus

The nature of radioactivity and matter were the research interests of a British physicist, Ernest Rutherford. In 1907, Rutherford was studying the scattering of alpha particles directed toward a thin sheet of metal. As shown in Figure 8.5, alpha particles from a radioactive source were allowed to move through a small

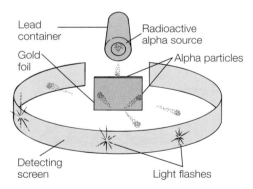

FIGURE 8.5

Rutherford and his coworkers studied alpha particle scattering from a thin metal foil. The alpha particles struck the detecting screen, producing a flash of visible light. Measurements of the angles between the flashes, the metal foil, and the source of the alpha particles showed that the particles were scattered in all directions, including straight back toward the source.

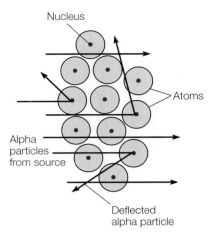

FIGURE 8.6

Rutherford's nuclear model of the atom explained the alpha scattering results as positive alpha particles experiencing a repulsive force from the positive nucleus. Measurements of the percentage of alpha particles passing straight through and of the various angles of scattering of those coming close to the nuclei gave Rutherford a means of estimating the size of the nucleus.

opening in a lead container, so only a narrow beam of the massive, fast-moving particles would penetrate a very thin sheet of gold. The alpha particles were detected by plates that produced small flashes of light when struck by alpha particles.

Rutherford found that most of the alpha particles went straight through the foil. However, he was astounded to find that some alpha particles were deflected at very large angles and some were even reflected backward. He could account for this only by assuming that the massive, positively charged alpha particles were repelled by a massive positive charge concentrated in a small region of the atom (Figure 8.6). He concluded that an atom must have a tiny, massive, and positively charged **nucleus** surrounded by electrons.

From measurements of the scattering, Rutherford estimated electrons must be moving around the nucleus at a distance 100,000 times the radius of the nucleus. This means the volume of an atom is mostly empty space. A few years later Rutherford was able to identify the discrete unit of positive charge which we now call a **proton.** Rutherford also speculated about the existence of a neutral particle in the nucleus, a neutron. The **neutron** was eventually identified in 1932 by James Chadwick.

Today, the number of protons in the nucleus of an atom is called the **atomic number.** All of the atoms of a particular element have the same number of protons in their nuclei, so all atoms of an element have the same atomic number. Hydrogen has an atomic number of 1, so any atom that has one proton in its nucleus is an atom of the element hydrogen. Today, scientists have identified 113 different kinds of elements, each with a different number of protons.

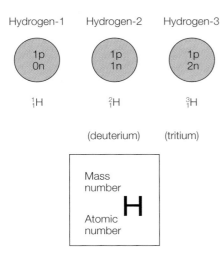

Hydrogen-1   Hydrogen-2   Hydrogen-3

1p   1p   1p
0n   1n   2n

$_1^1H$ $_1^2H$ $_1^3H$

(deuterium)   (tritium)

Mass number
Atomic number
H

FIGURE 8.7

The three isotopes of hydrogen have the same number of protons but different numbers of neutrons. Hydrogen-1 is the most common isotope. Hydrogen-2, with an additional neutron, is named *deuterium,* and hydrogen-3 is called *tritium.*

TABLE 8.1
Selected atomic weights calculated from mass and abundance of isotopes

Stable Isotopes	Mass of Isotope Compared to C-12	Abundance	Atomic Weight
$_1^1H$	1.007	99.985%	
$_1^2H$	2.0141	0.015%	1.0079
$_4^9Be$	9.01218	100.%	9.01218
$_7^{14}N$	14.00307	99.63%	
$_7^{15}N$	15.00011	0.37%	14.0067
$_8^{16}O$	15.99491	99.759%	
$_8^{17}O$	16.99914	0.037%	
$_8^{18}O$	17.00016	0.204%	15.9994
$_9^{19}F$	18.9984	100.%	18.9984
$_{10}^{20}Ne$	19.99244	90.92%	
$_{10}^{21}Ne$	20.99395	0.257%	
$_{10}^{22}Ne$	21.99138	8.82%	20.179
$_{13}^{27}Al$	26.9815	100.%	26.9815

The neutrons of the nucleus, along with the protons, contribute to the mass of an atom. Although all the atoms of an element must have the same number of protons in their nuclei, the number of neutrons may vary. Atoms of an element that have different numbers of neutrons are called **isotopes.** There are three isotopes of hydrogen illustrated in Figure 8.7. All three isotopes have the same number of protons and electrons, but one isotope has no neutrons, one isotope has one neutron (deuterium) and one isotope has two neutrons (tritium).

An atom is very tiny, and it is impossible to find the mass of a given atom. It is possible, however, to compare the mass of one atom to another. The mass of any atom is compared to the mass of an atom of a particular isotope of carbon. This particular carbon isotope is assigned a mass of exactly 12.00 units called **atomic mass units** (u). This isotope, called *carbon-12,* provides the standard to which the masses of all other isotopes are compared. The relative mass of any isotope is based on the mass of a carbon-12 isotope.

The relative mass of the hydrogen isotope without a neutron is 1.007 when compared to carbon-12. The relative mass of the hydrogen isotope with 1 neutron is 2.0141 when compared to carbon-12. Elements occur in nature as a mixture of isotopes, and the contribution of each is calculated in the atomic weight. **Atomic weight** for the atoms of an element is an average of the isotopes based on their mass compared to carbon-12, and their relative abundance in nature. Of all the hydrogen isotopes, for example, 99.985% occur as the isotope without a neutron and 0.015% are the isotope with one neutron (the other isotope is not considered because it is radioactive). The fractional part of occurrence is multiplied by the relative atomic mass for each isotope and the results

summed to obtain the atomic weight. Table 8.1 gives the atomic weight of hydrogen as 1.0079 as a result of this calculation.

The sum of the number of protons and neutrons in a nucleus of an atom is called the **mass number** of that atom. Mass numbers are used to identify isotopes. A hydrogen atom with 1 proton and 1 neutron has a mass number of $1 + 1$, or 2, and is referred to as hydrogen-2. A hydrogen atom with 1 proton and 2 neutrons has a mass number of $1 + 2$, or 3, and is referred to as hydrogen-3. Using symbols, hydrogen-3 is written as

$$_1^3H$$

where H is the chemical symbol for hydrogen, the subscript to the bottom left is the atomic number, and the superscript to the top left is the mass number.

How are the electrons moving around the nucleus? It might occur to you, as it did to Rutherford and others, that an atom might be similar to a miniature solar system. In this analogy, the nucleus is in the role of the sun, electrons in the role of moving planets in their orbits, and electrical attractions between the nucleus and electrons in the role of gravitational attraction. There are, however, big problems with this idea. If electrons were moving in circular orbits, they would continually change their direction of travel and would therefore be accelerating. According to the Maxwell model of electromagnetic radiation, an accelerating electric charge emits electromagnetic radiation such as light. If an electron gave off light, it would lose energy. The energy loss would mean that the electron could not maintain its orbit, and it would be pulled into the oppositely charged nucleus. The atom would collapse as electrons spiraled into the nucleus. Since atoms do not collapse like this, there is a significant problem with the solar system model of the atom.

The Bohr Model

Niels Bohr was a young Danish physicist who visited Rutherford's laboratory in 1912 and became very interested in questions about the solar system model of the atom. He wondered what determined the size of the electron orbits and the energies of the electrons. He wanted to know why orbiting electrons did not give off electromagnetic radiation. Seeking answers to questions such as these led Bohr to incorporate the *quantum concept* of Planck and Einstein with Rutherford's model to describe the electrons in the outer part of the atom. This quantum concept will be briefly reviewed before proceeding with the development of Bohr's model of the hydrogen atom.

The Quantum Concept

In the year 1900, Max Planck introduced the idea that matter emits and absorbs energy in discrete units that he called **quanta.** Planck had been trying to match data from spectroscopy experiments with data that could be predicted from the theory of electromagnetic radiation. In order to match the experimental findings with the theory, he had to assume that specific, discrete amounts of energy were associated with different frequencies of radiation. In 1905, Albert Einstein extended the quantum concept to light, stating that light consists of discrete units of energy that are now called **photons.** The energy of a photon is directly proportional to the frequency of vibration, and the higher the frequency of light, the greater the energy of the individual photons. In addition, the interaction of a photon with matter is an "all-or-none" affair,

that is, matter absorbs an entire photon or none of it. The relationship between frequency (f) and energy (E) is

$$E = hf$$

equation 8.1

where h is the proportionality constant known as *Planck's constant* (6.63×10^{-34} J·s). This relationship means that higher-frequency light, such as ultraviolet, has more energy than lower-frequency light, such as red light.

EXAMPLE 8.1

What is the energy of a photon of red light with a frequency of 4.60×10^{14} Hz?

Solution

$$f = 4.60 \times 10^{14} \text{ Hz}$$
$$h = 6.63 \times 10^{-34} \text{ J·s}$$
$$E = ?$$

$$E = hf$$
$$= (6.63 \times 10^{-34} \text{J·s})\left(4.60 \times 10^{14}\frac{1}{\text{s}}\right)$$
$$= (6.63 \times 10^{-34})(4.60 \times 10^{14}) \text{ J·s} \times \frac{1}{\text{s}}$$
$$= \boxed{3.05 \times 10^{-19} \text{ J}}$$

EXAMPLE 8.2

What is the energy of a photon of violet light with a frequency of 7.30×10^{14} Hz? (Answer: 4.84×10^{-19} J)

Atomic Spectra

Planck was concerned with hot solids that emit electromagnetic radiation. The nature of this radiation, called *blackbody radiation,* depends on the temperature of the source. When this light is passed through a prism, it is dispersed into a *continuous spectrum,* with one color gradually blending into the next as in a rainbow. Today, it is understood that a continuous spectrum comes from solids, liquids, and dense gases because the atoms interact, and all frequencies within a temperature-determined range are emitted. Light from an incandescent gas, on the other hand, is dispersed into a **line spectrum,** narrow lines of colors with no light between the lines (Figure 8.8). The atoms in the incandescent gas are able to emit certain characteristic frequencies, and each frequency is a line of color that represents a definite value of energy. The line spectra are specific for a substance, and increased or decreased temperature changes only the intensity

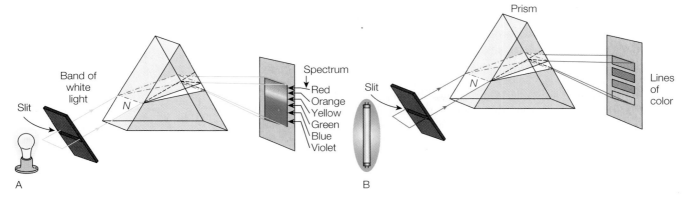

FIGURE 8.8

(A) Light from incandescent solids, liquids, or dense gases produces a continuous spectrum as atoms interact to emit all frequencies of visible light. *(B)* Light from an incandescent gas produces a line spectrum as atoms emit certain frequencies that are characteristic of each element.

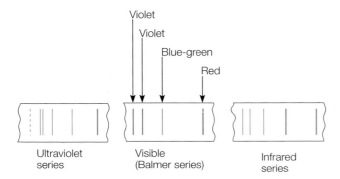

FIGURE 8.9

Atomic hydrogen produces a series of characteristic line spectra in the ultraviolet, visible, and infrared parts of the total spectrum. The visible light spectra always consist of two violet lines, a blue-green line, and a bright red line.

of the lines of colors. Thus hydrogen always produces the same colors of lines in the same position. Helium has its own specific set of lines, as do other substances. Line spectra are a kind of fingerprint that can be used to identify a gas. A line spectrum might also extend beyond visible light into ultraviolet, infrared, and other electromagnetic regions.

In 1885, a Swiss mathematics teacher named J. J. Balmer was studying the regularity of spacing of the hydrogen line spectra. Balmer was able to develop an equation that fit all the visible lines. By assigning values (*n*) of 3, 4, 5, and 6 to the four lines, he found the wavelengths fit the equation

$$\frac{1}{\lambda} = R\left(\frac{1}{2^2} - \frac{1}{n^2}\right)$$

equation 8.2

when *R* is a constant of 1.097×10^7 1/m.
Balmer's findings were:

Violet line	(*n* = 6)	$\lambda = 4.1 \times 10^{-7}$m
Violet line	(*n* = 5)	$\lambda = 4.3 \times 10^{-7}$m
Blue-green line	(*n* = 4)	$\lambda = 4.8 \times 10^{-7}$m
Red line	(*n* = 3)	$\lambda = 6.6 \times 10^{-7}$m

These four lines became known as the **Balmer series.** Other series were found later, outside the visible part of the spectrum (Figure 8.9). The equations of the other series were different only in the value of *n* and the number in the other denominator.

Such regularity of observable spectral lines must reflect some unseen regularity in the atom. At this time it was known that hydrogen had only one electron. How could one electron produce a series of spectral lines with such regularity?

EXAMPLE 8.3

Calculate the wavelength of the violet line (*n* = 6) in the hydrogen line spectra according to Balmer's equation.

Solution

$n = 6$

$R = 1.097 \times 10^7$ 1/m

$\lambda = ?$

$$\frac{1}{\lambda} = R\left(\frac{1}{2^2} - \frac{1}{n^2}\right)$$

$$= 1.097 \times 10^7 \frac{1}{m}\left(\frac{1}{2^2} - \frac{1}{6^2}\right)$$

$$= 1.097 \times 10^7\left(\frac{1}{4} - \frac{1}{36}\right)\frac{1}{m}$$

$$= 1.097 \times 10^7(0.222)\frac{1}{m}$$

$$\frac{1}{\lambda} = 2.44 \times 10^6 \frac{1}{m}$$

$$\lambda = \boxed{4.11 \times 10^{-7} \text{ m}}$$

Bohr's Theory

An acceptable model of the hydrogen atom would have to explain the characteristic line spectra and their regularity as described by Balmer. In fact, a successful model should be able to predict the occurrence of each color line as well as account for its origin. By 1913, Bohr was able to do this by applying the quantum concept to a solar system model of the atom. He began by considering the single hydrogen electron to be a single "planet" revolving in a circular orbit around the nucleus. There were three sets of rules that described this electron:

1. **Allowed Orbits.** An electron can revolve around an atom only in specific allowed orbits. Bohr considered the electron to be a particle with a known mass in motion around the nucleus, and used Newtonian mechanics to calculate the distances of the allowed orbits. According to the Bohr model, electrons can exist only in one of these allowed orbits and nowhere else.

2. **Radiationless Orbits.** An electron in an allowed orbit does not emit radiant energy as long as it remains in the orbit. According to Maxwell's theory of electromagnetic radiation, an accelerating electron should emit an electromagnetic wave, such as light, which would move off into space from the electron. Bohr recognized that electrons moving in a circular orbit are accelerating, since they are changing direction continuously. Yet, hydrogen atoms did not emit light in their normal state. Bohr decided that the situation must be different for orbiting electrons and that electrons could stay in their allowed orbits and not give off light. He postulated this rule as a way to make his theory consistent with other scientific theories.

3. **Quantum Leaps.** An electron gains or loses energy only by moving from one allowed orbit to another (Figure 8.10). In the Bohr model, the energy an electron has depends on which allowable orbit it occupies. The only way that an electron can change its energy is to jump from one allowed orbit to another in quantum "leaps." An electron must acquire energy to jump from a lower orbit to a higher one. Likewise, an electron gives up energy when jumping from a higher orbit to a lower one. Such jumps must be all at once, not part way and not gradual. An electron acquires energy from high temperatures or from electrical discharges to jump to a higher orbit. An electron jumping from a higher to a lower orbit gives up energy in the form of light. A single photon is emitted when a downward jump occurs and the energy of the photon is *exactly* equal to the difference in the energy level of the two orbits.

The energy level diagram in Figure 8.11 shows the energy states for the orbits of a hydrogen atom. The lowest energy state is the **ground state** (or normal state). The higher states are the **excited states**. The electron in a hydrogen atom would normally occupy the ground state, but high temperatures or electric discharge can give the electron sufficient energy to jump to one of the excited states. Once in an excited state, the electron immediately jumps back to a lower state, as shown by the arrows in the figure. The length of the arrow represents the frequency of the photon that the electron emits in the process. A hydrogen atom can give off only one photon at a time, and the many lines of a hydrogen line spectrum come from many atoms giving off many photons at the same time.

The reference level for the potential energy of an electron is considered to be zero when the electron is *removed* from an

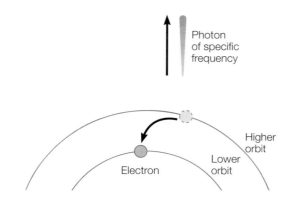

FIGURE 8.10

Each time an electron makes a "quantum leap," moving from a higher-energy orbit to a lower-energy orbit, it emits a photon of a specific frequency and energy value.

atom. The electron, therefore, has a lower and lower potential energy at closer and closer distances to the nucleus and has a negative value when it is in some allowed orbit. By way of analogy, you could consider ground level as a reference level where the potential energy of some object equals zero. But suppose there are two basement levels below the ground. An object on either basement level would have a gravitational potential energy less than zero, and work would have to be done on each object to bring it back to the zero level. Thus, each object would have a negative potential energy. The object on the lowest level would have the largest negative value of energy, since more

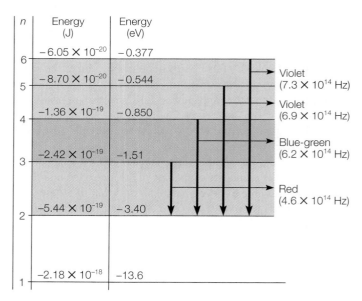

FIGURE 8.11

An energy level diagram for a hydrogen atom, not drawn to scale. The energy levels (n) are listed on the left side, followed by the energies of each level in J and eV. The color and frequency of the visible light photons emitted are listed on the right side, with the arrow showing the orbit moved from and to.

work would have to be done on it to bring it back to the zero level. Therefore, the object on the lowest level would have the *least* potential energy, and this would be expressed as the *largest negative value.*

Just as the objects on different basement levels have negative potential energy, the electron has a definite negative potential energy in each of the allowed orbits. Bohr calculated the energy of an electron in the orbit closest to the nucleus to be -2.18×10^{-18} J, which is called the energy of the lowest state. The energy of electrons can be expressed in units of the **electron volt** (eV). An electron volt is defined as the energy of an electron moving through a potential of one volt. Since this energy is charge times voltage (from $V = W/q$), 1.00 eV is equivalent to 1.60×10^{-19} J. Therefore the energy of an electron in the innermost orbit is its energy in joules divided by 1.60×10^{-19} J/eV, or -13.6 eV.

Bohr found that the energy of each of the allowed orbits could be found from the simple relationship of

$$E_n = \frac{E_1}{n^2}$$

equation 8.3

where E_1 is the energy of the innermost orbit (-13.6 eV), and n is the quantum number for an orbit, or 1, 2, 3, and so on. Thus, the energy for the second orbit ($n = 2$) is $E_2 = -13.6$ eV/4 $= -3.40$ eV. The energy for the third orbit out ($n = 3$) is $E_3 = -13.6$ eV/9 $= -1.51$ eV, and so forth (Figure 8.11). Thus, the energy of each orbit is *quantized,* occurring only as a definite value.

In the Bohr model, the energy of the electron is determined by which allowable orbit it occupies. The only way that an electron can change its energy is to jump from one allowed orbit to another in quantum "jumps." An electron must *acquire* energy to jump from a lower orbit to a higher one. Likewise an electron *gives up* energy when jumping from a higher orbit to a lower one. Such jumps must be all at once, not part way and not gradual. By way of analogy, this is very much like the gravitational potential energy that you have on the steps of a staircase. You have the lowest potential on the bottom step and the greatest amount on the top step. Your potential energy is quantized because you can increase or decrease it by going up or down a number of steps, but you cannot stop between the steps.

An electron acquires energy from high temperatures or from electrical discharges to jump to a higher orbit. An electron jumping from a higher to a lower orbit gives up energy in the form of light. A single photon is emitted when a downward jump occurs, and the *energy of the photon is exactly equal to the difference in the energy level* of the two orbits. If E_L represents the lower-energy level (closest to the nucleus) and E_H represents a higher-energy level (farthest from the nucleus), the energy of the emitted photon is

$$hf = E_H - E_L$$

equation 8.4

where h is Planck's constant, and f is the frequency of the emitted light (Figure 8.12).

FIGURE 8.12

These fluorescent lights emit light as electrons of mercury atoms inside the tubes gain energy from the electric current. As soon as they can, the electrons drop back to a lower-energy orbit, emitting photons with ultraviolet frequencies. Ultraviolet radiation strikes the fluorescent chemical coating inside the tube, stimulating the emission of visible light.

As you can see, the energy level diagram in Figure 8.11 shows how the change of known energy levels from known orbits results in the exact energies of the color lines in the Balmer series. Bohr's theory did offer an explanation for the lines in the hydrogen spectrum with a remarkable degree of accuracy. However, the model did not have much success with larger atoms. Larger atoms had spectra lines that could not be explained by the Bohr model with its single quantum number. A German physicist, A. Sommerfeld, tried to modify Bohr's model by adding elliptical orbits in addition to Bohr's circular orbits. It soon became apparent that the "patched up" model, too, was not adequate. Bohr had made the rule that there were radiationless orbits without an explanation, and he did not have an explanation for the quantized orbits. There was something fundamentally incomplete about the model.

EXAMPLE **8.4**

An electron in a hydrogen atom jumps from the excited energy level $n = 4$ to $n = 2$. What is the frequency of the emitted photon?

Solution

The frequency of an emitted photon can be calculated from equation 8.4, $hf = E_H - E_L$. The values for the two energy levels can be obtained from Figure 8.11. (Note: E_H and E_L must be in joules. If the values are in electron volts, they can be converted to joules by multiplying by the ratio of joules per electron volt, or (eV)(1.60×10^{-19} J/eV) = joules.)

$E_H = -1.36 \times 10^{-19}$ J
$E_L = -5.44 \times 10^{-19}$ J
$h = 6.63 \times 10^{-34}$ J·s
$f = ?$

$$hf = E_H - E_L \quad \therefore \quad f = \frac{E_H - E_L}{h}$$

$$f = \frac{(-1.36 \times 10^{19} \text{ J}) - (-5.44 \times 10^{-19} \text{J})}{6.63 \times 10^{-34} \text{ J·s}}$$

$$= \frac{4.08 \times 10^{-19}}{6.63 \times 10^{-34}} \frac{\cancel{\text{J}}}{\cancel{\text{J}}\text{·s}}$$

$$= 6.15 \times 10^{14} \frac{1}{s}$$

$$= \boxed{6.15 \times 10^{14} \text{ Hz}}$$

This is approximately the blue-green line in the hydrogen line spectrum.

Quantum Mechanics

The Bohr model of the atom successfully accounted for the line spectrum of hydrogen and provided an understandable mechanism for the emission of photons by atoms. However, the model did not predict the spectra of any atom larger than hydrogen, and there were other limitations. A new, better theory was needed. The roots of a new theory would again come from experiments with light. Experiments with light had established that sometimes light behaves like a stream of particles, and at other times it behaves like a wave (see chapter 7). Eventually scientists began to accept that light has both wave properties and particle properties, which is now referred to as the *wave-particle duality of light*. This dual nature of light was recognized in 1905 when Einstein applied Planck's quantum concept to the energy of a photon with the relationship found in equation 8.1, $E = hf$, where E is the energy of a photon particle, f is the frequency of the associated wave, and h is Planck's constant.

Matter Waves

In 1923, Louis de Broglie, a French physicist, reasoned that symmetry is usually found in nature, so if a particle of light has a dual nature, then particles such as electrons should too. De Broglie reasoned further that if this is true, an electron in its circular path around the nucleus would have to have a particular wavelength that would fit into the circumference of the orbit (Figure 8.13). De Broglie derived a relationship from equations concerning light and energy, which was

$$\lambda = \frac{h}{mv}$$

equation 8.5

where λ is the wavelength, m is mass, v is velocity, and h is again Planck's constant. This equation means that any moving particle has a wavelength that is associated with its mass and velocity.

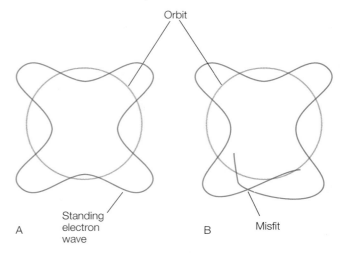

A — Standing electron wave
B — Misfit
Orbit

FIGURE 8.13
(A) Schematic of de Broglie wave, where the standing wave pattern will just fit in the circumference of an orbit. This is an allowed orbit. (B) This orbit does not have a circumference that will match a whole number of wavelengths; it is not an allowed orbit.

In other words, de Broglie was proposing a wave-particle duality of matter, the existence of **matter waves.** According to equation 8.5, *any* moving object should exhibit wave properties. However, an ordinary-sized object would have wavelengths so small that they could not be observed. This is different for electrons because they have such a tiny mass.

EXAMPLE 8.5

What is the de Broglie wavelength associated with a 0.150 kg baseball with a velocity of 50.0 m/s?

Solution

$m = 0.150$ kg
$v = 50.0$ m/s
$h = 6.63 \times 10^{-34}$ J·s
$\lambda = ?$

$$\lambda = \frac{h}{mv}$$

$$= \frac{6.63 \times 10^{-34} \text{ J·s}}{(0.150 \text{ kg})\left(50.0\frac{\text{m}}{\text{s}}\right)}$$

$$= \frac{6.63 \times 10^{-34}}{(0.150)(50.0)} \frac{\text{J·s}}{\text{kg} \times \dfrac{\text{m}}{\text{s}}}$$

$$= \frac{6.63 \times 10^{-34}}{7.50} \frac{\dfrac{\text{kg·m}^2}{\text{s}^2}\text{·s}}{\dfrac{\text{kg·m}}{\text{s}}}$$

$$= \boxed{8.84 \times 10^{-35} \text{ m}}$$

What is the de Broglie wavelength associated with an electron with a velocity of 6.00×10^6 m/s?

Solution

$m = 9.11 \times 10^{-31}$ kg

$v = 6.00 \times 10^6$ m/s

$h = 6.63 \times 10^{-34}$ J·s

$\lambda = ?$

$$\lambda = \frac{h}{mv}$$

$$= \frac{6.63 \times 10^{-34} \text{ J·s}}{(9.11 \times 10^{-31} \text{ kg})\left(6.00 \times 10^6 \dfrac{\text{m}}{\text{s}}\right)}$$

$$= \frac{6.63 \times 10^{-34}}{5.47 \times 10^{-24}} \frac{\text{J·s}}{\text{kg} \times \dfrac{\text{m}}{\text{s}}}$$

$$= 1.21 \times 10^{-10} \frac{\dfrac{\text{kg·m}^2}{\text{s}^2} \cdot \text{s}}{\dfrac{\text{kg·m}}{\text{s}}}$$

$$= \boxed{1.21 \times 10^{-10} \text{ m}}$$

The baseball wavelength of 8.84×10^{-35} m is much too small to be detected or measured. The electron wavelength of 1.21×10^{-10} m, on the other hand, is comparable to the distances between atoms in a crystal, so a beam of electrons through a crystal should produce diffraction.

The idea of matter waves was soon tested after de Broglie published his theory. Experiments with a beam of light passing by the edge of a sharp-edged obstacle produced interference patterns. This was part of the evidence for the wave nature of light, since such results could only be explained by waves, not particles. When similar experiments were performed with a beam of electrons, *identical* wave property behaviors were observed. This and many related experiments showed without doubt that electrons have both wave properties and particle properties. And, as was the case with light waves, measurements of the electron interference patterns provided a means to measure the wavelength of electron waves.

Recall that waves confined on a fixed string establish resonant modes of vibration called *standing waves* (see chapter 5). Only certain fundamental frequencies and harmonics can exist on a string, and the combination of the fundamental and overtones gives the stringed instrument its particular quality. The same result of resonant modes of vibrations is observed in *any* situation where waves are confined to a fixed space. Characteristic standing wave patterns depend on the wavelength and wave velocity for waves formed on strings, in enclosed columns of air, or for any kind of wave in a confined space. Electrons are confined to the space near a nucleus, and electrons have wave properties, so an electron in an atom must be a confined wave. Does an electron form a characteristic wave pattern? This was the question being asked in about 1925 when Heisenberg, Schrödinger, Dirac, and others applied the wave nature of the electron to develop a new model of the atom based on the mechanics of electron waves. The new theory is now called **wave mechanics,** or **quantum mechanics.**

Wave Mechanics

Erwin Schrödinger, an Austrian physicist, treated the atom as a three-dimensional system of waves to derive what is now called the *Schrödinger equation.* Instead of the simple circular planetary orbits of the Bohr model, solving the Schrödinger equation results in a description of three-dimensional shapes of the patterns that develop when electron waves are confined by a nucleus. Schrödinger first considered the hydrogen atom, calculating the states of vibration that would be possible for an electron wave confined by a nucleus. He found that the frequency of these vibrations, when multiplied by Planck's constant, matched exactly, to the last decimal point, the observed energies of the quantum states of the hydrogen atom ($E = hf$). The conclusion is that the wave nature of the electron is the important property to consider for a successful model of the atom.

The quantum mechanics theory of the atom proved to be very successful; it confirmed all the known experimental facts and predicted new discoveries. The theory does have some of the same quantum ideas as the Bohr model; for example, an electron emits a photon when jumping from a higher state to a lower one. The Bohr model, however, considered the particle nature of an electron moving in a circular orbit with a definitely assigned position at a given time. Quantum mechanics considers the wave nature, with the electron as a confined wave with well-defined shapes and frequencies. A wave is not localized like a particle and is spread out in space. The quantum mechanics model is, therefore, a series of orbitlike smears, or fuzzy statistical representations, of where the electron might be found.

The Quantum Mechanics Model

The quantum mechanics model is a highly mathematical treatment of the mechanics of matter waves. In addition, the wave properties are considered as three-dimensional problems, and three quantum numbers are needed to describe the fuzzy electron cloud. The mathematical detail will not be presented here. The following is a qualitative description of the main ideas in the quantum mechanics model. It will describe the results of the mathematics and will provide a mental visualization of what it all means.

First, understand that the quantum mechanical theory is not an extension or refinement of the Bohr model. The Bohr model considered electrons as particles in circular orbits that could be only certain distances from the nucleus. The quantum mechanical model, on the other hand, considers the electron as a wave and considers the energy of its harmonics, or modes, of standing waves. In the Bohr model, the location of an electron was certain—in an orbit. In the quantum mechanical model, the electron is a spread-out wave.

Quantum mechanics describes the energy state of an electron wave with four *quantum numbers:*

1. **Distance from the Nucleus.** The *principal quantum number* describes the *main energy level* of an electron in terms of its most probable distance from the nucleus. The lowest energy state possible is closest to the nucleus and is assigned the principal quantum number of 1 ($n = 1$). Higher states are assigned progressively higher positive whole numbers of $n = 2$, $n = 3$, $n = 4$, and so on. Electrons with higher principal quantum numbers have higher energies and are located farther from the nucleus.

2. **Energy Sublevel.** The *angular momentum quantum number* defines energy sublevels within the main energy levels. Each sublevel is identified with a letter. The first four of these letters, in order of increasing energy, are s, p, d, and f. The letter s represents the lowest sublevel, and the letter f represents the highest sublevel. A principal quantum number and a letter indicating the angular momentum quantum number are combined to identify the main energy state and energy sublevel of an electron. For an electron in the lowest main energy level, $n = 1$ and in the lowest sublevel, s, the number and letter are 1s (read as "one-s"). Thus 1s indicates an electron that is as close to the nucleus as possible in the lowest energy sublevel possible.

The Bohr model considered the location of an electron as certain, like a tiny shrunken marble in an orbit. The quantum mechanical model considers the electron as a wave, and knowledge of its location is very uncertain. The **Heisenberg uncertainty principle** states that you cannot measure the exact position of a wave because a wave is spread out. One cannot specify the position and the momentum of a spread-out electron. The location of the electron can only be described in terms of *probabilities* of where it might be at a given instant. The probability of location is described by a fuzzy region of space called an **orbital**. An orbital defines the space where an electron is likely to be found. Orbitals have characteristic three-dimensional shapes and sizes and are identified with electrons of characteristic energy levels (Figure 8.14). An orbital shape

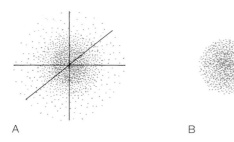

FIGURE 8.14

(*A*) An electron distribution sketch representing probability regions where an electron is most likely to be found. (*B*) A boundary surface, or contour, that encloses about 90 percent of the electron distribution shown in (*A*). This three-dimensional space around the nucleus, where there is the greatest probability of finding an electron, is called an *orbital*.

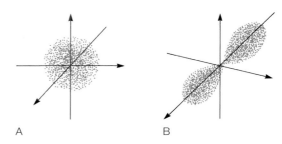

FIGURE 8.15

(*A*) A contour representation of an s orbital. (*B*) A contour representation of a p orbital.

represents where an electron could probably be located at any particular instant. This "probability cloud" could likewise have any particular orientation in space, and the direction of this orientation is uncertain.

3. **Orientation in Space.** An external magnetic field applied to an atom produces different energy levels that are related to the orientation of the orbital to the magnetic field. The orientation of an orbital in space is described by the *magnetic quantum number.* This number is related to the energies of orbitals as they are oriented in space relative to an external magnetic field, a kind of energy sub-sublevel. In general, the lowest-energy sublevel (s) has only one orbital orientation. The next higher-energy sublevel (p) can have three orbital orientations (Figure 8.15). The d sublevel can have five orbital orientations, and the highest sublevel, f, can have a total of seven different orientations (Table 8.2).

4. **Direction of Spin.** Detailed studies have shown that an electron spinning one way (say clockwise) in an external magnetic field would have a different energy than one spinning the other way (say counterclockwise). The *spin quantum number* describes these two spin orientations (Figure 8.16).

Electron spin is an important property of electrons that helps determine the electronic structure of an atom. As it turns

TABLE 8.2			
Quantum numbers and electron distribution to $n = 4$			
Main Energy Level	**Energy Sublevels**	**Maximum Number of Electrons**	**Maximum Number of Electrons per Main Energy Level**
$n = 1$	s	2	2
$n = 2$	s	2	
	p	6	8
$n = 3$	s	2	
	p	6	
	d	10	18
$n = 4$	s	2	
	p	6	
	d	10	
	f	14	32

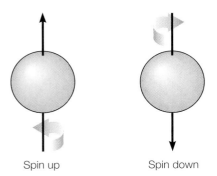

Spin up   Spin down

FIGURE 8.16

Experimental evidence supports the concept that electrons can be considered to spin one way or the other as they move about an orbital under an external magnetic field.

TABLE 8.3		
Electron configuration for the first twenty elements		
Atomic Number	**Element**	**Electron Configuration**
1	Hydrogen	$1s^1$
2	Helium	$1s^2$
3	Lithium	$1s^2 2s^1$
4	Beryllium	$1s^2 2s^2$
5	Boron	$1s^2 2s^2 2p^1$
6	Carbon	$1s^2 2s^2 2p^2$
7	Nitrogen	$1s^2 2s^2 2p^3$
8	Oxygen	$1s^2 2s^2 2p^4$
9	Fluorine	$1s^2 2s^2 2p^5$
10	Neon	$1s^2 2s^2 2p^6$
11	Sodium	$1s^2 2s^2 2p^6 3s^1$
12	Magnesium	$1s^2 2s^2 2p^6 3s^2$
13	Aluminum	$1s^2 2s^2 2p^6 3s^2 3p^1$
14	Silicon	$1s^2 2s^2 2p^6 3s^2 3p^2$
15	Phosphorus	$1s^2 2s^2 2p^6 3s^2 3p^3$
16	Sulfur	$1s^2 2s^2 2p^6 3s^2 3p^4$
17	Chlorine	$1s^2 2s^2 2p^6 3s^2 3p^5$
18	Argon	$1s^2 2s^2 2p^6 3s^2 3p^6$
19	Potassium	$1s^2 2s^2 2p^6 3s^2 3p^6 4s^1$
20	Calcium	$1s^2 2s^2 2p^6 3s^2 3p^6 4s^2$

out, two electrons spinning in opposite directions, called an **electron pair,** can occupy the same orbital. This was summarized in 1924 by Wolfgang Pauli, a German physicist. His summary, now known as the **Pauli exclusion principle,** states that *no two electrons in an atom can have the same four quantum numbers.* This provides the key for understanding the electron structure of atoms.

Electron Configuration

The arrangement of electrons in orbitals is called the *electron configuration.* Before you can describe the electron arrangement, you need to know how many electrons are present in an atom. An atom is electrically neutral, so the number of protons (positive charge) must equal the number of electrons (negative charge). The atomic number therefore identifies the number of electrons as well as the number of protons. Now that you have a means of finding the number of electrons, consider the various energy levels to see how the electron configuration is determined.

According to the Pauli exclusion principle, no two electrons in an atom can have all four quantum numbers the same. As it works out, this means there can only be *a maximum of two electrons in any given orbital.* There are four things to consider: (1) the main energy level, (2) the energy sublevel, (3) the number of orbital orientations, and (4) the electron spin. Recall that the lowest-energy level is $n = 1$, and successive numbers identify progressively higher-energy levels. Recall also that the energy sublevels, in order of increasing energy, are s, p, d, and f. This electron configuration is written in shorthand, with 1s standing for the lowest-energy sublevel of the first energy level. A superscript gives the number of electrons present in a sublevel. Thus, the electron configuration for a helium atom, which has two electrons, is written as $1s^2$. This combination of symbols has the following meaning: The symbols mean an atom with two electrons in the s sublevel of the first main energy level.

Table 8.3 gives the electron configurations for the first twenty elements. The configurations of the p energy sublevel have been condensed in this table. There are three possible orientations of the p orbital, each with two electrons (Figure 8.17). This is shown as p^6, which designates the number of electrons in all of the three possible p orientations. Note that the sum of the electrons in all the orbitals equals the atomic number. Note also that as you proceed from a lower atomic number to a higher one, the higher element has the same configuration as the element before it with the addition of one more electron. In general, it is then possible to begin with the simplest atom, hydrogen, and add one electron at a time to the order of energy sublevels and obtain the electron configuration for all the elements. The exclusion principle limits the number of electrons in any orbital, and allowances will need to be made for the more complex behavior of atoms with many electrons.

The energies of the orbital are different for each element, and there are several factors that influence their energies. The first orbitals are filled in a straightforward 1s, 2s, 2p, 3s, then 3p order. Then the order becomes contrary to what you might expect. One useful way of figuring out the order in which orbitals are filled is illustrated in Figure 8.18. Each row of this matrix represents a principal energy level with possible energy sublevels increasing from left to right. The order of filling is indicated by the diagonal arrows. There are exceptions to the order of filling shown by the matrix, but it works for most of the elements.

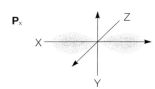

P_x

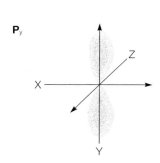

P_y

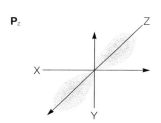

P_z

FIGURE 8.17

There are three possible orientations of the p orbital, and these are called p_x, p_y, and p_z. Each orbital can hold two electrons, so a total of six electrons are possible in the three orientations; thus the notation p^6.

Concepts Applied

Firework Configuration

Certain strontium (atomic number 38) chemicals are used to add the pure red color to flares and fireworks. Write the electron configuration of strontium and do this before looking at the solution that follows.

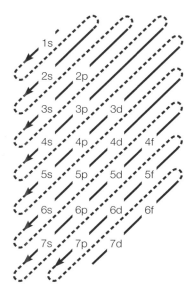

FIGURE 8.18

A matrix showing the order in which the orbitals are filled. Start at the top left, then move from the head of each arrow to the tail of the one immediately below it. This sequence moves from the lowest-energy level to the next higher level for each orbital.

First, note that an atomic number of 38 means a total of thirty-eight electrons. Second, refer to the order of filling matrix in Figure 8.18. Remember that only two electrons can occupy an orbital, but there are three orientations of the p orbital, for a total of six electrons. There are likewise five possible orientations of the d orbital, for a total of ten electrons. Starting at the lowest energy level, two electrons go in 1s, making $1s^2$; then two go in 2s, making $2s^2$. That is a total of four electrons so far. Next $2p^6$ and $3s^2$ use eight more electrons, for a total of twelve so far. The $3p^6$, $4s^2$, $3d^{10}$, and $4p^6$ use up twenty-four more electrons, for a total of thirty-six. The remaining two go into the next sublevel, $5s^2$, and the complete answer is

Strontium: $1s^2\ 2s^2\ 2p^6\ 3s^2\ 3p^6\ 4s^2\ 3d^{10}\ 4p^6\ 5s^2$

The Periodic Table

The periodic table is made up of rows and columns of cells, with each element having its own cell in a specific location. The cells are not arranged symmetrically. The arrangement has a meaning, both about atomic structure and about chemical behaviors. The key to meaningful, satisfying use of the table is to understand the code of this structure. The following explains some of what the code means. It will facilitate your understanding of the code if you refer frequently to a periodic table during the following discussion (Figure 8.19).

Periodic Table of the Elements

Period

IA (1)				
Alkali Metals	Alkaline Earth Metals			

1
Hydrogen
1 **H** 1.008

IIA (2)

2
Lithium 3 **Li** 6.941 — Beryllium 4 **Be** 9.012

3
Sodium 11 **Na** 22.99 — Magnesium 12 **Mg** 24.31

4
Potassium 19 **K** 39.10 — Calcium 20 **Ca** 40.08

5
Rubidium 37 **Rb** 85.47 — Strontium 38 **Sr** 87.62

6
Cesium 55 **Cs** 132.9 — Barium 56 **Ba** 137.3

7
Francium 87 **Fr** (223) — Radium 88 **Ra** (226)

Transition Elements

IIIB (3)	IVB (4)	VB (5)	VIB (6)	VIIB (7)	VIIIB (8)	VIIIB (9)	VIIIB (10)	IB (11)	IIB (12)
Scanium 21 **Sc** 44.96	Titanium 22 **Ti** 47.88	Vanadium 23 **V** 50.94	Chromium 24 **Cr** 52.00	Maganese 25 **Mn** 54.94	Iron 26 **Fe** 55.85	Cobalt 27 **Co** 58.93	Nickel 28 **Ni** 58.69	Copper 29 **Cu** 63.55	Zinc 30 **Zn** 65.39
Yttrium 39 **Y** 88.91	Zirconium 40 **Zr** 91.22	Niobium 41 **Nb** 92.91	Molybdenum 42 **Mo** 95.94	Technetium 43 **Tc** (98)	Ruthenium 44 **Ru** 101.1	Rhodium 45 **Rh** 102.9	Palladium 46 **Pd** 106.4	Silver 47 **Ag** 107.9	Cadmium 48 **Cd** 112.4
Lanthanum 57 **La** 138.9 †	Hafnium 72 **Hf** 178.5	Tantalum 73 **Ta** 180.9	Tungsten 74 **W** 183.8	Rhenium 75 **Re** 186.2	Osmium 76 **Os** 190.2	Iridium 77 **Ir** 192.2	Platinum 78 **Pt** 195.1	Gold 79 **Au** 197.0	Mercury 80 **Hg** 200.6
Actinium 89 **Ac** (227) ‡	Rutherfordium 104 **Rf** (261)	Dubnium 105 **Db** (262)	Seaborgium 106 **Sg** (266)	Bohrium 107 **Bh** (264)	Hassium 108 **Hs** (277)	Meitnerium 109 **Mt** (268)	Darmstadtium 110 **Ds** (281)	Unununium 111 **Uuu** (272)	Ununbium 112 **Uub** (285)

IIIA (13)	IVA (14)	VA (15)	VIA (16)	VIIA (17)	VIIIA (18)
				Halogens	Noble gases
					Helium 2 **He** 4.003
Boton 5 **B** 10.81	Carbon 6 **C** 12.01	Nitrogen 7 **N** 14.01	Oxygen 8 **O** 16.00	Fluorine 9 **F** 19.00	Neon 10 **Ne** 20.18
Aluminum 13 **Al** 26.98	Silicon 14 **Si** 28.09	Phosphorus 15 **P** 30.97	Sulfur 16 **S** 32.07	Chlorine 17 **Cl** 35.45	Argon 18 **Ar** 39.95
Gallium 31 **Ga** 69.72	Germanium 32 **Ge** 72.61	Arsenic 33 **As** 74.92	Selenium 34 **Se** 78.96	Bromine 35 **Br** 79.90	Krypton 36 **Kr** 83.80
Indium 49 **In** 114.8	Tin 50 **Sn** 118.7	Antimony 51 **Sb** 121.8	Tellurium 52 **Te** 127.6	Iodine 53 **I** 126.9	Xenon 54 **Xe** 131.3
Thallium 81 **Tl** 204.4	Lead 82 **Pb** 207.2	Bismuth 83 **Bi** 209.0	Polonium 84 **Po** (209)	Astatine 85 **At** (210)	Radon 86 **Rn** (222)
	Ununquadium 114 **Uuq** (289)				

	Metals
	Semiconductors
	Nonmetals

Inner Transition Elements

†Lanthanides 6

Cerium 58 **Ce** 140.1	Praseodymium 59 **Pr** 140.9	Neodymium 60 **Nd** 144.2	Promethium 61 **Pm** (145)	Samarium 62 **Sm** 150.4	Europium 63 **Eu** 152.0	Gadolinium 64 **Gd** 157.3	Terbium 65 **Tb** 158.9	Dysprosium 66 **Dy** 162.5	Holmium 67 **Ho** 164.9	Erbium 68 **Er** 167.3	Thulium 69 **Tm** 168.9	Ytterbium 70 **Yb** 173.0	Lutetium 71 **Lu** 175.0

‡Actinides 7

Thorium 90 **Th** 232.0	Protactinium 91 **Pa** 231.0	Uranium 92 **U** 238.0	Neptunium 93 **Np** (237)	Plutonium 94 **Pu** (244)	Americium 95 **Am** (243)	Curium 96 **Cm** (247)	Berkelium 97 **Bk** (247)	Californium 98 **Cf** (251)	Einsteinium 99 **Es** (252)	Fermium 100 **Fm** (257)	Mendelevium 101 **Md** (258)	Nobelium 102 **No** (259)	Lawrencium 103 **Lr** (262)

Values in parentheses are the mass numbers of the most stable or best-known isotopes.

Names and symbols for elements 111–114 are under review.

Key

element name —— Hydrogen
symbol of element —— 1 — atomic number
H
1.008 — atomic weight

FIGURE 8.19

The periodic table of the elements.

TABLE 8.4
Electron structures of the alkali metal family

Element	Electron Configuration	Number of Electrons in Main Energy Level						
		1st	2nd	3rd	4th	5th	6th	7th
Lithium (Li)	[He] $2s^1$	2	1	—	—	—	—	—
Sodium (Na)	[Ne] $3s^1$	2	8	1	—	—	—	—
Potassium (K)	[Ar] $4s^1$	2	8	8	1	—	—	—
Rubidium (Rb)	[Kr] $5s^1$	2	8	18	8	1	—	—
Cesium (Cs)	[Xe] $6s^1$	2	8	18	18	8	1	—
Francium (Fr)	[Rn] $7s^1$	2	8	18	32	18	8	1

TABLE 8.5
Electron structures of the noble gas family

Element	Electron Configuration	Number of Electrons in Main Energy Level						
		1st	2nd	3rd	4th	5th	6th	7th
Helium (He)	$1s^2$	2	—	—	—	—	—	—
Neon (Ne)	[He] $2s^2 2p^6$	2	8	—	—	—	—	—
Argon (Ar)	[Ne] $3s^2 3p^6$	2	8	8	—	—	—	—
Krypton (Kr)	[Ar] $4s^2 3d^{10} 4p^6$	2	8	18	8	—	—	—
Xenon (Xe)	[Kr] $5s^2 4d^{10} 5p^6$	2	8	18	18	8	—	—
Radon (Rn)	[Xe] $6s^2 4f^{14} 5d^{10} 6p^6$	2	8	18	32	18	8	—

An element is identified in each cell with its chemical symbol. The number above the symbol is the atomic number of the element, and the number below the symbol is the rounded atomic weight of the element. Horizontal rows of elements run from left to right with increasing atomic numbers. Each row is called a *period*. The periods are numbered from 1 to 7 on the left side. A vertical column of elements is called a *family* (or group) of elements. Elements in families have similar properties, but this is more true of some families than others.

As shown in Table 8.4, all of the elements in the first column have an outside electron configuration of one electron. With the exception of hydrogen, the elements of the first column are shiny, low-density metals that are so soft you can cut them easily with a knife. These metals are called the *alkali metals* because they react violently with water to form an alkaline solution. The alkali metals do not occur in nature as free elements because they are so reactive. Hydrogen is a unique element in the periodic table. It is not an alkali metal and is placed in the group because it seems to fit there because it has one electron in its outer s orbital.

The elements in the second column all have an outside configuration of two electrons and are called the *alkaline earth metals*. The alkaline earth metals are soft, reactive metals but not as reactive or soft as the alkali metals. Calcium and magnesium, in the form of numerous compounds, are familiar examples of this group.

The elements in group VIIA (17) all have an outside configuration of seven electrons, needing only one more electron to completely fill the outer (p) orbitals. These elements are called the *halogens*. The halogens are very reactive nonmetals. The halogens fluorine and chlorine are greenish-colored gases. Bromine is a reddish-brown liquid and iodine is a dark purple solid. Halogens are used as disinfectants, bleaches, and combined with a metal as a source of light in halogen lights. Halogens react with metals to form a group of chemicals called *salts*, such as sodium chloride. In fact, the word *halogen* is Greek, meaning "salt former."

As shown in Table 8.5, the elements in group VIIIA (18) have orbitals that are filled to capacity. These elements are colorless, odorless gases that almost never react with other elements to form compounds. Sometimes they are called the noble gases because they are chemically inert, perhaps indicating they are above the other elements. They have also been called the *rare gases* because of their scarcity and *inert gases* because they are mostly chemically inert, not forming compounds. The noble gases are inert because they have filled outer electron configurations, a particularly stable condition.

Each period *begins* with a single electron in a new orbital. Second, each period *ends* with the filling of an orbital, completing the maximum number of electrons that can occupy that main energy level. Since the first A family is identified as IA, this means that all the atoms of elements in this family have one electron in their outer orbitals. All the atoms of elements in family IIA have two electrons in their outer orbitals. This pattern continues on to family VIIIA, in which all the atoms of elements have eight electrons in their outer orbitals except helium. Thus, the number identifying the A families *also identifies the number of electrons in the outer orbitals,* with the exception of helium. Helium is nonetheless similar to the other elements in this family, since all have filled outer orbitals. The electron theory of chemical bonding, which is discussed in chapter 9, states that only the electrons in the outermost orbitals of an atom are involved in chemical reactions. Thus, *the outer orbital electrons are mostly responsible for the chemical properties of an element.* Since the members of a family all have similar outer configurations, you would expect them to have similar chemical behaviors, and they do.

The members of the A-group families are called the *main-group* or **representative elements.** The members of the B-group families are called the **transition elements** (or metals).

Concepts Applied

Spiral Origins

Make a periodic table. List the elements in boxes on a roll of adding machine tape. Have someone hold the tape up by the hydrogen end while you make a spiral with the noble gases appearing one below the other. Use cellophane tape to hold the noble gases family together, and then cut the tape just to the right of this group. Spread this cut spiral flat and a long form of the periodic table will be the result. Moving the inner transition elements shown in Figure 8.19 (58 to 71 and 90 to 103) will produce the familiar short form of the periodic table.

Periodic Practice

Identify the periodic table period and family of the element silicon. Write your answer before reading the solution in the next paragraph.

According to the list of elements on the inside back cover of this text, silicon has the symbol Si and an atomic number of 14. The square with the symbol Si and the atomic number 14 is located in the third period (third row) and in the column identified as IVA (14).

Now, can you identify the period and family of the element iron? Compare your answer with a classmate's to check.

Metals, Nonmetals, and Semiconductors

As indicated earlier, chemical behavior is mostly concerned with the outer orbital electrons. The outer orbital electrons, that is, the highest energy level electrons, are conveniently represented with an **electron dot notation,** made by writing the chemical symbol with dots around it indicating the number of outer orbital electrons. Electron dot notations are shown for the representative elements in Figure 8.20. Again, note the pattern in Figure 8.20—all the noble gases are in group VIIIA, and all (except helium) have eight outer electrons. All the group IA elements (alkali metals) have one dot, all the IIA elements have two dots, and so on. This pattern will explain the difference in metals, nonmetals, and a third group of in-between elements called semiconductors.

One way to group substances is according to the physical properties of metals and nonmetals—luster, conductivity, malleability, and ductility. Metals and nonmetals also have certain

chemical properties that are related to their positions in the periodic table. Figure 8.21 shows where the *metals, nonmetals,* and *semiconductors* are located. Note that about 80 percent of all the elements are metals.

The noble gases have completely filled outer orbitals in their highest energy levels, and this is a particularly stable arrangement. Other elements react chemically, either *gaining or losing electrons to attain a filled outermost energy level like the noble gases.* When an atom loses or gains electrons, it acquires an unbalanced electron charge and is called an **ion.** An atom of lithium, for example, has three protons (plus charges) and three electrons (negative charges). If it loses the outermost electron, it now has an outer filled orbital structure like helium, a noble gas. It is also now an ion, since it has three protons (3+) and two electrons (2−), for a net charge of 1+. A lithium ion thus has a 1+ charge.

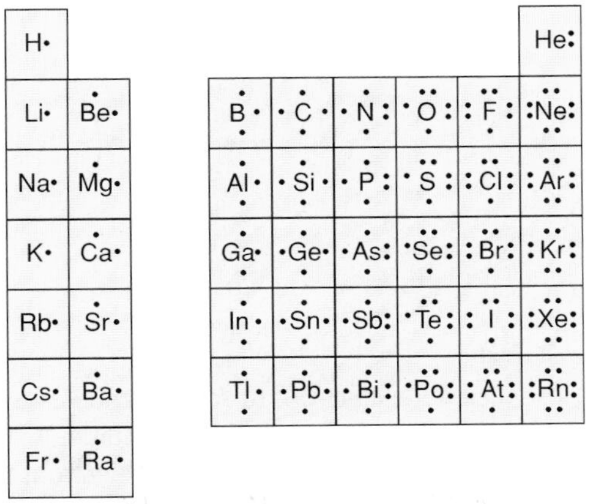

Metals and Charge

Is strontium a metal, nonmetal, or semiconductor? What is the charge on a strontium ion?

The list of elements inside the back cover identifies the symbol for strontium as Sr (atomic number 38). In the periodic table, Sr is located in family IIA, which means that an atom of strontium has two electrons in its outer orbital. For several reasons, you know that strontium is a metal: (1) An atom of strontium has two electrons in its outer orbital and atoms with one, two, or three outer electrons are identified as metals; (2) strontium is located in the IIA family, the alkaline earth metals; and (3) strontium is located on the left side of the periodic table and, in general, elements located in the left two-thirds of the table are metals.

Elements with one, two, or three outer electrons tend to lose electrons to form positive ions. Since strontium has an atomic number of 38, you know that it has thirty-eight protons (38+) and thirty-eight electrons (38−). When it loses its two outer orbital electrons, it has 38+ and 36− for a charge of 2+.

FIGURE 8.20
Electron dot notation for the representative elements.

Elements with one, two, or three outer electrons tend to lose electrons to form positive ions. The metals lose electrons like this, and the *metals are elements that lose electrons to form positive ions* (Figure 8.22). Nonmetals, on the other hand, are elements with five to seven outer electrons that tend to acquire electrons to fill their outer orbitals. *Nonmetals are elements that gain electrons to form negative ions.* In general, elements located in the left two-thirds or so of the periodic table are metals. The nonmetals are on the right side of the table (Figure 8.21).

1 H																	2 He
3 Li	4 Be											5 B	6 C	7 N	8 O	9 F	10 Ne
11 Na	12 Mg											13 Al	14 Si	15 P	16 S	17 Cl	18 Ar
19 K	20 Ca	21 Sc	22 Ti	23 V	24 Cr	25 Mn	26 Fe	27 Co	28 Ni	29 Cu	30 Zn	31 Ga	32 Ge	33 As	34 Se	35 Br	36 Kr
37 Rb	38 Sr	39 Y	40 Zr	41 Nb	42 Mo	43 Tc	44 Ru	45 Rh	46 Pd	47 Ag	48 Cd	49 In	50 Sn	51 Sb	52 Te	53 I	54 Xe
55 Cs	56 Ba	57 La	72 Hf	73 Ta	74 W	75 Re	76 Os	77 Ir	78 Pt	79 Au	80 Hg	81 Tl	82 Pb	83 Bi	84 Po	85 At	86 Rn
87 Fr	88 Ra	89 Ac	104 Rf	105 Db	106 Sg	107 Bh	108 Hs	109 Mt	110 Ds	111	112		114				

Metals

Nonmetals

Semiconductors

FIGURE 8.21
The location of metals, nonmetals, and semiconductors in the periodic table.

Concepts Applied

Outer Orbitals

How many outer orbital electrons are found in an atom of
(a) oxygen, (b) calcium, and (c) aluminum? Write your answers
before reading the answers in the next paragraph.

(a) According to the list of elements on the inside back
cover of this text, oxygen has the symbol O and an atomic
number of 8. The square with the symbol O and the atomic
number 8 is located in the column identified as VIA. Since the A
family number is the same as the number of electrons in the outer
orbital, oxygen has six outer orbital electrons. (b) Calcium has the
symbol Ca (atomic number 20) and is located in column IIA, so a
calcium atom has two outer orbital electrons. (c) Aluminum has
the symbol Al (atomic number 13) and is located in column IIIA, so
an aluminum atom has three outer orbital electrons.

The dividing line between the metals and nonmetals is a
steplike line from the left top of group IIIA down to the bottom
left of group VIIA. This is not a line of sharp separation between
the metals and nonmetals, and elements *along* this line sometimes act like metals, sometimes like nonmetals, and sometimes
like both. These hard-to-classify elements are called **semiconductors** (or *metalloids*). Silicon, germanium, and arsenic have
physical properties of nonmetals; for example, they are brittle
materials that cannot be hammered into a new shape. Yet these
elements conduct electric currents under certain conditions.
The ability to conduct an electric current is a property of a
metal, and nonmalleability is a property of nonmetals, so as you
can see, these semiconductors have the properties of both metals and nonmetals.

The transition elements, which are all metals, are located in
the B-group families. Unlike the representative elements, which
form vertical families of similar properties, the transition
elements tend to form horizontal groups of elements with
similar properties. Iron (Fe), cobalt (Co), and nickel (Ni) in
group VIIIB, for example, are three horizontally arranged
metallic elements that show magnetic properties.

A family of representative elements all form ions with the
same charge. Alkali metals, for example, all lose an electron to
form a 1+ ion. The transition elements have *variable charges*.
Some transition elements, for example, lose their one outer

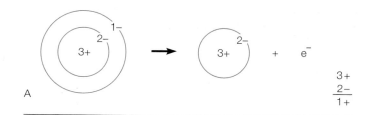

A

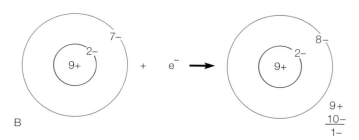

B

FIGURE 8.22
(A) Metals lose their outer electrons to acquire a noble gas
structure and become positive ions. Lithium becomes a 1+ ion as it
loses its one outer electron. (B) Nonmetals gain electrons to acquire
an outer noble gas structure and become negative ions. Fluorine
gains a single electron to become a 1− ion.

Compounds of the rare earths were first identified when they were isolated from uncommon minerals in the late 1700s. The elements are very reactive and have similar chemical properties, so they were not recognized as elements until some fifty years later. Thus, they were first recognized as earths, that is, nonmetal substances, when in fact they are metallic elements. They were also considered to be rare since, at that time, they were known to occur only in uncommon minerals. Today, these metallic elements are known to be more abundant in the earth than gold, silver, mercury, or tungsten. The rarest of the rare earths, thulium, is twice as abundant as silver. The rare earth elements are neither rare nor earths, and they are important materials in glass, electronic, and metallurgical industries.

You can identify the rare earths in the two lowest rows of the periodic table. These rows contain two series of elements that actually belong in periods 6 and 7, but they are moved below so that the entire table is not so wide. Together, the two series are called the inner transition elements. The top series is fourteen elements wide from elements 58 through 71. Since this series belongs next to element 57, lanthanum, it is sometimes called the *lanthanide series*. This series is also known as the rare earths. The second series of fourteen elements is called the *actinide series*. These are mostly the artificially prepared elements that do not occur naturally.

You may never have heard of the rare earth elements, but they are key materials in many advanced or high-technology products. Lanthanum, for example, gives glass special refractive properties and is used in optic fibers and expensive camera lenses. Samarium, neodymium, and dysprosium are used to manufacture crystals used in lasers. Samarium, ytterbium, and terbium have special magnetic properties that have made possible new electric motor designs, magnetic-optical devices in computers, and the creation of a ceramic superconductor. Other rare earth metals are also being researched for use in possible high-temperature superconductivity materials. Many rare earths are also used in metal alloys; for example, an alloy of cerium is used to make heat-resistant jet-engine parts. Erbium is also used in high-performance metal alloys. Dysprosium and holmium have neutron-absorbing properties and are used in control rods to control nuclear fission. Europium should be mentioned because of its role in making the red color of color television screens. The rare earths are relatively abundant metallic elements that play a key role in many common and high-technology applications. They may also play a key role in superconductivity research.

People Behind the Science

Erwin Schrödinger (1887–1961)

Erwin Schrödinger was an Austrian physicist who founded wave mechanics with the formulation of the Schrödinger wave equation to describe the behavior of electrons in atoms. For this achievement, he was awarded the 1993 Nobel Peace Prize with Paul Dirac (1902–1984) and Werner Heisenberg (1901–1976), who also made important advances in the theory of atomic structure.

Schrödinger was born in Vienna on August 12, 1887. His father was an oilcloth manufacturer who had studied chemistry and his mother was the daughter of a chemistry professor. Apart from a few weeks when he attended an elementary school in Innsbruck, Schrödinger received his early education from a private tutor. In 1898 he entered the Gymnasium in Vienna where he enjoyed mathematics, physics and ancient languages. He then attended the University of Vienna, specializing in physics. Schrödinger obtained his doctorate in 1910 and a year later he became an assistant in the University's Second Physics Institute. His early research ranged over many topics in experimental and theoretical physics.

During World War I, Schrödinger served as an artillery officer and then returned to his previous post at Vienna. Conditions were difficult in Austria after the war and Schrödinger decided to go to Germany in 1920. After a series of short-lived posts at Jena, Stuttgart, and Breslau, he became Professor of Physics at Zurich in 1921.

Schrödinger's most productive work was done at Zurich and it resulted in his succeeding Max Planck (1858–1947) as Professor of Theoretical Physics at Berlin in 1927. He remained there until the rise of the Nazis in 1933, when Schrödinger went to Oxford, England, where he became a fellow of Magdalen College. Homesick, he returned to Austria in 1936 to take up a post at Graz, but the Nazi takeover of Austria in 1938 placed Schrödinger in danger. The

—Continued top of next page

intervention of the Prime Minister of Ireland, Eamon de Valera (1882–1975), led to his appointment in 1939 to a post at the Institute for Advanced Studies in Dublin. Schrödinger continued work in theoretical physics there until 1956, when he returned to Austria to a chair at the University of Vienna. In the following year Schrödinger suffered a severe illness from which he never fully recovered. He died in Vienna on January 4, 1961.

The origin of Schrödinger's great discovery of wave mechanics began with the work of Louis de Broglie (1892–1960) who, in 1924, using ideas from Einstein's special theory of relativity, showed that an electron or any other particle has a wave associated with it. The fundamental result was that

$$\lambda = h/p$$

where λ is the wavelength of the associated wave, h is Planck's constant and p is the momentum of the particle. An immediate deduction from this discovery was that if particles, and particularly electrons, have waves then their behavior should be capable of description by a particular type of partial differential equation known as a wave equation in the same way as sound and other kinds of waves. These ideas were taken up by both de Broglie and Schrödinger and in 1926 each published the same wave equation which, when written in relativistic terms, is:

$$\frac{1}{c^2}\frac{\delta^2\psi}{\delta t^2} = \frac{\delta^2\psi}{\delta x^2} + \frac{\delta^2\psi}{\delta y^2} + \frac{\delta^2\psi}{\delta z^2} - \frac{4\pi m^2 c^2}{h^2}\psi$$

where ψ is the wave function, t is the time, m is the mass of the electron, c is the velocity of light, h is Planck's constant and x, y, and z represent the position of the electron in Cartesian coordinates. Unfortunately, while the equation is true, it was of very little help in developing further facts and explanations.

Later in the same year, however, Schrödinger used a new approach. After spending some time studying the mathematics of partial differential equations and using the Hamiltonian function, a powerful idea in mechanics due to William Rowan Hamilton (1805–1865), he formulated an equation in terms of the energies of the electron and the field in which it was situated. His new equation was:

$$\frac{\delta^2\psi}{\delta x^2} + \frac{\delta^2\psi}{\delta y^2} + \frac{\delta^2\psi}{\delta z^2} - \frac{8\pi^2 m}{h^2}(E - V)\psi = 0$$

where E is the total energy of the electron and V is the potential of the field in which the electron is moving. This equation neglects the small effects of special relativity. Partial differential equations have many solutions and very stringent conditions had to be fulfilled by the individual solutions of this equation in order for it to be useful in describing the electron. Among other things, they had to be finite and possess only one value. These solutions were associated with special values of E, known as proper values or eigenvalues. Schrödinger solved the equation for the hydrogen atom, where:

$$V = -e^2/r$$

e being the electron's charge and r its distance from the nucleus, and found that the values of E corresponded with those of the energy levels given in the older theory of Niels Bohr (1885–1962). Also, to each value of E there corresponded a finite number of particular solutions for the wave function ψ, and these could be associated with lines in the spectrum of atomic hydrogen. In the hydrogen atom the wave function describes where we can expect to find the electron, and it turns out that while it is most likely to be where Bohr predicted it to be, it does not follow a circular orbit but is described by the more complicated notion of an orbital, a region in space where the electron can be found with varying degrees of probability.

Atoms other than hydrogen and also molecules and ions can be described by Schrödinger's wave equation but such cases are very difficult to solve. In certain cases approximations have been used, usually with the numerical work being carried out on a computer.

Schrödinger's mathematical description of electron waves found immediate acceptance because these waves could be visualized as standing waves around the nucleus. In 1925, a year before Schrödinger published his results, a mathematical system called matrix mechanics, developed by Max Born (1882–1970) and Werner Heisenberg (1901–1976) had also succeeded in describing the structure of the atom but it was totally theoretical and gave no picture of the atom. Schrödinger's vindication of de Broglie's picture of electron waves immediately overturned matrix mechanics, though it was later shown that wave mechanics is equivalent to matrix mechanics.

During his later years, Schrödinger became increasingly worried by the way quantum mechanics, of which wave mechanics is a part, was interpreted, in particular with the probabilistic nature of the wave function. Schrödinger believed he had given a great description to the atom in the same way that Newton's laws described mechanics and Maxwell's equations described electrodynamics, only to find that the structure of the atom became increasingly more difficult to describe explicitly with each new discovery. Much of his later work was concerned with philosophy, particularly as applied to physics and the atom.

Schrödinger made a fundamental contribution to physics in finally producing a solid mathematical explanation of the quantum theory first advanced by Planck in 1900, and the subsequent structures of the atom formulated by Bohr and de Broglie.

From the Hutchinson *Dictionary of Scientific Biography*. © Research Machines plc [2003] All Rights Reserved. Helicon Publishing is a division of Research Machines.

224   **Chapter Eight:** Atoms and Periodic Properties   *8-20*

electron to form 1+ ions (copper, silver). Copper, because of its special configuration, can also lose an additional electron to form a 2+ ion. Thus, copper can form either a 1+ ion or a 2+ ion. Most transition elements have two outer s orbital electrons and lose them both to form 2+ ions (iron, cobalt, nickel), but some of these elements also have special configurations that permit them to lose more of their electrons. Thus, iron and cobalt, for example, can form either a 2+ ion or a 3+ ion. Much more can be interpreted from the periodic table, and more generalizations will be made as the table is used in the following chapters.

Summary

Attempts at understanding matter date back to ancient Greek philosophers, who viewed matter as being composed of *elements,* or simpler substances. Two models were developed that considered matter to be (1) *continuous,* or infinitely divisible, or (2) *discontinuous,* made up of particles called *atoms.*

In the early 1800s, Dalton published an *atomic theory,* reasoning that matter was composed of hard, indivisible atoms that were joined together or dissociated during chemical change.

When a good air pump to provide a vacuum was invented in 1885, *cathode rays* were observed to move from the negative terminal in an evacuated glass tube. The nature of cathode rays was a mystery. The mystery was solved in 1897 when Thomson discovered they were negatively charged particles now known as *electrons.* Thomson had discovered the first elementary particle of which atoms are made and measured their charge-to-mass ratio.

Rutherford developed a solar system model based on experiments with alpha particles scattered from a thin sheet of metal. This model had a small, massive, and positively charged *nucleus* surrounded by moving electrons. These electrons were calculated to be at a distance from the nucleus of 100,000 times the radius of the nucleus, so the volume of an atom is mostly empty space. Later, Rutherford proposed that the nucleus contained two elementary particles: *protons* with a positive charge and *neutrons* with no charge. The *atomic number* is the number of protons in an atom. Atoms of elements with different numbers of neutrons are called *isotopes.* The mass of each isotope is compared to the mass of carbon-12, which is assigned a mass of exactly 12.00 *atomic mass units.* The mass contribution of the isotopes of an element according to their abundance is called the *atomic weight* of an element. Isotopes are identified by their *mass number,* which is the sum of the number of protons and neutrons in the nucleus. Isotopes are identified by their chemical symbol with the atomic number as a subscript and the mass number as a superscript.

Bohr developed a model of the hydrogen atom to explain the characteristic *line spectra* emitted by hydrogen. His model specified that (1) electrons can move only in allowed orbits, (2) electrons do not emit radiant energy when they remain in an orbit, and (3) electrons move from one allowed orbit to another when they gain or lose energy. When an electron jumps from a higher orbit to a lower one, it gives up energy in the form of a single photon. The energy of the photon corresponds to the difference in energy between the two levels. The Bohr model worked well for hydrogen but not for other atoms.

De Broglie proposed that moving particles of matter (electrons) should have wave properties like moving particles of light (photons).

His derived equation, $\lambda = h/mv$, showed that these *matter waves* were only measurable for very small particles such as electrons. De Broglie's proposal was tested experimentally, and the experiments confirmed that electrons do have wave properties.

Schrödinger and others used the wave nature of the electron to develop a new model of the atom called *wave mechanics,* or *quantum mechanics.* This model was found to confirm exactly all the experimental data as well as predict new data. The quantum mechanical model describes the energy state of the electron in terms of quantum numbers based on the wave nature of the electron. The quantum numbers defined the *probability* of the location of an electron in terms of fuzzy regions of space called *orbitals.*

The *periodic table* has horizontal rows of elements called *periods* and vertical columns of elements called *families.* Families have the same outer orbital electron configurations and it is the electron configuration that is mostly responsible for the chemical properties of an element.

Summary of Equations

8.1

$$\text{energy} = (\text{Planck's constant})(\text{frequency})$$

$$E = hf$$

$$\text{where } h = 6.63 \times 10^{-34} \text{ J·s}$$

8.2

$$\frac{1}{\text{wavelength}} = \text{constant}\left(\frac{1}{2^2} - \frac{1}{\text{number}^2}\right)$$

$$\frac{1}{\lambda} = R\left(\frac{1}{2^2} - \frac{1}{n^2}\right)$$

$$\text{where } R = 1.097 \times 10^7 \text{ l/m}$$

8.3

$$\text{energy state of orbit number} = \frac{\text{energy state of innermost orbit}}{\text{number squared}}$$

$$E_n = \frac{E_1}{n^2}$$

$$\text{where } E_1 = -13.6 \text{ eV, and } n = 1, 2, 3, \ldots$$

8.4

$$\begin{pmatrix}\text{energy} \\ \text{of} \\ \text{photon}\end{pmatrix} = \begin{pmatrix}\text{energy state} \\ \text{of} \\ \text{higher orbit}\end{pmatrix} - \begin{pmatrix}\text{energy state} \\ \text{of} \\ \text{lower orbit}\end{pmatrix}$$

$$hf = E_H - E_L$$

$$\text{where } h = 6.63 \times 10^{-34} \text{ J·s; } E_H \text{ and } E_L \text{ must be in joules}$$

8.5

$$\text{wavelength} = \frac{\text{Planck's constant}}{(\text{mass})(\text{velocity})}$$

$$\lambda = \frac{h}{mv}$$

$$\text{where } h = 6.63 \times 10^{-34} \text{ J·s}$$

atomic mass units (p. 209)

atomic number (p. 208)

atomic weight (p. 209)

Balmer series (p. 211)

electron (p. 207)

electron dot notation (p. 221)

electron pair (p. 217)

electron volt (p. 213)

excited states (p. 212)

ground state (p. 212)

Heisenberg uncertainty principle (p. 216)

ion (p. 221)

isotopes (p. 209)

line spectrum (p. 210)

mass number (p. 209)

matter waves (p. 214)

neutron (p. 208)

nucleus (p. 208)

orbital (p. 216)

Pauli exclusion principle (p. 217)

photons (p. 210)

proton (p. 208)

quanta (p. 210)

quantum mechanics (p. 215)

representative elements (p. 220)

semiconductors (p. 222)

transition elements (p. 220)

wave mechanics (p. 215)

APPLYING THE CONCEPTS

1. The electron was discovered through experiments with
   a. radioactivity.
   b. light.
   c. matter waves.
   d. electricity.

2. Thomson was convinced that he had discovered a subatomic particle, the electron, from the evidence that
   a. the charge-to-mass ratio was the same for all materials.
   b. cathode rays could move through a vacuum.
   c. electrons were attracted toward a negatively charged plate.
   d. the charge was always 1.60×10^{-19} coulomb.

3. The existence of a tiny, massive, and positively charged nucleus was deduced from the observation that
   a. fast, massive, and positively charged alpha particles all move straight through metal foil.
   b. alpha particles were deflected by a magnetic field.
   c. some alpha particles were deflected by metal foil.
   d. None of the above is correct.

4. According to Rutherford's calculations, the volume of an atom is mostly
   a. occupied by protons and neutrons.
   b. filled with electrons.
   c. occupied by tightly bound protons, electrons, and neutrons.
   d. empty space.

5. The atomic number is the number of
   a. protons.
   b. protons plus neutrons.
   c. protons plus electrons.
   d. protons, neutrons, and electrons in an atom.

6. All neutral atoms of an element have the same
   a. atomic number.
   b. number of electrons.
   c. number of protons.
   d. All of the above are correct.

7. The main problem with a solar system model of the atom is that
   a. electrons move in circular, not elliptical orbits.
   b. the electrons should lose energy since they are accelerating.
   c. opposite charges should attract one another.
   d. the mass ratio of the nucleus to the electrons is wrong.

8. The energy of a photon
   a. varies inversely with the frequency.
   b. is directly proportional to the frequency.
   c. varies directly with the velocity, not the frequency.
   d. is inversely proportional to the velocity.

9. The frequency of a particular color of light is equal to
   a. Eh.
   b. h/E.
   c. E/h.
   d. $Eh/2$.

10. A photon of which of the following has the most energy?
    a. red light
    b. orange light
    c. green light
    d. blue light

11. The lines of color in a line spectrum from a given element
    a. change colors with changes in the temperature.
    b. are always the same, with a regular spacing pattern.
    c. are randomly spaced, having no particular pattern.
    d. have the same colors, with a spacing pattern that varies with the temperature.

12. Hydrogen, with its one electron, produces a line spectrum in the visible light range with
    a. one color line.
    b. two color lines.
    c. three color lines.
    d. four color lines.

13. Using the laws of motion for moving particles and the laws of electrical attraction, Bohr calculated that electrons could
    a. move only in orbits of certain allowed radii.
    b. move, as do the planets, in orbits at any distance from the nucleus.
    c. move in orbits at distances from the nucleus that matched the distances between colors in the line spectrum.
    d. move in orbits at variable distances from the nucleus that are directly proportional to the velocity of the electrons.

14. According to the Bohr model, an electron gains or loses energy only by
    a. moving faster or slower in an allowed orbit.
    b. jumping from one allowed orbit to another.
    c. being completely removed from an atom.
    d. jumping from one atom to another atom.

15. When an electron in a hydrogen atom jumps from an orbit farther from the nucleus to an orbit closer to the nucleus, it
    a. emits a single photon with an energy equal to the energy difference of the two orbits.
    b. emits four photons, one for each of the color lines observed in the line spectrum of hydrogen.
    c. emits a number of photons dependent on the number of orbit levels jumped over.
    d. None of the above is correct.

16. The Bohr model of the atom
    a. explained the color lines in the hydrogen spectrum.
    b. could not explain the line spectrum of atoms larger than hydrogen.
    c. had some made-up rules without explanations.
    d. All of the above are correct.

17. The proposal that matter, like light, has wave properties in addition to particle properties was
    a. verified by diffraction experiments with a beam of electrons.
    b. never tested, since it was known to be impossible.
    c. tested mathematically, but not by actual experiments.
    d. verified by physical measurement of a moving baseball.

18. The quantum mechanics model of the atom is based on
    a. the quanta, or measured amounts of energy of a moving particle.
    b. the energy of a standing electron wave that can fit into an orbit.
    c. calculations of the energy of the three-dimensional shape of a circular orbit of an electron particle.
    d. Newton's laws of motion, but scaled down to the size of electron particles.

19. The Bohr model of the atom described the energy state of electrons with one quantum number. The quantum mechanics model uses how many quantum numbers to describe the energy state of an electron?
    a. one
    b. two
    c. four
    d. ten

20. An electron in the second main energy level and the second sublevel is described by the symbols
    a. 1s.
    b. 2s.
    c. 1p.
    d. 2p.

21. The space in which it is probable that an electron will be found is described by a(an)
    a. circular orbit.
    b. elliptical orbit.
    c. orbital.
    d. geocentric orbit.

22. Two electrons can occupy the same orbital because they have different
    a. principal quantum numbers.
    b. angular momentum quantum numbers.
    c. magnetic quantum numbers.
    d. spin quantum numbers.

23. Two isotopes of the same element have
    a. the same number of protons, neutrons, and electrons.
    b. the same number of protons and neutrons, but different numbers of electrons.
    c. the same number of protons and electrons, but different numbers of neutrons.
    d. the same number of neutrons and electrons, but different numbers of protons.

24. Atomic weight is
    a. the weight of an atom in grams.
    b. the average atomic mass of the isotopes as they occur in nature.
    c. the number of protons and neutrons in the nucleus.
    d. all of the above.

25. The mass of any isotope is based on the mass of
    a. hydrogen, which is assigned the number 1 since it is the lightest element.
    b. oxygen, which is assigned a mass of 16.
    c. an isotope of carbon, which is assigned a mass of 12.
    d. its most abundant isotope as found in nature.

26. The isotopes of a given element always have
    a. the same mass and the same chemical behavior.
    b. the same mass and a different chemical behavior.
    c. different masses and different chemical behaviors.
    d. different masses and the same chemical behavior.

27. If you want to know the number of protons in an atom of a given element, you would look up the
    a. mass number.
    b. atomic number.
    c. atomic weight.
    d. abundance of isotopes compared to the mass number.

28. If you want to know the number of neutrons in an atom of a given element, you would
    a. round the atomic weight to the nearest whole number.
    b. add the mass number and the atomic number.
    c. subtract the atomic number from the mass number.
    d. add the mass number and the atomic number, then divide by two.

29. Which of the following is always a whole number?
    a. atomic mass of an isotope
    b. mass number of an isotope
    c. atomic weight of an element
    d. None of the above is correct.

30. The chemical family of elements called the noble gases is found in what column of the periodic table?
    a. IA
    b. IIA
    c. VIIA
    d. VIIIA

31. A particular element is located in column IVA of the periodic table. How many dots would be placed around the symbol of this element in its electron dot notation?

    a. 1

    b. 3

    c. 4

    d. 8

Answers

1. d 2. a 3. c 4. d 5. a 6. d 7. b 8. b 9. c 10. d 11. b 12. d 13. a 14. b 15. a
16. d 17. a 18. b 19. c 20. d 21. c 22. d 23. c 24. b 25. c 26. d 27. b 28. c
29. b 30. d 31. c

QUESTIONS FOR THOUGHT

1. Describe the experimental evidence that led Rutherford to the concept of a nucleus in an atom.
2. What is the main problem with a solar system model of the atom?
3. Compare the size of an atom to the size of its nucleus.
4. An atom has 11 protons in the nucleus. What is the atomic number? What is the name of this element? What is the electron configuration of this atom?
5. Why do the energies of electrons in an atom have negative values? (*Hint:* It is *not* because of the charge of the electron.)
6. What is similar about the Bohr model of the atom and the quantum mechanical model? What are the fundamental differences?
7. What is the difference between a hydrogen atom in the ground state and one in the excited state?
8. Which of the following are whole numbers, and which are not whole numbers? Explain why for each.

    (a) atomic number

    (b) isotope mass

    (c) mass number

    (d) atomic weight

9. Why does the carbon-12 isotope have a whole-number mass but not the other isotopes?
10. What do the members of the noble gas family have in common? What are their differences?
11. How are the isotopes of an element similar? How are they different?
12. What patterns are noted in the electron structures of elements found in a period and in a family in the periodic table?

PARALLEL EXERCISES

The exercises in groups A and B cover the same concepts. Solutions to group A exercises are located in appendix D.

Group A

1. A neutron with a mass of 1.68×10^{-27} kg moves from a nuclear reactor with a velocity of 3.22×10^3 m/s. What is the de Broglie wavelength of the neutron?
2. Calculate the energy (a) in eV and (b) in joules for the sixth energy level ($n = 6$) of a hydrogen atom.
3. How much energy is needed to move an electron in a hydrogen atom from $n = 2$ to $n = 6$? Give the answer (a) in joules and (b) in eV. (See Figure 8.11 for needed values.)
4. What frequency of light is emitted when an electron in a hydrogen atom jumps from $n = 6$ to $n = 2$? What color would you see?
5. How much energy is needed to completely remove the electron from a hydrogen atom in the ground state?
6. Thomson determined the charge-to-mass ratio of the electron to be -1.76×10^{11} coulomb/kilogram. Millikan determined the charge on the electron to be -1.60×10^{-19} coulomb. According to these findings, what is the mass of an electron?
7. Assume that an electron wave making a standing wave in a hydrogen atom has a wavelength of 1.67×10^{-10} m. Considering the mass of an electron to be 9.11×10^{-31} kg, use the de Broglie equation to calculate the velocity of an electron in this orbit.
8. Using any reference you wish, write the complete electron configurations for (a) boron, (b) aluminum, and (c) potassium.
9. Explain how you know that you have the correct *total* number of electrons in your answers for 8a, 8b, and 8c.

Group B

1. An electron with a mass of 9.11×10^{-31} kg has a velocity of 4.3×10^6 m/s in the innermost orbit of a hydrogen atom. What is the de Broglie wavelength of the electron?
2. Calculate the energy (a) in eV and (b) in joules of the third energy level ($n = 3$) of a hydrogen atom.
3. How much energy is needed to move an electron in a hydrogen atom from the ground state ($n = 1$) to $n = 3$? Give the answer (a) in joules and (b) in eV.
4. What frequency of light is emitted when an electron in a hydrogen atom jumps from $n = 2$ to the ground state ($n = 1$)?
5. How much energy is needed to completely remove an electron from $n = 2$ in a hydrogen atom?
6. If the charge-to-mass ratio of a proton is 9.58×10^7 coulomb/kilogram and the charge is 1.60×10^{-19} coulomb, what is the mass of the proton?
7. An electron wave making a standing wave in a hydrogen atom has a wavelength of 8.33×10^{-11} m. If the mass of the electron is 9.11×10^{-31} kg, what is the velocity of the electron according to the de Broglie equation?
8. Using any reference you wish, write the complete electron configurations for (a) nitrogen, (b) phosphorus, and (c) chlorine.
9. Explain how you know that you have the correct *total* number of electrons in your answers for 8a, 8b, and 8c.

10. Refer to Figure 8.18 *only*, and write the complete electron configurations for (a) argon, (b) zinc, and (c) bromine.

11. Lithium has two naturally occurring isotopes: lithium-6 and lithium-7. Lithium-6 has a mass of 6.01512 relative to carbon-12 and makes up 7.42 percent of all naturally occurring lithium. Lithium-7 has a mass of 7.016 compared to carbon-12 and makes up the remaining 92.58 percent. According to this information, what is the atomic weight of lithium?

12. Identify the number of protons, neutrons, and electrons in the following isotopes:
    (a) $^{12}_{6}C$
    (b) $^{1}_{1}H$
    (c) $^{40}_{18}Ar$
    (d) $^{2}_{1}H$
    (e) $^{197}_{79}Au$
    (f) $^{235}_{92}U$

13. Identify the period and the family in the periodic table for the following elements:
    (a) Radon
    (b) Sodium
    (c) Copper
    (d) Neon
    (e) Iodine
    (f) Lead

14. How many outer-orbital electrons are found in an atom of
    (a) Li
    (b) N
    (c) F
    (d) Cl
    (e) Ra
    (f) Be

15. Write electron dot notations for the following elements:
    (a) Boron
    (b) Bromine
    (c) Calcium
    (d) Potassium
    (e) Oxygen
    (f) Sulfur

16. Identify the charge on the following ions:
    (a) Boron
    (b) Bromine
    (c) Calcium
    (d) Potassium
    (e) Oxygen
    (f) Nitrogen

17. Use the periodic table to identify if the following are metals, nonmetals, or semiconductors:
    (a) Krypton
    (b) Cesium
    (c) Silicon
    (d) Sulfur
    (e) Molybdenum
    (f) Plutonium

10. Referring to Figure 8.18 *only*, write the complete electron configuration for (a) neon, (b) sulfur, and (c) calcium.

11. Boron has two naturally occurring isotopes, boron-10 and boron-11. Boron-10 has a mass of 10.0129 relative to carbon-12 and makes up 19.78 percent of all naturally occurring boron. Boron-11 has a mass of 11.00931 compared to carbon-12 and makes up the remaining 80.22 percent. What is the atomic weight of boron?

12. Identify the number of protons, neutrons, and electrons in the following isotopes:
    (a) $^{14}_{7}N$
    (b) $^{7}_{3}Li$
    (c) $^{35}_{17}Cl$
    (d) $^{48}_{20}Ca$
    (e) $^{63}_{29}Cu$
    (f) $^{230}_{92}U$

13. Identify the period and the family in the periodic table for the following elements:
    (a) Xenon
    (b) Potassium
    (c) Chromium
    (d) Argon
    (e) Bromine
    (f) Barium

14. How many outer-orbital electrons are found in an atom of
    (a) Na
    (b) P
    (c) Br
    (d) I
    (e) Te
    (f) Sr

15. Write electron dot notations for the following elements:
    (a) Aluminum
    (b) Fluorine
    (c) Magnesium
    (d) Sodium
    (e) Carbon
    (f) Chlorine

16. Identify the charge on the following ions:
    (a) Aluminum
    (b) Chlorine
    (c) Magnesium
    (d) Sodium
    (e) Sulfur
    (f) Hydrogen

17. Use the periodic table to identify if the following are metals, nonmetals, or semiconductors:
    (a) Radon
    (b) Francium
    (c) Arsenic
    (d) Phosphorus
    (e) Hafnium
    (f) Uranium

18. From their charges, predict the periodic table family number for the following ions:
    (a) Br^{-1}
    (b) K^{+1}
    (c) Al^{+3}
    (d) S^{-2}
    (e) Ba^{+2}
    (f) O^{-2}

19. Use chemical symbols and numbers to identify the following isotopes:
    (a) Oxygen-16
    (b) Sodium-23
    (c) Hydrogen-3 $^{3}_{1}H = 1p, 2n$
    (d) Chlorine-35

18. From their charges, predict the periodic table family number for the following ions:
    (a) F^{-1}
    (b) Li^{+1}
    (c) B^{+3}
    (d) O^{-2}
    (e) Be^{+2}
    (f) Si^{+4}

19. Use chemical symbols and numbers to identify the following isotopes:
    (a) Potassium-39
    (b) Neon-22
    (c) Tungsten-184
    (d) Iodine-127

A chemical change occurs when iron rusts, and rust is a different substance with different physical and chemical properties than iron. This rusted anchor makes a colorful display on the bow of a grain ship.

CHAPTER

9

Chemical Bonds

In the previous chapter, you learned how the modern atomic theory is used to describe the structures of atoms of different elements. The electron structures of different atoms successfully account for the position of elements in the periodic table as well as for groups of elements with similar properties. On a large scale, all metals were found to have a similarity in electron structure, as were nonmetals. On a smaller scale, chemical families such as the alkali metals were found to have the same outer electron configurations. Thus, the modern atomic theory accounts for observed similarities between elements in terms of atomic structure.

So far, only individual, isolated atoms have been discussed; we have not considered how atoms of elements join together to produce chemical compounds. There is a relationship between the electron structure of atoms and the reactions they undergo to produce specific compounds. Understanding this relationship will explain the changes that matter itself undergoes. For example, hydrogen is a highly flammable, gaseous element that burns with an explosive reaction. Oxygen, on the other hand, is a gaseous element that supports burning. As you know, hydrogen and oxygen combine to form water. Water is a liquid that neither burns nor supports burning. What happens when atoms of elements such as hydrogen and oxygen join to form molecules such as water? Why do such atoms join and why do they stay together? Why does water have different properties from the elements that combine to produce it? And finally, why is water H_2O and not H_3O or H_4O?

Answers to questions about why and how atoms join together in certain numbers are provided by considering the electronic structures of the atoms. Chemical substances are formed from the interactions of electrons as their structures merge, forming new patterns that result in molecules with new properties. It is the new electron pattern of the water molecule that gives water different properties than the oxygen or hydrogen from which it formed (Figure 9.1). Understanding how electron structures of atoms merge to form new patterns is understanding the changes that matter itself undergoes, the topic of this chapter.

Compounds and Chemical Change

There are more than 100 elements listed in the periodic table, and all matter on the earth is made of these elements. However, very few pure elements are found in your surroundings. The air you breathe, the liquids you drink, and all the other things around you are mostly *compounds,* substances made up of combinations of elements. Water, sugar, gasoline, and chalk are examples of compounds and each can be broken down into the elements that make it up. Examples of elements are hydrogen, carbon, and calcium. Why and how these elements join together in different ways to form the different compounds that make up your surroundings is the subject of this chapter.

You have already learned that elements are made up of atoms that can be described by the modern atomic theory. You can also consider an **atom** to be *the smallest unit of an element that can exist alone or in combination with other elements.* Compounds are formed when atoms are held together by an attractive force called a *chemical bond.* The chemical bond binds individual atoms together in a compound. A molecule is generally thought of as a tightly bound group of atoms that maintains its identity. More specifically, a **molecule** is defined as *the smallest particle of a compound, or a gaseous element, that can exist and still retain the characteristic chemical properties of a substance.* Compounds with one type of chemical bond, as you will see, have molecules that are electrically neutral groups of atoms held together strongly enough to be considered independent units. For example, water is a

compound. The smallest unit of water that can exist alone is an electrically neutral unit made up of two hydrogen atoms and one oxygen atom held together by chemical bonds. The concept of a molecule will be expanded as chemical bonds are discussed.

Compounds occur naturally as gases, liquids, and solids. Many common gases occur naturally as molecules made up of two or more atoms. For example, at ordinary temperatures hydrogen gas occurs as molecules of two hydrogen atoms bound together. Oxygen gas also usually occurs as molecules of two oxygen atoms bound together. Both hydrogen and oxygen occur naturally as *diatomic molecules* ("di-" means "two"). Oxygen sometimes occurs as molecules of three oxygen atoms bound together. These *triatomic molecules* ("tri-" means "three") are called *ozone.* The noble gases are unique, occurring as single atoms called *monatomic* ("mon-" or "mono-" means "one") (Figure 9.2). These monatomic particles are sometimes called *monatomic molecules* since they are the smallest units of the noble gases that can exist alone. Helium and neon are examples of the monatomic noble gases.

When molecules of any size are formed or broken down into simpler substances, new materials with new properties are produced. This kind of a change in matter is called a chemical change, and the process is called a chemical reaction. A **chemical reaction** is defined as

a change in matter in which different chemical substances are created by forming or breaking chemical bonds.

FIGURE 9.1

Water is the most abundant liquid on the earth and is necessary for all life. Because of water's great dissolving properties, any sample is a solution containing solids, other liquids, and gases from the environment. This stream also carries suspended, ground-up rocks, called *rock flour,* from a nearby glacier.

In general, chemical bonds are formed when atoms of elements are bound together to form compounds. Chemical bonds are broken when a compound is decomposed into simpler substances. Chemical bonds are electrical in nature, formed by electrical attractions, as discussed in chapter 6.

Chemical reactions happen all the time, all around you. A growing plant, burning fuels, and your body's utilization of food all involve chemical reactions. These reactions produce different chemical substances with greater or smaller amounts of internal potential energy (see chapter 4 for a discussion of internal potential energy). Energy is *absorbed* to produce new chemical substances with more internal potential energy.

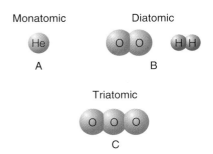

FIGURE 9.2

(*A*) The noble gases are monatomic, occurring as single atoms. (*B*) Many gases, such as hydrogen and oxygen, are diatomic, with two atoms per molecule. (*C*) Ozone is a form of oxygen that is triatomic, occurring with three atoms per molecule.

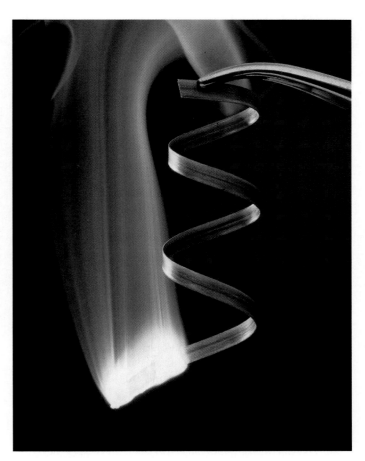

FIGURE 9.3

Magnesium is an alkaline earth metal that burns brightly in air, releasing heat and light. As chemical energy is released, new chemical substances are formed. One new chemical material produced here is magnesium oxide, a soft, powdery material that forms an alkaline solution in water (called *milk of magnesia*).

Energy is *released* when new chemical substances are produced with less internal potential energy (Figure 9.3). In general, changes in internal potential energy are called **chemical energy.** For example, new chemical substances are produced in green plants through the process called *photosynthesis*. A green plant uses radiant energy (sunlight), carbon dioxide, and water to produce new chemical materials and oxygen. These new chemical materials, the stuff that leaves, roots, and wood are made of, contain more chemical energy than the carbon dioxide and water they were made from.

A **chemical equation** is a way of describing what happens in a chemical reaction. Later, you will learn how to use formulas in a chemical reaction. For now, the chemical reaction of photosynthesis will be described by using words in an equation:

energy (sunlight)	+	carbon dioxide molecules	+	water molecules	→	plant material molecules	+	oxygen molecules

The substances that are changed are on the left side of the word equation and are called *reactants.* The reactants are carbon dioxide molecules and water molecules. The equation also indicates that energy is absorbed, since the term *energy* appears on the left side. The arrow means *yields*. The new chemical substances are on the right side of the word equation and are called *products*. Reading the photosynthesis reaction as a sentence you would say, "Carbon dioxide and water use energy to react, yielding plant materials and oxygen."

The plant materials produced by the reaction have more internal potential energy, also known as *chemical energy,* than the reactants. You know this from the equation because the term *energy* appears on the left side but not the right. This means that the energy on the left went into internal potential energy on the right. You also know this because the reaction can be reversed to release the stored energy (Figure 9.4). When plant materials (such as wood) are burned, the materials react with oxygen, and chemical energy is released in the form of radiant energy (light) and high kinetic energy of the newly formed gases and vapors. In words,

plant
material + oxygen → carbon + water + energy
molecules   molecules   dioxide   molecules
             molecules

If you compare the two equations, you will see that burning is the opposite of the process of photosynthesis! The energy released in burning is exactly the same amount of solar energy that was stored as internal potential energy by the plant. Such chemical changes, in which chemical energy is stored in one reaction and released by another reaction, are the result of the making, then the breaking, of chemical bonds. Chemical bonds were formed by utilizing energy to produce new chemical

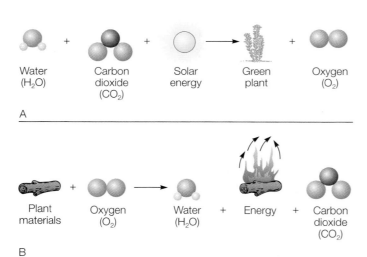

Water (H$_2$O) + Carbon dioxide (CO$_2$) + Solar energy → Green plant + Oxygen (O$_2$)

A

Plant materials + Oxygen (O$_2$) → Water (H$_2$O) + Energy + Carbon dioxide (CO$_2$)

B

FIGURE 9.4

(*A*) New chemical bonds are formed as a green plant makes new materials and stores solar energy through the photosynthesis process. (*B*) The chemical bonds are later broken, and the same amount of energy and the same original materials are released. The same energy and the same materials are released rapidly when the plant materials burn, and they are released slowly when the plant decomposes.

substances. Energy was released when these bonds were broken to produce the original substances. In this example, chemical reactions and energy flow can be explained by the making and breaking of chemical bonds. Chemical bonds can be explained in terms of changes in the electron structures of atoms. Thus, the place to start in seeking understanding about chemical reactions is the electron structure of the atoms themselves.

Valence Electrons and Ions

As discussed in chapter 8, it is the number of electrons in the outermost orbital that usually determines the chemical properties of an atom. These outer electrons are called **valence electrons,** and it is the valence electrons that participate in chemical bonding. The inner electrons are in stable, fully occupied orbitals and do not participate in chemical bonds. The representative elements (the A-group families) have valence electrons in the outermost orbitals, which contain from one to eight valence electrons. Recall that you can easily find the number of valence electrons by referring to a periodic table. The number at the top of each representative family is the same as the number of outer orbital electrons (with the exception of helium).

The noble gases have filled outer orbitals and do not normally form compounds. Apparently, half-filled and filled orbitals are particularly stable arrangements. Atoms have a tendency to seek such a stable, filled outer orbital arrangement such as the one found in the noble gases. For the representative elements, this tendency is called the **octet rule.** The octet rule states that *atoms attempt to acquire an outer orbital with eight electrons* through chemical reactions. This rule is a generalization, and a few elements do not meet the requirement of eight electrons but do seek the same general trend of stability. There are a few other exceptions, and the octet rule should be considered a generalization that helps keep track of the valence electrons in most representative elements.

The family number of the representative element in the periodic table tells you the number of valence electrons and what the atom must do to reach the stability suggested by the octet rule. For example, consider sodium (Na). Sodium is in family IA, so it has one valence electron. If the sodium atom can get rid of this outer valence electron through a chemical reaction, it will have the same outer electron configuration as an atom of the noble gas neon (Ne) (compare Figure 9.5B and 9.5C).

When a sodium atom (Na) loses an electron to form a sodium ion (Na$^+$), it has the same, stable outer electron configuration as a neon atom (Ne). The sodium ion (Na$^+$) is still a form of sodium since it still has eleven protons. But it is now a sodium *ion*, not a sodium *atom*, since it has eleven protons (eleven positive charges) and now has ten electrons (ten negative charges) for a total of

$$\begin{array}{r} 11 + \text{(protons)} \\ \underline{10 - \text{(electrons)}} \\ 1 + \text{(net charge on sodium ion)} \end{array}$$

This charge is shown on the chemical symbol of Na$^+$ for the *sodium ion*. Note that the sodium nucleus and the inner orbitals

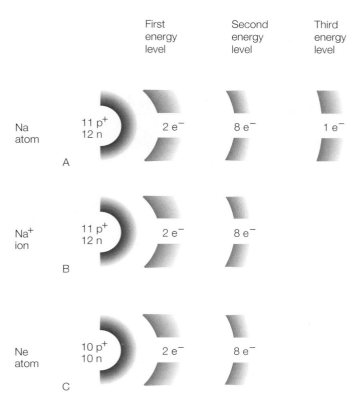

| First energy level | Second energy level | Third energy level |

Na atom 11 p^+ 12 n 2 e^- 8 e^- 1 e^-

A

Na^+ ion 11 p^+ 12 n 2 e^- 8 e^-

B

Ne atom 10 p^+ 10 n 2 e^- 8 e^-

C

FIGURE 9.5

(*A*) A sodium atom has two electrons in the first energy level, eight in the second energy level, and one in the third level. (*B*) When it loses its one outer, or valence, electron, it becomes a sodium ion with the same electron structure as an atom of neon (*C*).

do not change when the sodium atom is ionized. The sodium ion is formed when a sodium atom loses its valence electron, and the process can be described by

$$\text{energy} \quad + \quad \text{Na} \cdot \quad \longrightarrow \quad Na^+ \quad + \quad e^-$$

where Na · is the electron dot symbol for sodium, and the e^- is the electron that has been pulled off the sodium atom.

EXAMPLE 9.1

What is the symbol and charge for a calcium ion?

Solution

From the list of elements on the inside back cover, the symbol for calcium is Ca, and the atomic number is 20. The periodic table tells you that Ca is in family IIA, which means that calcium has 2 valence electrons. According to the octet rule, the calcium ion must lose 2 electrons to acquire the stable outer arrangement of the noble gases. Since the atomic number is 20, a calcium atom has 20

protons (20+) and 20 electrons (20−). When it is ionized, the calcium ion will lose 2 electrons for a total charge of (20+) + (18−), or 2+. The calcium ion is represented by the chemical symbol for calcium and the charge shown as a superscript: Ca^{2+}.

EXAMPLE 9.2

What is the symbol and charge for an aluminum ion? (Answer: Al^{3+})

Chemical Bonds

Atoms gain or lose electrons through a chemical reaction to achieve a state of lower energy, the stable electron arrangement of the noble gas atoms. Such a reaction results in a **chemical bond**, an *attractive force that holds atoms together in a compound.* There are three general classes of chemical bonds: (1) ionic bonds, (2) covalent bonds, and (3) metallic bonds.

Ionic bonds are formed when atoms *transfer* electrons to achieve the noble gas electron arrangement. Electrons are given up or acquired in the transfer, forming positive and negative ions. The electrostatic attraction between oppositely charged ions forms ionic bonds, and ionic compounds are the result. In general, ionic compounds are formed when a metal from the left side of the periodic table reacts with a nonmetal from the right side.

Covalent bonds result when atoms achieve the noble gas electron structure by *sharing* electrons. Covalent bonds are generally formed between the nonmetallic elements on the right side of the periodic table.

Metallic bonds are formed in solid metals such as iron, copper, and the other metallic elements that make up about 80 percent of all the elements. The atoms of metals are closely packed and share many electrons in a "sea" that is free to move throughout the metal, from one metal atom to the next. Metallic bonding accounts for metallic properties such as high electrical conductivity.

Ionic, covalent, and metallic bonds are attractive forces that hold atoms or ions together in molecules and crystals. There are two ways to describe what happens to the electrons when one of these bonds is formed, by considering (1) the new patterns formed when atomic orbitals overlap to form a combined orbital called a *molecular orbital* or (2) the atoms in a molecule as *isolated atoms* with changes in their outer shell arrangements. The molecular orbital description considers that the electrons belong to the whole molecule and form a molecular orbital with its own shape, orientation, and energy levels. The isolated atom description considers the electron energy levels as if the atoms in the molecule were alone, isolated from the molecule. The isolated atom description is less accurate than the molecular orbital description, but it is less complex and more easily understood. Thus, the following details about chemical bonding will mostly consider individual atoms and ions in compounds.

Ionic Bonds

An **ionic bond** is defined as the *chemical bond of electrostatic attraction* between negative and positive ions. Ionic bonding occurs when an atom of a metal reacts with an atom of a nonmetal. The reaction results in a transfer of one or more valence electrons from the metal atom to the valence shell of the nonmetal atom. The atom that loses electrons becomes a positive ion, and the atom that gains electrons becomes a negative ion. Oppositely charged ions attract one another, and when pulled together, they form an ionic solid with the ions arranged in an orderly geometric structure (Figure 9.6). This results in a crystalline solid that is typical of salts such as sodium chloride (Figure 9.7).

As an example of ionic bonding, consider the reaction of sodium (a soft reactive metal) with chlorine (a pale yellow-green gas). When an atom of sodium and an atom of chlorine collide, they react violently as the valence electron is transferred from the sodium to the chlorine atom. This produces a sodium ion and a chlorine ion. The reaction can be illustrated with electron dot symbols as follows:

$$Na \cdot \ + \ \cdot \ddot{\underset{..}{Cl}} : \ \longrightarrow \ Na^+ \ (: \ddot{\underset{..}{Cl}} :)^-$$

As you can see, the sodium ion transferred its valence electron, and the resulting ion now has a stable electron configuration. The chlorine atom accepted the electron in its outer orbital

FIGURE 9.7
You can clearly see the cubic structure of these ordinary table salt crystals because they have been magnified about ten times.

to acquire a stable electron configuration. Thus, a stable positive ion and a stable negative ion are formed. Because of opposite electrical charges, the ions attract each other to produce an ionic bond. When many ions are involved, each Na^+ ion is surrounded by six Cl^- ions, and each Cl^- ion is surrounded by six Na^+ ions. This gives the resulting solid NaCl its crystalline cubic structure, as shown in Figure 9.7. In the solid state, all the sodium ions and all the chlorine ions are bound together in one giant unit. Thus, the term *molecule* is not really appropriate for ionic solids such as sodium chloride. But the term is sometimes used anyway, since any given sample will have the same number of Na^+ ions as Cl^- ions.

Energy and Electrons in Ionic Bonding

The sodium ions and chlorine ions in a crystal of sodium chloride can be formed from separated sodium and chlorine atoms. The energy involved in such a sodium-chlorine reaction can be assumed to consist of three separate reactions:

1. $\quad$ energy $\ + \ Na \cdot \longrightarrow Na^+ \ + \ e^-$

2. $\quad \cdot \ddot{\underset{..}{Cl}} : \ + \ e^- \ \longrightarrow \ (: \ddot{\underset{..}{Cl}} :)^- + \ energy$

3. $\quad Na^+ \ + \ (: \ddot{\underset{..}{Cl}} :)^- \longrightarrow Na^+ \ (: \ddot{\underset{..}{Cl}} :)^- + \ energy$

The overall effect is that energy is released and an ionic bond is formed. The energy released is called the **heat of formation.** It is also the amount of energy required to decompose the compound (sodium chloride) into its elements. The reaction does not take place in steps as described, however, but occurs all at once. Note again, as in the photosynthesis-burning reactions described earlier, that the total amount of chemical energy is conserved. The energy released by the formation of the sodium chloride compound is the *same* amount of energy needed to decompose the compound.

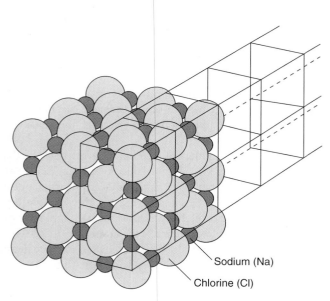

Sodium (Na)
Chlorine (Cl)

FIGURE 9.6
Sodium chloride crystals are composed of sodium and chlorine ions held together by electrostatic attraction. A crystal builds up, giving the sodium chloride crystal a cubic structure.

Ionic bonds are formed by electron transfer, and electrons are conserved in the process. This means that electrons are not created or destroyed in a chemical reaction. The same total number of electrons exists after a reaction that existed before the reaction. There are two rules you can use for keeping track of electrons in ionic bonding reactions:

1. Ions are formed as atoms gain or lose valence electrons to achieve the stable noble gas structure.
2. There must be a balance between the number of electrons lost and the number of electrons gained by atoms in the reaction.

The sodium-chlorine reaction follows these two rules. The loss of one valence electron from a sodium atom formed a stable sodium ion. The gain of one valence electron by the chlorine atom formed a stable chlorine ion. Thus, both ions have noble gas configurations (rule 1), and one electron was lost and one was gained, so there is a balance in the number of electrons lost and the number gained (rule 2).

Ionic Compounds and Formulas

The **formula** of a compound *describes what elements are in the compound and in what proportions.* Sodium chloride contains one positive sodium ion for each negative chlorine ion. The formula of the compound sodium chloride is NaCl. If there are no subscripts at the lower right part of each symbol, it is understood that the symbol has a number "1." Thus, NaCl indicates a compound made up of the elements sodium and chlorine, and there is one sodium atom for each chlorine atom.

Calcium (Ca) is an alkaline metal in family IIA, and fluorine (F) is a halogen in family VIIA. Since calcium is a metal and fluorine is a nonmetal, you would expect calcium and fluorine atoms to react, forming a compound with ionic bonds. Calcium must lose two valence electrons to acquire a noble gas configuration. Fluorine needs one valence electron to acquire a noble gas configuration. So calcium needs to lose two electrons and fluorine needs to gain one electron to achieve a stable configuration (rule 1). Two fluorine atoms, each acquiring one electron, are needed to balance the number of electrons lost and the number of electrons gained. The compound formed from the reaction, calcium fluoride, will therefore have a calcium ion with a charge of plus two for every fluorine ion with a charge of minus one. Recalling that electron dot symbols show only the outer valence electrons, you can see that the reaction is

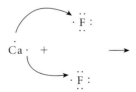

which shows that a calcium atom transfers two electrons, one each to two fluorine atoms. Now showing the results of the reaction, a calcium ion is formed from the loss of two electrons (charge $2+$) and two fluorine ions are formed by gaining one electron each (charge $1-$):

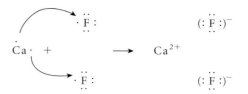

The formula of the compound is therefore CaF_2, with the subscript 2 for fluorine and the understood subscript 1 for calcium. This means that there are two fluorine atoms for each calcium atom in the compound.

Sodium chloride (NaCl) and magnesium fluoride (MgF_2) are examples of compounds held together by ionic bonds. Such compounds are called **ionic compounds.** Ionic compounds of the representative elements are generally white, crystalline solids that form colorless solutions. Sodium chloride, the most common example, is common table salt. Many of the transition elements form colored compounds that make colored solutions. Ionic compounds dissolve in water, producing a solution of ions that can conduct an electric current.

In general, the elements in families IA and IIA of the periodic table tend to form positive ions by losing electrons. The ion charge for these elements equals the family number of these elements. The elements in families VIA and VIIA tend to form negative ions by gaining electrons. The ion charge for these elements equals their family number minus 8. The elements in families IIIA and VA have less of a tendency to form ionic compounds, except for those in higher periods. Common ions of representative elements are given in Table 9.1. The transition elements form positive ions of several different charges. Some common ions of the transition elements are listed in Table 9.2.

The single-charge representative elements and the variable-charge transition elements form single, monatomic negative ions. There are also many polyatomic (*poly* means "many")

TABLE 9.1
Common ions of some representative elements

Element	Symbol	Ion
Lithium	Li	1+
Sodium	Na	1+
Potassium	K	1+
Magnesium	Mg	2+
Calcium	Ca	2+
Barium	Ba	2+
Aluminum	Al	3+
Oxygen	O	2−
Sulfur	S	2−
Hydrogen	H	1+,1−
Fluorine	F	1−
Chlorine	Cl	1−
Bromine	Br	1−
Iodine	I	1−

TABLE 9.2
Common ions of some transition elements

Single-Charge Ions

Element	Symbol	Charge
Zinc	Zn	2+
Tungsten	W	6+
Silver	Ag	1+
Cadmium	Cd	2+

Variable-Charge Ions

Element	Symbol	Charge
Chromium	Cr	2+,3+,6+
Manganese	Mn	2+,4+,7+
Iron	Fe	2+,3+
Cobalt	Co	2+,3+
Nickel	Ni	2+,3+
Copper	Cu	1+,2+
Tin	Sn	2+,4+
Gold	Au	1+,3+
Mercury	Hg	1+,2+
Lead	Pb	2+,4+

TABLE 9.3
Some common polyatomic ions

Ion Name	Formula
Acetate	$(C_2H_3O_2)^-$
Ammonium	$(NH_4)^+$
Borate	$(BO_3)^{3-}$
Carbonate	$(CO_3)^{2-}$
Chlorate	$(ClO_3)^-$
Chromate	$(CrO_4)^{2-}$
Cyanide	$(CN)^-$
Dichromate	$(Cr_2O_7)^{2-}$
Hydrogen carbonate (or bicarbonate)	$(HCO_3)^-$
Hydrogen sulfate (or bisulfate)	$(HSO_4)^-$
Hydroxide	$(OH)^-$
Hypochlorite	$(ClO)^-$
Nitrate	$(NO_3)^-$
Nitrite	$(NO_2)^-$
Perchlorate	$(ClO_4)^-$
Permanganate	$(MnO_4)^-$
Phosphate	$(PO_4)^{3-}$
Phosphite	$(PO_3)^{3-}$
Sulfate	$(SO_4)^{2-}$
Sulfite	$(SO_3)^{2-}$

negative ions, charged groups of atoms that act like a single unit in ionic compounds. Polyatomic ions are listed in Table 9.3.

EXAMPLE 9.3

Use electron dot notation to predict the formula of a compound formed when aluminum (Al) combines with fluorine (F).

Solution

Aluminum, atomic number 13, is in family IIIA so it has three valence electrons and an electron dot notation of

$$\dot{A}l\cdot$$

According to the octet rule, the aluminum atom would need to lose three electrons to acquire the stable noble gas configuration. Fluorine, atomic number 9, is in family VIIA so it has seven valence electrons and an electron dot notation of

$$\cdot\ddot{\underset{..}{F}}:$$

Fluorine would acquire a noble gas configuration by accepting one electron. Three fluorine atoms, each acquiring one electron, are needed to balance the three electrons lost by aluminum. The reaction can be represented as

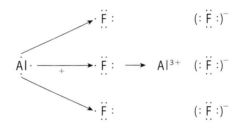

The ratio of aluminum atoms to fluorine atoms in the compound is 1:3. The formula for aluminum fluoride is therefore AlF_3.

EXAMPLE 9.4

Predict the formula of the compound formed between aluminum and oxygen using electron dot notation. (Answer: Al_2O_3).

Covalent Bonds

Most substances do not have the properties of ionic compounds since they are not composed of ions. Most substances are molecular, composed of electrically neutral groups of atoms that are tightly bound together. As noted earlier, many gases are diatomic, occurring naturally as two atoms bound together as an electrically neutral molecule. Hydrogen, for example, occurs as molecules of H_2 and no ions are involved. The hydrogen atoms are held together by a covalent bond. A **covalent bond** is a *chemical bond formed by the sharing of a pair of electrons.* In the diatomic hydrogen molecule each hydrogen atom contributes a single electron to the shared pair. Both hydrogen atoms count the shared pair of electrons in achieving their noble gas configuration. Hydrogen atoms both share one pair of electrons, but other elements might share more than one pair to achieve a noble gas structure.

Consider how the covalent bond forms between two hydrogen atoms by imagining two hydrogen atoms moving toward one another. Each atom has a single electron. As the atoms move closer and closer together, their orbitals begin to overlap. Each electron is attracted to the oppositely charged nucleus of the other atom and the overlap tightens. Then the repulsive forces from the like-charged nuclei will halt the merger. A state of stability is reached between the two nuclei and two electrons, and an H_2 molecule has been formed. The two electrons are now shared by both atoms, and the attraction of one nucleus for the other electron and vice versa holds the atoms together (Figure 9.8).

Covalent Compounds and Formulas

Electron dot notation can be used to represent the formation of covalent bonds. For example, the joining of two hydrogen atoms to form an H_2 molecule can be represented as

$$H \cdot \ + \ H \cdot \longrightarrow H : H$$

Since an electron pair is *shared* in a covalent bond, the two electrons move throughout the entire molecular orbital. Since each

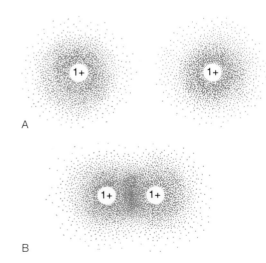

A

B

FIGURE 9.8

(*A*) Two hydrogen atoms, each with its own probability distribution of electrons about the nucleus. (*B*) When the hydrogen atoms bond, a new electron distribution pattern forms around the entire molecule, and both electrons occupy the molecular orbital.

hydrogen atom now has both electrons on an equal basis, each can be considered to now have the noble gas configuration of helium. A dashed circle around each symbol shows that both atoms have two electrons:

$$H \cdot \ + \ H \cdot \longrightarrow \left(H \!:\! H \right)$$

Hydrogen and fluorine react to form a covalent molecule (how this is known will be discussed shortly), and this bond can be represented with electron dots. Fluorine is in the VIIA family, so you know an atom of fluorine has seven valence electrons in the outermost energy level. The reaction is

$$H \cdot \ + \ \cdot \ddot{\ddot{F}} : \longrightarrow \left(H \!:\! \ddot{\ddot{F}} : \right)$$

Each atom shares a pair of electrons to achieve a noble gas configuration. Hydrogen achieves the helium configuration, and fluorine achieves the neon configuration. All the halogens have seven valence electrons and all need to gain one electron (ionic bond) or share an electron pair (covalent bond) to achieve a noble gas configuration. This also explains why the halogen gases occur as diatomic molecules. Two fluorine atoms can achieve a noble gas configuration by sharing a pair of electrons:

$$\cdot \ddot{\ddot{F}} : \ + \ \cdot \ddot{\ddot{F}} : \longrightarrow \left(: \ddot{\ddot{F}} \!:\! \ddot{\ddot{F}} : \right)$$

Each fluorine atom thus achieves the neon configuration by bonding together. Note that there are two types of electron pairs: (1) orbital pairs and (2) bonding pairs. Orbital pairs are not shared, since they are the two electrons in an orbital, each with a separate spin. Orbital pairs are also called *lone pairs,* since they are not shared. *Bonding pairs,* as the name implies, are the electron pairs shared between two atoms. Considering again the F_2 molecule,

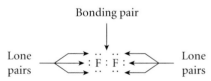

Often, the number of bonding pairs that are formed by an atom is the same as the number of single, *unpaired* electrons in the atomic electron dot notation. For example, hydrogen has one unpaired electron, and oxygen has two unpaired electrons. Hydrogen and oxygen combine to form an H_2O molecule, as follows:

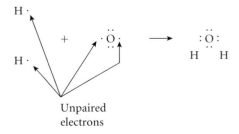

TABLE 9.4
Structures and compounds of nonmetallic elements combined with hydrogen

Nonmetallic Elements	Element (E Represents Any Element of Family)	Compound
Family IVA: C, Si, Ge	·Ė·	H H:E:H H
Family VA: N, P, As, Sb	·Ė·	H:E:H H
Family VIA: O, S, Se, Te	·Ë·	H:E:H
Family VIIA: F, Cl, Br, I	·Ë·	H:E:

The diatomic hydrogen (H_2) and fluorine (F_2), hydrogen fluoride (HF), and water (H_2O) are examples of compounds held together by covalent bonds. A compound held together by covalent bonds is called a **covalent compound.** In general, covalent compounds form from nonmetallic elements on the right side of the periodic table. For elements in families IVA through VIIA, the number of unpaired electrons (and thus the number of covalent bonds formed) is eight minus the family number. You can get a lot of information from the periodic table from generalizations like this one. For another generalization, compare Table 9.4 with the periodic table. The table gives the structures of nonmetals combined with hydrogen and the resulting compounds.

Multiple Bonds

Two dots can represent a lone pair of valence electrons, or they can represent a bonding pair, a single pair of electrons being shared by two atoms. Bonding pairs of electrons are often represented by a simple line between two atoms. For example,

H : H   is shown as   H — H

and

:Ö: is shown as   O
H   H   H   H

Note that the line between the two hydrogen atoms represents an electron pair, so each hydrogen atom has two electrons in the outer orbital, as does helium. In the water molecule, each hydrogen atom has two electrons as before. The oxygen atom has two lone pairs (a total of four electrons) and two bonding pairs (a total of four electrons) for a total of eight electrons. Thus, oxygen has acquired a stable octet of electrons.

A covalent bond in which a single pair of electrons is shared by two atoms is called a *single covalent bond* or simply a **single bond.** Some atoms have two unpaired electrons and can share more than one electron pair. A **double bond** is a covalent bond formed when *two pairs* of electrons are shared by two atoms. This happens mostly in compounds involving atoms of the elements C, N, O, and S. Ethylene, for example, is a gas given off from ripening fruit. The electron dot formula for ethylene is

The ethylene molecule has a double bond between two carbon atoms. Since each line represents two electrons, you can simply count the lines around each symbol to see if the octet rule has been satisfied. Each H has one line, so each H atom is sharing two electrons. Each C has four lines so each C atom has eight electrons, satisfying the octet rule.

A **triple bond** is a covalent bond formed when *three pairs* of electrons are shared by two atoms. Triple bonds occur mostly in compounds with atoms of the elements C and N. Acetylene, for example, is a gas often used in welding torches (Figure 9.9). The electron dot formula for acetylene is

H : C ⦂⦂⦂ C : H   or   H — C ≡ C — H

FIGURE 9.9

Acetylene is a hydrocarbon consisting of two carbon atoms and two hydrogen atoms held together by a triple covalent bond between the two carbon atoms. When mixed with oxygen gas, the resulting flame is hot enough to cut through most metals.

The acetylene molecule has a triple bond between two carbon atoms. Again, note that each line represents two electrons. Each C atom has four lines, so the octet rule is satisfied.

Bond Polarity

How do you know if a bond between two atoms will be ionic or covalent? In general, ionic bonds form between metal atoms and nonmetal atoms, especially those from the opposite sides of the periodic table. Also in general, covalent bonds form between the atoms of nonmetals. If an atom has a much greater electron-pulling ability than another atom, the electron is pulled completely away from the atom with lesser pulling ability, and an ionic bond is the result. If the electron-pulling ability is more even between the two atoms, the electron is shared, and a covalent bond results. As you can imagine, all kinds of reactions are possible between atoms with different combinations of electron-pulling abilities. The result is that it is possible to form many gradations of bonding between completely ionic and completely covalent bonding. Which type of bonding will result can be found by comparing the electronegativity of the elements involved. **Electronegativity** is the *comparative ability of atoms of an element to attract bonding electrons*. The assigned numerical values for electronegativities are given in Figure 9.10. Elements with higher

TABLE 9.5

The meaning of absolute differences in electronegativity

Absolute Difference	→	Type of Bond Expected
1.7 or greater	means	ionic bond
between 0.5 and 1.7	means	polar covalent bond
0.5 or less	means	covalent bond

values have the greatest attraction for bonding electrons, and elements with the lowest values have the least attraction for bonding electrons.

The absolute ("absolute" means without plus or minus signs) difference in the electronegativity of two bonded atoms can be used to predict if a bond is ionic or covalent (Table 9.5). A large difference means that one element has a much greater attraction for bonding electrons than the other element. *If the absolute difference in electronegativity is 1.7 or more,* one atom pulls the bonding electron completely away and *an ionic bond results.* For example, sodium (Na) has an electronegativity of 0.9. Chlorine (Cl) has an electronegativity of 3.0. The difference is 2.1, so you can expect sodium and chloride to form ionic bonds. *If the absolute difference in electronegativity is 0.5 or less,* both atoms have about the same ability to attract bonding electrons. The result is that the electron is shared, and *a covalent*

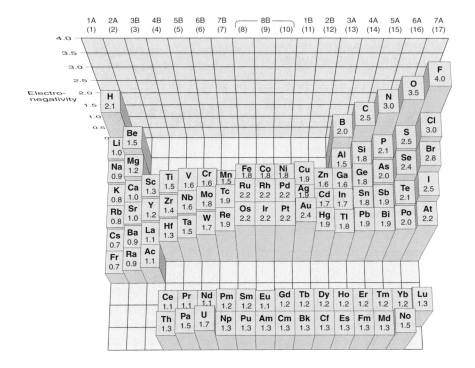

FIGURE 9.10

Elements with the highest electronegativity values have the strongest attraction for the electrons within a chemical bond. Note that the electronegativity of atoms in a group decreases moving down the periodic table, and electronegativity of atoms in a period increases from left to right.

bond results. A given hydrogen atom (H) has an electronegativity of another hydrogen atom, so the difference is 0. Zero is less than 0.5 so you can expect a molecule of hydrogen gas to have a covalent bond.

An ionic bond can be expected when the difference in electronegativity is 1.7 or more, and a covalent bond can be expected when the difference is less than 0.5. What happens when the difference is between 0.5 and 1.7? A covalent bond is formed, but there is an inequality since one atom has a greater bonding electron attraction than the other atom. Thus, the bonding electrons are shared unequally. A **polar covalent bond** is *a covalent bond in which there is an unequal sharing of bonding electrons.* Thus, the bonding electrons spend more time around one atom than the other. The term "polar" means "poles," and that is what forms in a polar molecule. Since the bonding electrons spend more time around one atom than the other, one end of the molecule will have a negative pole, and the other end will have a positive pole. Since there are two poles, the molecule is sometimes called a *dipole.* Note that the molecule as a whole still contains an equal number of electrons and protons, so it is overall electrically neutral. The poles are created by an uneven charge distribution, not an imbalance of electrons and protons. Figure 9.11 shows this uneven charge distribution for a polar covalent compound. The bonding electrons spend more time near the atom on the right, giving this side of the molecule a negative pole.

Figure 9.11 also shows a molecule that has an even charge distribution. The electron distribution around one atom is just like the charge distribution around the other. This molecule is thus a *covalent molecule.* Thus, a polar bond can be viewed as an intermediate type of bond between a covalent bond and an ionic bond. Many gradations are possible between the transition from a purely covalent bond and a purely ionic bond.

EXAMPLE 9.5

Predict if the following bonds are covalent, polar covalent, or ionic: (a) H-O; (b) C-Br; and (c) K-Cl

Solution

From the electronegativity values in Figure 9.10, the absolute differences are

- **(a)** H-O, 1.4
- **(b)** C-Br, 0.3
- **(c)** K-Cl, 2.2

Since an absolute difference of less than 0.5 means covalent, between 0.5 and 1.7 means polar covalent, and greater than 1.7 means ionic, then

- **(a)** H-O, polar covalent
- **(b)** C-Br, covalent
- **(c)** K-Cl, ionic

EXAMPLE 9.6

Predict if the following bonds are covalent, polar covalent, or ionic: (a) Ca-O; (b) H-Cl; and, (c) C-O. Answer: (a) ionic; (b) polar covalent; (c) polar covalent.

Composition of Compounds

As you can imagine, there are literally millions of different chemical compounds from all the possible combinations of over ninety natural elements held together by ionic or covalent bonds. Each of these compounds has its own name, so there are millions of names and formulas for all the compounds. In the early days, compounds were given *common names* according to how they were used, where they came from, or some other means of identifying them. Thus, sodium carbonate was called soda, and closely associated compounds were called baking soda (sodium bicarbonate), washing soda (sodium carbonate), caustic soda (sodium hydroxide), and the bubbly drink made by reacting soda with acid was called soda water, later called soda pop (Figure 9.12). Potassium carbonate was extracted from charcoal by soaking in water and came to be called potash. Such common names are colorful, and some are descriptive, but it was impossible to keep up with the names as the number of known compounds grew. So a systematic set of rules was developed to determine the name and formula of each compound. Once you know the rules, you can write the

Electron distribution and kinds of bonding

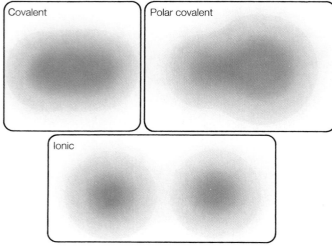

FIGURE 9.11
The absolute difference in electronegativities determines the kind of bond formed.

FIGURE 9.12
These substances are made up of sodium and some form of a carbonate ion. All have common names with the term "soda" for this reason. Soda water (or "soda pop") was first made by reacting soda (sodium carbonate) with an acid, so it was called "soda water."

TABLE 9.6
Modern names of some variable-charge ions

Ion	Name of Ion
Fe^{2+}	Iron(II) ion
Fe^{3+}	Iron(III) ion
Cu^+	Copper(I) ion
Cu^{2+}	Copper(II) ion
Pb^{2+}	Lead(II) ion
Pb^{4+}	Lead(IV) ion
Sn^{2+}	Tin(II) ion
Sn^{4+}	Tin(IV) ion
Cr^{2+}	Chromium(II) ion
Cr^{3+}	Chromium(III) ion
Cr^{6+}	Chromium(VI) ion

formula when you hear the name. Conversely, seeing the formula will tell you the systematic name of the compound. This can be an interesting intellectual activity and can also be important when reading the list of ingredients to understand the composition of a product.

There is a different set of systematic rules to be used with ionic compounds and covalent compounds, but there are a few rules in common. For example, a compound made of only two different elements always ends with the suffix "-ide." So when you hear the name of a compound ending with "-ide" you automatically know that the compound is made up of only two elements. Sodium chlor*ide* is an ionic compound made up of sodium and chlorine ions. Carbon diox*ide* is a covalent compound with carbon and oxygen atoms. Thus, the systematic name tells you what elements are present in a compound with an "-ide" ending.

Ionic Compound Names

Ionic compounds formed by representative metal ions are named by stating the name of the metal (positive ion) first, then the name of the nonmetal (negative ion). Ionic compounds formed by variable-charge ions of the transition elements have an additional rule to identify which variable-charge ion is involved. There was an old way of identifying the charge on the ion by adding either "-ic" or "-ous" to the name of the metal. The suffix "-ic" meant the higher of two possible charges, and the suffix "-ous" meant the lower of two possible charges. For example, iron has two possible charges, 2+ or 3+. The old system used the Latin name for the root. The Latin name for iron is ferrum, so a higher charged iron ion (3+) was named a ferric ion. The lower charged iron ion (2+) was called a ferrous ion.

You still hear the old names sometimes, but chemists now have a better way to identify the variable-charge ion. The newer system uses the English name of the metal with Roman numerals in parentheses to indicate the charge number. Thus, an iron ion with a charge of 2+ is called an iron(II) ion and an iron ion with a charge of 3+ is an iron(III) ion. Table 9.6 gives some of the modern names for variable-charge ions. These names are used with the name of a nonmetal ending in "-ide," just like the single-charge ions in ionic compounds made up of two different elements.

Some ionic compounds contain three or more elements, and so are more complex than a combination of a metal ion and a nonmetal ion. This is possible because they have *polyatomic ions,* groups of two or more atoms that are bound together tightly and behave very much like a single monatomic ion. For example, the OH^- ion is an oxygen atom bound to a hydrogen atom with a net charge of 1−. This polyatomic ion is called a *hydroxide ion.* The hydroxide compounds make up one of the main groups of ionic compounds, the *metal hydroxides.* A metal hydroxide is an ionic compound consisting of a metal with the hydroxide ion. Another main group consists of the salts with polyatomic ions.

The metal hydroxides are named by identifying the metal first and the term *hydroxide* second. Thus, NaOH is named sodium hydroxide and KOH is potassium hydroxide. The salts are similarly named, with the metal (or ammonium ion) identified first, then the name of the polyatomic ion. Thus, $NaNO_3$ is named sodium nitrate and $NaNO_2$ is sodium nitrite. Note that the suffix "-ate" means the polyatomic ion with one more oxygen atom than the "-ite" ion. For example, the chlor*ate* ion is $(ClO_3)^-$, and the chlor*ite* ion is $(ClO_2)^-$. Sometimes more than two possibilities exist, and more oxygen atoms are identified with the prefix "per-", and less with the prefix "hypo-". Thus, the *per*chlor*ate* ion is $(ClO_4)^-$ and the *hypo*chlor*ite* ion is $(ClO)^-$.

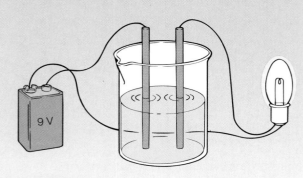
Ionic Compound Formulas

The formulas for ionic compounds are easy to write. There are two rules:

1. The symbols—write the symbol for the positive element first, followed by the symbol for the negative element (same order as in the name).
2. The subscripts—add subscripts to indicate the numbers of ions needed to produce an electrically neutral compound.

As an example, let us write the formula for the compound calcium chloride. The name tells you that this compound consists of positive calcium ions and negative chlorine ions. The suffix "-ide" tells you there are only two elements present. Following rule 1, the symbols would be CaCl.

For rule 2 note the calcium ion is Ca^{2+}, and the chlorine ion is Cl^-. You know the calcium is $+2$ and chlorine is -1 by applying the atomic theory, knowing their positions in the periodic table, or by using a table of ions and their charges. To be electrically neutral, the compound must have an equal number of pluses and minuses. Thus, you will need two negative chlorine ions for every calcium ion with its $2+$ charge. Therefore, the formula is $CaCl_2$. The total charge of two chlorines is thus $2-$, which balances the $2+$ charge on the calcium ion.

One easy way to write a formula showing that a compound is electrically neutral is to cross over the absolute charge numbers (without plus or minus signs) and use them as subscripts. For example, the symbols for the calcium ion and the chlorine ion are

$$Ca^{2+}Cl^{1-}$$

Crossing the absolute numbers as subscripts, as follows

and then dropping the charge numbers gives

$$Ca_1 Cl_2$$

No subscript is written for 1; it is understood. The formula for calcium chloride is thus

$$CaCl_2$$

The crossover technique works because ionic bonding results from a transfer of electrons, and the net charge is conserved. A calcium ion has a $2+$ charge because the atom lost two electrons and two chlorine atoms gain one electron each, for a total of two electrons gained. Two electrons lost equals two electrons gained, and the net charge on calcium chloride is zero, as it has to be. When using the crossover technique it is sometimes necessary to reduce the ratio to the lowest common multiple. Thus, Mg_2O_2 means an equal ratio of magnesium and oxygen ions, so the correct formula is MgO.

The formulas for variable-charge ions are easy to write, since the Roman numeral tells you the charge number. The formula for tin(II) fluoride is written by crossing over the charge numbers (Sn^{2+}, F^{1-}), and the formula is SnF_2.

EXAMPLE 9.7

Name the following compounds: (a) LiF and (b) PbF_2. Write the formulas for the following compounds: (c) potassium bromide and (d) copper(I) sulfide.

Solution

(a) The formula LiF means that the positive metal ions are lithium, the negative nonmetal ions are fluorine, and there are only two elements in the compound. Lithium ions are Li^{1+} (family IA), and fluorine ions are F^{1-} (family VIIA). The name is lithium fluoride.

(b) Lead is a variable-charge transition element (Table 9.6), and fluorine ions are F^{1-}. The lead ion must be Pb^{2+} because the compound PbF_2 is electrically neutral. Therefore, the name is lead(II) fluoride.

(c) The ions are K^{1+} and Br^{1-}. Crossing over the charge numbers and dropping the signs gives the formula KBr.

(d) The Roman numeral tells you the charge on the copper ion, so the ions are Cu^{1+} and S^{2-}. The formula is Cu_2S.

(c) The ions are Ca^{2+} and $(CO_3)^{2-}$. Crossing over the charge numbers and dropping the signs gives the formula $Ca_2(CO_3)_2$. Reducing the ratio to the lowest common multiple gives the correct formula of $CaCO_3$.

(d) The ions are Ca^{2+} and $(PO_4)^{3-}$ (from Table 9.3). Using the crossover technique gives the formula $Ca_3(PO_4)_2$. The parentheses indicate that the entire phosphate unit is taken twice.

The formulas for ionic compounds with polyatomic ions are written from combinations of positive metal ions or the ammonium ion with the polyatomic ions, as listed in Table 9.3. Since the polyatomic ion is a group of atoms that has a charge and stays together in a unit, it is sometimes necessary to indicate this with parentheses. For example, magnesium hydroxide is composed of Mg^{2+} ions and $(OH)^{1-}$ ions. Using the crossover technique to write the formula, you get

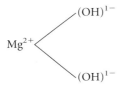

$$Mg^{2+} \quad OH_2^{1-} \quad \text{or} \quad Mg(OH)_2$$

The parentheses are used and the subscript is written *outside* the parenthesis to show that the entire hydroxide unit is taken twice. The formula $Mg(OH)_2$ means

$$Mg^{2+} \Big\langle \begin{array}{l} (OH)^{1-} \\ \\ (OH)^{1-} \end{array}$$

which shows that the pluses equal the minuses. Parentheses are not used, however, when only one polyatomic ion is present. Sodium hydroxide is NaOH, not $Na(OH)_1$.

EXAMPLE 9.8

Name the following compounds: (a) Na_2SO_4 and (b) $Cu(OH)_2$. Write formulas for the following compounds: (c) calcium carbonate and (d) calcium phosphate.

Solution

(a) The ions are Na^+ (sodium ion) and $(SO_4)^{2-}$ (sulfate ion). The name of the compound is sodium sulfate.

(b) Copper is a variable-charge transition element (Table 9.6), and the hydroxide ion $(OH)^{1-}$ has a charge of $1-$. Since the compound $Cu(OH)_2$ must be electrically neutral, the copper ion must be Cu^{2+}. The name is copper(II) hydroxide.

Covalent Compound Names

Covalent compounds are molecular, and the molecules are composed of two *nonmetals*, as opposed to the metal and nonmetal elements that make up ionic compounds. The combinations of nonmetals alone do not present simple names as the ionic compounds did, so a different set of rules for naming and formula writing is needed.

Ionic compounds were named by stating the name of the positive metal ion, then the name of the negative nonmetal ion with an "-ide" ending. This system is not adequate for naming the covalent compounds. To begin, covalent compounds are composed of two or more nonmetal atoms that form a molecule. It is possible for some atoms to form single, double, or even triple bonds with other atoms, including atoms of the same element, and coordinate covalent bonding is also possible in some compounds. The net result is that the same two elements can form more than one kind of covalent compound. Carbon and oxygen, for example, can combine to form the gas released from burning and respiration, carbon dioxide (CO_2). Under certain conditions the very same elements combine to produce a different gas, the poisonous carbon monoxide (CO). Similarly, sulfur and oxygen can combine differently to produce two different covalent compounds. A successful system for naming covalent compounds must therefore provide a means of identifying different compounds made of the same elements. This is accomplished by using a system of Greek prefixes (see Table 9.7). The rules are as follows:

1. The first element in the formula is named first with a prefix indicating the number of atoms if the number is greater than 1.
2. The stem name of the second element in the formula is next. A prefix is used with the stem if two elements form more than one compound. The suffix "-ide" is again used to indicate a compound of only two elements.

For example, CO is carbon monoxide and CO_2 is carbon dioxide. The compound BF_3 is boron trifluoride and N_2O_4 is dinitrogen tetroxide. Knowing the formula and the prefix and stem information in Table 9.7, you can write the name of any covalent compound made up of two elements by ending it with "-ide." Conversely, the name will tell you the formula. However, there are a few polyatomic ions with "-ide" endings that are compounds made up of more than just two elements (hydroxide and cyanide). Compounds formed with ammonium will

A microwave oven rapidly cooks foods that contain water, but paper, glass, and plastic products remain cool in the oven. If they are warmed at all, it is from the heat conducted from the food. The explanation of how the microwave oven heats water, but not most other substances, begins with the nature of the chemical bond.

A chemical bond acts much like a stiff spring, resisting both compression and stretching as it maintains an equilibrium distance between the atoms. As a result, a molecule tends to vibrate when energized or buffeted by other molecules. The rate of vibration depends on the "stiffness" of the spring, which is determined by the bond strength, and the mass of the atoms making up the molecule. Each kind of molecule therefore has its own set of characteristic vibrations, a characteristic natural frequency.

Disturbances with a wide range of frequencies can impact a vibrating system. When the frequency of a disturbance matches the natural frequency, energy is transferred very efficiently, and the system undergoes a large increase in amplitude. Such a frequency match is called resonance. When the disturbance is visible light or some other form of radiant energy, a resonant match results in absorption of the radiant energy and an increase in the molecular kinetic energy of vibration. Thus, a resonant match results in a temperature increase.

The natural frequency of a water molecule matches the frequency of infrared radiation, so resonant heating occurs when infrared radiation strikes water molecules. It is the water molecules in your skin that absorb infrared radiation from the sun, a fire, or some hot object, resulting in the warmth that you feel. Because of this match between the frequency of infrared radiation and the natural frequency of a water molecule, infrared is often called "heat radiation." Since infrared radiation is absorbed by water molecules, it is mostly absorbed on the surface of an object, penetrating only a short distance.

BOX TABLE 9.1
Approximate ranges of visible light, infrared radiation, and microwave radiation

Radiation	Frequency Range (Hz)
Visible light	4×10^{14} to 8×10^{14}
Infrared radiation	3×10^{11} to 4×10^{14}
Microwave radiation	1×10^{9} to 3×10^{11}

The frequency ranges of visible light, infrared radiation, and microwave radiation are given in Box Table 9.1. Most microwave ovens operate at the lower end of the microwave frequency range, at 2.45 gigahertz. This frequency is too low for a resonant match with water molecules, so something else must transfer energy from the microwaves to heat the water. This something else is a result of another characteristic of the water molecule, the type of covalent bond holding the molecule together.

The difference in electronegativity between a hydrogen and oxygen atom is 1.4, meaning the water molecule is held together by a polar covalent bond. The electrons are strongly shifted toward the oxygen end of the molecule, creating a negative pole at the oxygen end and a positive pole at the hydrogen end. The water molecule is thus a dipole, as shown in Box Figure 9.2A.

The dipole of water molecules has two effects: (1) the molecule can be rotated by the electric field of a microwave (see Box Figure 9.2B) and (2) groups of individual molecules are held together by an electrostatic attraction between the positive hydrogen ends of a water molecule and the negative oxygen end of another molecule (see Box Figure 9.2C).

One model to explain how microwaves heat water involves a particular group of three molecules, arranged so that the end

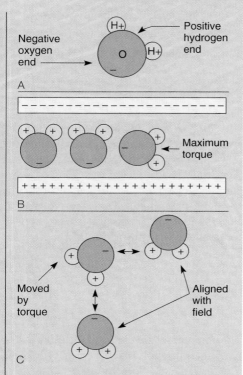

BOX FIGURE 9.2

(A) A water molecule is polar, with a negative pole on the oxygen end and positive poles on the hydrogen end. (B) An electric field aligns the water dipoles, applying a maximum torque at right angles to the dipole vector. (C) Electrostatic attraction between the dipoles holds groups of water molecules together.

molecules of the group are aligned with the microwave electric field, with the center molecule not aligned. The microwave torques the center molecule, breaking its hydrogen bond. The energy of the microwave goes into doing the work of breaking the hydrogen bond, and the molecule now has increased potential energy as a consequence. The detached water molecule reestablishes its hydrogen bond, giving up its potential energy, which goes into the vibration of the group of molecules.

—Continued top of next page

Continued—

Thus, the energy of the microwaves is converted into a temperature increase of the water. The temperature increase is high enough to heat and cook most foods.

Microwave cooking is different from conventional cooking because the heating results from energy transfer in polar water molecules, not conduction and convection. The surface of the food never reaches a temperature over the boiling point of water, so a microwave oven does not brown food (a conventional oven may reach temperatures almost twice as high). Large food items continue to cook for a period of time after being in a microwave oven as the energy is conducted from the water molecules to the food. Most recipes allow for this continued cooking by specifying a waiting period after removing the food from the oven.

Microwave ovens are able to defrost frozen foods because ice always has a thin layer of liquid water (which is what makes it slippery). To avoid "spot cooking" of small pockets of liquid water, many microwave ovens cycle on and off in the defrost cycle. The electrons in metals, like the dipole water molecules, are affected by the electric field of a microwave. A piece of metal near the wall of a microwave oven can result in sparking, which can ignite paper. Metals also reflect microwaves, which can damage the radio tube that produces the microwaves.

TABLE 9.7
Prefixes and element stem names

Prefixes		Stem Names	
Prefix	Meaning	Element	Stem
Mono-	1	Hydrogen	Hydr-
Di-	2	Carbon	Carb-
Tri-	3	Nitrogen	Nitr-
Tetra-	4	Oxygen	Ox-
Penta-	5	Fluorine	Fluor-
Hexa-	6	Phosphorus	Phosph-
Hepta-	7	Sulfur	Sulf-
Octa-	8	Chlorine	Chlor-
Nona-	9	Bromine	Brom-
Deca-	10	Iodine	Iod-

Note: the *a* or *o* ending on the prefix is often dropped if the stem name begins with a vowel, e.g., "tetroxide," not "tetraoxide."

also have an "-ide" ending, and these are also made up of more than two elements.

Covalent Compound Formulas

The systematic name tells you the formula for a covalent compound. The gas that dentists use as an anesthetic, for example, is dinitrogen monoxide. This tells you there are two nitrogen atoms and one oxygen atom in the molecule, so the formula is N_2O. A different molecule composed of the very same elements is nitrogen dioxide. Nitrogen dioxide is the pollutant responsible for the brownish haze of smog. The formula for nitrogen dioxide is NO_2. Other examples of formulas from systematic names are carbon dioxide (CO_2) and carbon tetrachloride (CCl_4).

Formulas of covalent compounds indicate a pattern of how many atoms of one element combine with atoms of another. Carbon, for example, combines with no more than two oxygen atoms to form carbon dioxide. Carbon combines with no more than four chlorine atoms to form carbon tetrachloride. Electron dot formulas show these two molecules as

$$\ddot{O}::C::\ddot{O} \qquad\qquad :\overset{\ddots}{\underset{\ddot{Cl}}{\overset{:\ddot{Cl}:}{\underset{}{Cl}}}}:\ddot{C}:\ddot{Cl}:$$

Using a dash to represent bonding pairs, we have

$$O=C=O \qquad\qquad \overset{Cl}{\underset{Cl}{\overset{|}{Cl-\overset{|}{C}-Cl}}}$$

In both of these compounds, the carbon atom forms four covalent bonds with another atom. The number of covalent bonds that an atom can form is called its **valence.** Carbon has a valence of four and can form single, double, or triple bonds. Here are the possibilities for a single carbon atom (combining elements not shown):

$$-\overset{|}{\underset{|}{C}}- \qquad -\overset{|}{C}= \qquad =C= \qquad -C\equiv$$

Hydrogen has only one unshared electron, so the hydrogen atom has a valence of one. Oxygen has a valence of two and nitrogen has a valence of three. Here are the possibilities for hydrogen, oxygen, and nitrogen:

$$H- \qquad -\ddot{O}- \qquad :\ddot{O}=$$

$$-\ddot{N}- \qquad -\ddot{N}= \qquad :N\equiv$$

Robert Wilhelm Bunsen (1811–1899)

Robert Bunsen was a German chemist who pioneered the use of the spectroscope to analyze chemical compounds. Using the technique, he discovered two new elements, rubidium and cesium. He also devised several pieces of laboratory apparatus, although he probably played only a minor part (if any) in the invention of the Bunsen burner.

Bunsen was born on March 31, 1811, at Göttingen, son of a librarian and linguistics professor at the local university. He studied chemistry there and at Paris, Berlin, and Vienna, gaining his Ph.D. in 1830. He was appointed professor at the Polytechnic Institute of Kassel in 1836, and subsequently held chairs at Marburg (1838) and Breslau (1851) before becoming Professor of Experimental Chemistry at Heidelberg in 1852. He remained there until he retired. Bunsen never married, and ten years after retiring he died, on August 16, 1899.

Bunsen's first significant work, begun in 1837, was on cacodyl compounds, unpleasant and dangerous organic compounds of arsenic. A laboratory explosion cost Bunsen the sight of one eye and he nearly died of arsenic poisoning. He did, however, stimulate later researches into organometallic compounds by his student, the British chemist Edward Frankland. In 1841 he devised the Bunsen cell, 1.9-volt carbon-zinc primary cell, which he used to produce an extremely bright electric arc light. He then (1844) invented a grease-spot photometer to measure brightness (by comparing a light source of known bright-

ness with that being investigated). His contribution to the improvement of laboratory instruments and techniques gave rise also to the Bunsen ice calorimeter, which he developed in 1870 to measure the heat capacities of substances that were available in only small quantities.

Bunsen's first work in inorganic chemistry made use of his primary cell. Using electrolysis, he was the first to isolate metallic magnesium and to demonstrate the intense light produced when the metal is burned in air. But his major contribution was the analysis of the spectra produced when metal salts (particularly chlorides) are heated to incandescence in a flame, a technique first advocated by the U.S. physicist David Alter (1807–1891). Working with Gustav Kirchhoff (1824–1887) in about 1860, Bunsen observed "new" lines in the spectra of minerals which represented the elements rubidium (which has a prominent red line) and cesium (blue line). Other workers using the same technique soon discovered several other new elements.

The Bunsen burner (Box Figure 9.3), probably used to heat the materials for spectroscopic analysis, seems to have been designed by Peter Desdega, Bunsen's technician. Gas (originally coal gas, but any inflammable gas can be used) is released from a jet at the base of a chimney. A hole or holes at the base of the chimney are encircled by a movable collar, which also has holes. Rotation of the collar controls the amount of air admitted at the base of the chimney; the air-gas mixture burns at the top. With the air holes

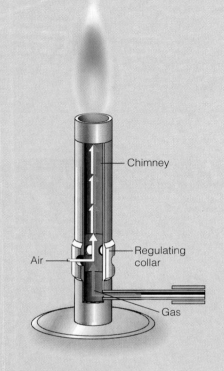

BOX FIGURE 9.3

The Bunsen burner, used for heating laboratory equipment and chemicals. The flame can reach temperatures of 1,500°C/ 2,732°F and is at its hottest when the collar is open.

closed, the gas burns with a luminous, sooty flame. With the air holes open, the air-gas mixture burns with a hot, nonluminous flame (and makes a characteristic roaring sound).

Summary

Elements are basic substances that cannot be broken down into anything simpler, and an *atom* is the smallest unit of an element. *Compounds* are combinations of two or more elements and can be broken down into simpler substances. Compounds are formed when atoms are held together by an attractive force called a *chemical bond*. A *molecule* is the smallest unit of a compound, or a gaseous element, that can exist and still retain the characteristic properties of a substance.

A *chemical change* produces new substances with new properties, and the new materials are created by making or breaking chemical bonds. The process of chemical change in which different chemical substances are created by forming or breaking chemical bonds is called a *chemical reaction*. During a chemical reaction, different chemical substances with greater or lesser amounts of internal potential energy are produced. *Chemical energy* is the change of internal potential energy during a chemical reaction, and other reactions absorb energy. A *chemical equation* is a shorthand way of describing a chemical reaction. An equation shows the substances that are changed, the *reactants*, on the left side, and the new substances produced, the *products*, on the right side.

Chemical reactions involve *valence electrons*, the electrons in the outermost orbital of an atom. Atoms tend to lose or acquire electrons to achieve the configuration of the noble gases with stable, filled outer orbitals. This tendency is generalized as the *octet rule*, that atoms lose or gain electrons to acquire the noble gas structure of eight electrons in the outer orbital. Atoms form negative or positive *ions* in the process.

A chemical bond is an attractive force that holds atoms together in a compound. Chemical bonds that are formed when atoms transfer electrons to become ions are *ionic bonds*. An ionic bond is an electrostatic attraction between oppositely charged ions. Chemical bonds formed when ions share electrons are *covalent bonds*.

Ionic bonds result in *ionic compounds* with a crystalline structure. The energy released when an ionic compound is formed is called the *heat of formation*. It is the same amount of energy that is required to decompose the compound into its elements. A *formula* of a compound uses symbols to tell what elements are in a compound and in what proportions. Ions of representative elements have a single, fixed charge, but many transition elements have variable charges. Electrons are conserved when ionic compounds are formed, and the ionic compound is electrically neutral. The formula shows this overall balance of charges.

Covalent compounds are molecular, composed of electrically neutral groups of atoms bound together by *covalent bonds*. A single *covalent bond* is formed by the sharing of a pair of electrons, with each atom contributing a single electron to the shared pair. Covalent bonds formed when two pairs of electrons are shared are called *double bonds*. A *triple bond* is the sharing of three pairs of electrons.

The electron-pulling ability of an atom in a bond is compared with arbitrary values of *electronegativity*. A high electronegative value means a greater attraction for bonding electrons. If the absolute difference in electronegativity of two bonded atoms is 1.7 or more, one atom pulls the bonding electron away, and an ionic bond results. If the difference is less than 0.5, the electrons are equally shared in a covalent bond. Between 0.5 and 1.7, the electrons are shared unequally in a *polar covalent bond*. A polar covalent bond results in electrons spending more time around the atom or atoms with the greater pulling ability, creating a negative pole at one end and a positive pole at the other. Such a molecule is called a *dipole*, since it has two poles, or centers, of charge.

Compounds are named with systematic rules for ionic and covalent compounds. Both ionic and covalent compounds that are made up of only two different elements always end with an *-ide* suffix, but there are a few *-ide* names for compounds that have more than just two elements.

The modern systematic system for naming variable-charge ions states the English name and gives the charge with Roman numerals in parentheses. Ionic compounds are electrically neutral, and formulas must show a balance of charge. The *crossover technique* is an easy way to write formulas that show a balance of charge.

Covalent compounds are molecules of two or more nonmetal atoms held together by a covalent bond. The system for naming covalent compounds uses Greek prefixes to identify the numbers of atoms, since more than one compound can form from the same two elements (CO and CO_2, for example).

KEY TERMS

atom (p.232)
chemical bond (p.235)
chemical energy (p. 233)
chemical equation (p. 233)
chemical reaction (p. 232)
covalent bond (p. 239)
covalent compound (p. 240)
double bond (p. 240)
electronegativity (p. 241)
formula (p. 237)

heat of formation (p. 236)
ionic bond (p. 236)
ionic compounds (p. 237)
molecule (p. 232)
octet rule (p. 234)
polar covalent bond (p. 242)
single bond (p. 240)
triple bond (p. 240)
valence (p. 247)
valence electrons (p. 234)

APPLYING THE CONCEPTS

1. The smallest unit of an element that can exist alone or in combination with other elements is the
   a. electron.
   b. atom.
   c. molecule.
   d. chemical bond.

2. The smallest unit of a covalent compound that can exist while retaining the chemical properties of the compound is the
   a. electron.
   b. atom.
   c. molecule.
   d. ionic bond.

3. You know that a chemical reaction is taking place if
   a. the temperature of a substance increases.
   b. electrons move in a steady current.
   c. chemical bonds are formed or broken.
   d. All of the above are correct.

4. Chemical reactions that involve changes in the internal potential energy of molecules always involve changes of
   a. the mass of the reactants as compared to the products.
   b. chemical energy.
   c. radiant energy.
   d. the weight of the reactants.

5. The energy released in burning materials produced by photosynthesis has what relationship to the solar energy that was absorbed in making the materials? It is
   a. less than the solar energy absorbed.
   b. the same as the solar energy absorbed.
   c. more than the solar energy absorbed.
   d. variable, having no relationship to the energy absorbed.

6. The electrons that participate in chemical bonding are (the)
   a. valence electrons.
   b. electrons in fully occupied orbitals.
   c. stable inner electrons.
   d. all of the above.

7. Atoms of the representative elements have a tendency to seek stability through
   a. acquiring the noble gas structure.
   b. filling or emptying their outer orbitals.
   c. any situation that will satisfy the octet rule.
   d. all of the above.

8. An ion is formed when an atom of a representative element
   a. gains or loses protons.
   b. shares electrons to achieve stability.
   c. loses or gains electrons to satisfy the octet rule.
   d. All of the above are correct.

9. An atom of an element that is in family VIA will have what charge when it is ionized?
   a. 2+
   b. 6+
   c. 6−
   d. 2−

10. Which type of chemical bond is formed by a transfer of electrons?
    a. ionic
    b. covalent
    c. metallic
    d. All of the above are correct.

11. Which type of chemical bond is formed between two atoms by the sharing of two electrons, with one electron from each atom?
    a. ionic
    b. covalent
    c. metallic
    d. any of the above.

12. Salts, such as sodium chloride, are what type of compounds?
    a. ionic compounds
    b. covalent compounds
    c. polar compounds
    d. Any of the above are correct.

13. If there are two bromide ions for each barium ion in a compound, the chemical formula is
    a. $_2Br_1Ba$.
    b. Ba_2Br.
    c. $BaBr_2$.
    d. none of the above.

14. Which combination of elements forms crystalline solids that will dissolve in water, producing a solution of ions that can conduct an electric current?
    a. metal and metal
    b. metal and nonmetal
    c. nonmetal and nonmetal
    d. All of the above are correct.

15. In a single covalent bond between two atoms,
    a. a single electron from one of the atoms is shared.
    b. a pair of electrons from one of the atoms is shared.
    c. a pair of electrons, one from each atom, is shared.
    d. a single electron is transferred from one atom.

16. Sulfur and oxygen are both in the VIA family of the periodic table. If element X combines with oxygen to form the compound X_2O, element X will combine with sulfur to form the compound
    a. XS_2.
    b. X_2S.
    c. X_2S_2.
    d. It is impossible to say without more information.

17. One element is in the IA family of the periodic table, and a second is in the VIIA family. What type of compound will the two elements form?
    a. ionic
    b. covalent
    c. They will not form a compound.
    d. More information is needed to answer this question.

18. One element is in the VA family of the periodic table, and a second is in the VIA family. What type of compound will these two elements form?
    a. ionic
    b. covalent
    c. They will not form a compound.
    d. More information is needed to answer this question.

19. A covalent bond in which there is an unequal sharing of bonding electrons is a
    a. single covalent bond.
    b. double covalent bond.
    c. triple covalent bond.
    d. polar covalent bond.

20. An inorganic compound made of only two different elements has a systematic name that always ends with the suffix
    a. -ite.
    b. -ate.
    c. -ide.
    d. -ous.

21. Dihydrogen monoxide is the systematic name for a compound that has the common name of
    a. laughing gas.
    b. water.
    c. smog.
    d. rocket fuel.

Answers

1. b 2. c 3. c 4. b 5. b 6. a 7. d 8. c 9. d 10. a 11. b 12. a 13. c 14. b 15. c 16. b 17. a 18. b 19. d 20. c 21. b

QUESTIONS FOR THOUGHT

1. Describe how the following are alike and how they are different: (a) a sodium atom and a sodium ion, and (b) a sodium ion and a neon atom.

2. What is the difference between a polar covalent bond and a nonpolar covalent bond?

3. What is the difference between an ionic and covalent bond? What do atoms forming the two bond types have in common?

4. What is the octet rule?

5. Is there a relationship between the number of valence electrons and how many covalent bonds an atom can form? Explain.

6. Write electron dot formulas for molecules formed when hydrogen combines with (a) chlorine, (b) oxygen, and (c) carbon.

7. Sodium fluoride is often added to water supplies to strengthen teeth. Is sodium fluoride ionic, covalent, or polar covalent? Explain the basis of your answer.

8. What is the modern systematic name of a compound with the formula (a) SnF_2? (b) PbS?

9. What kinds of elements are found in (a) ionic compounds with a name ending with an "-ide" suffix? (b) covalent compounds with a name ending with an "-ide" suffix?

10. Why is it necessary to use a system of Greek prefixes to name binary covalent compounds?

11. What are variable-charge ions? Explain how variable-charge ions are identified in the modern systematic system of naming compounds.

12. What is a polyatomic ion? Give the names and formulas for several common polyatomic ions.

13. Write the formula for magnesium hydroxide. Explain what the parentheses mean.

14. What is a double bond? A triple bond?

PARALLEL EXERCISES

The exercises in groups A and B cover the same concepts. Solutions to group A exercises are located in appendix D.

Group A

1. Use electron dot symbols in equations to predict the formula of the ionic compound formed from the following:
   (a) K and I
   (b) Sr and S
   (c) Na and O
   (d) Al and O

2. Name the following ionic compounds formed from variable-charge transition elements:
   (a) CuS
   (b) Fe_2O_3
   (c) CrO
   (d) PbS

3. Name the following polyatomic ions:
   (a) $(OH)^-$
   (b) $(SO_3)^{2-}$
   (c) $(ClO)^-$
   (d) $(NO_3)^-$
   (e) $(CO_3)^{2-}$
   (f) $(ClO_4)^-$

4. Use the crossover technique to write formulas for the following compounds: ← charge
   (a) Iron(III) hydroxide
   (b) Lead(II) phosphate — $Pb_3(PO4)_2$
   (c) Zinc carbonate
   (d) Ammonium nitrate
   (e) Potassium hydrogen carbonate
   (f) Potassium sulfite

5. Write formulas for the following covalent compounds:
   (a) Carbon tetrachloride
   (b) Dihydrogen monoxide
   (c) Manganese dioxide
   (d) Sulfur trioxide
   (e) Dinitrogen pentoxide
   (f) Diarsenic pentasulfide

Group B

1. Use electron dot symbols in equations to predict the formulas of the ionic compounds formed between the following:
   (a) Li and F
   (b) Be and S
   (c) Li and O
   (d) Al and S

2. Name the following ionic compounds formed from variable-charge transition elements:
   (a) $PbCl_2$
   (b) FeO
   (c) Cr_2O_3
   (d) PbO

3. Name the following polyatomic ions:
   (a) $(C_2H_3O_2)^-$
   (b) $(HCO_3)^-$
   (c) $(SO_4)^{2-}$
   (d) $(NO_2)^-$
   (e) $(MnO_4)^-$
   (f) $(CO_3)^{2-}$

4. Use the crossover technique to write formulas for the following compounds:
   (a) Aluminum hydroxide
   (b) Sodium phosphate
   (c) Copper(II) chloride
   (d) Ammonium sulfate
   (e) Sodium hydrogen carbonate
   (f) Cobalt(II) chloride

5. Write formulas for the following covalent compounds:
   (a) Silicon dioxide
   (b) Dihydrogen sulfide
   (c) Boron trifluoride
   (d) Dihydrogen dioxide
   (e) Carbon tetrafluoride
   (f) Nitrogen trihydride

6. Name the following covalent compounds:
   (a) CO
   (b) CO_2
   (c) CS_2
   (d) N_2O
   (e) P_4S_3
   (f) N_2O_3

7. Predict if the bonds formed between the following pairs of elements will be ionic, polar covalent, or covalent:
   (a) Si and O
   (b) O and O
   (c) H and Te
   (d) C and H
   (e) Li and F
   (f) Ba and S

6. Name the following covalent compounds:
   (a) N_2O
   (b) SO_2
   (c) SiC
   (d) PF_5
   (e) $SeCl_6$
   (f) N_2O_4

7. Predict if the bonds formed between the following pairs of elements will be ionic, polar covalent, or covalent:
   (a) Si and C
   (b) Cl and Cl
   (c) S and O
   (d) Sr and F
   (e) O and H
   (f) K and F

A clear solution of potassium iodide reacts with a clear solution of lead nitrate, producing a yellow precipitate of lead iodide. This is an example of one of the four types of chemical reactions.

CHAPTER

10

Chemical Reactions

We live in a chemical world that has been partly manufactured through controlled chemical change. Consider all of the synthetic fibers and plastics that are used in clothing, housing, and cars. Consider all the synthetic flavors and additives in foods, how these foods are packaged, and how they are preserved. Consider also the synthetic drugs and vitamins that keep you healthy. There are millions of such familiar products that are the direct result of chemical research. Most of these products simply did not exist sixty years ago.

Many of the products of chemical research have remarkably improved the human condition. For example, synthetic fertilizers have made it possible to supply food in quantities that would not otherwise be possible. Chemists learned how to take nitrogen from the air and convert it into fertilizers on an enormous scale. Other chemical research resulted in products such as weed killers, insecticides, and mold and fungus inhibitors. The fertilizers and these products have made it possible to supply food for millions of people who would otherwise have starved (Figure 10.1).

Yet, we also live in a world with concerns about chemical pollutants, the greenhouse effect, acid rain, and a disappearing ozone shield. The very nitrogen fertilizers that have increased food supplies also wash into rivers, polluting the waterways and bays. Such dilemmas require an understanding of chemical products and the benefits and hazards of possible alternatives. Understanding requires a knowledge of chemistry, since the benefits, and risks, are chemical in nature.

The previous chapters were about the modern atomic theory and how it explains elements and how compounds are formed in chemical change. This chapter is concerned with describing chemical changes and the different kinds of chemical reactions that occur. These reactions are explained with balanced chemical equations, which are concise descriptions of reactions that produce the products used in our chemical world.

Chemical Formulas

In chapter 9 you learned how to name and write formulas for ionic and covalent compounds, including the ionic compound of table salt and the covalent compound of ordinary water. Recall that a formula is a shorthand way of describing the elements or ions that make up a compound. There are basically three kinds of formulas that describe compounds: (1) *empirical* formulas, (2) *molecular* formulas, and (3) *structural* formulas. Empirical and molecular formulas, and their use, will be considered in this chapter. Structural formulas will be considered in chapter 12.

An **empirical formula** identifies the elements present in a compound and describes the *simplest whole number ratio* of atoms of these elements with subscripts. For example, the empirical formula for ordinary table salt is NaCl. This tells you that the elements sodium and chlorine make up this compound, and there is one atom of sodium for each chlorine atom. The empirical formula for water is H_2O, meaning there are two atoms of hydrogen for each atom of oxygen.

Covalent compounds exist as molecules. A chemical formula that identifies the *actual numbers* of atoms in a molecule is known as a **molecular formula.** Figure 10.2 shows the structure of some common molecules and their molecular formulas. Note that each formula identifies the elements and numbers of atoms in each molecule. The figure also indicates

how molecular formulas can be written to show how the atoms are arranged in the molecule. Formulas that show the relative arrangements are called *structural formulas.* Compare the structural formulas in the illustration with the three-dimensional representations and the molecular formulas.

How do you know if a formula is empirical or molecular? First, you need to know if the compound is ionic or covalent. You know that ionic compounds are usually composed of metal and nonmetal atoms with an electronegativity difference greater than 1.7. Formulas for ionic compounds are *always* empirical formulas. Ionic compounds are composed of many positive and negative ions arranged in an electrically neutral array. There is no discrete unit, or molecule, in an ionic compound, so it is only possible to identify ratios of atoms with an empirical formula.

Covalent compounds are generally nonmetal atoms bonded to nonmetal atoms in a molecule. You could therefore assume that a formula for a covalent compound is a molecular formula unless it is specified otherwise. You can be certain it is a molecular formula if it is not the simplest whole-number ratio. Glucose, for example, is a simple sugar (also known as dextrose) with the formula $C_6H_{12}O_6$. This formula is divisible by six, yielding a formula with the simplest whole-number ratio of CH_2O. Therefore, CH_2O is the empirical formula for glucose, and $C_6H_{12}O_6$ is the molecular formula.

FIGURE 10.1

The products of chemical research have substantially increased food supplies but have also increased the possibilities of pollution. Balancing the benefits and hazards of the use of chemicals requires a knowledge of chemistry and a knowledge of the alternatives.

Name	Molecular formula	Sketch	Structural formula
Water	H_2O		O / H H
Ammonia	NH_3		N / H H H
Hydrogen peroxide	H_2O_2		H / $O - O$ / H
Carbon dioxide	CO_2		$O = C = O$

FIGURE 10.2

The name, molecular formula, sketch, and structural formula of some common molecules. Compare the kinds and numbers of atoms making up each molecule in the sketch to the molecular formula.

Molecular and Formula Weights

The **formula weight** of a compound is the sum of the atomic weights of all the atoms in a chemical formula. For example, the formula for water is H_2O. Hydrogen and oxygen are both non-

metals, so the formula means that one atom of oxygen is bound to two hydrogen atoms in a molecule. From the periodic table, you know that the approximate (rounded) atomic weight of hydrogen is 1.0 u and oxygen is 16.0 u. Adding the atomic weights for all the atoms,

Atoms	Atomic Weight		Totals
2 of H	2×1.0 u	=	2.0 u
1 of O	1×16.0 u	=	16.0 u
	Formula weight	=	18.0 u

Thus, the formula weight of a water molecule is 18.0 u.

The formula weight of an ionic compound is found in the same way, by adding the rounded atomic weights of atoms (or ions) making up the compound. Sodium chloride is NaCl, so the formula weight is 23.0 u plus 35.5 u, or 58.5 u. The *formula weight* can be calculated for an ionic or molecular substance. The **molecular weight** is the formula weight of a molecular substance. The term *molecular weight* is sometimes used for all substances, whether or not they have molecules. Since ionic substances such as NaCl do not occur as molecules, this is not strictly correct. Both molecular and formula weights are calculated in the same way, but formula weight is a more general term.

EXAMPLE **10.1**

What is the formula weight of table sugar (sucrose), which has the formula $C_{12}H_{22}O_{11}$?

Solution

The formula identifies the numbers of each atom, and the atomic weights are from a periodic table:

Atoms	Atomic Weight		Totals
12 of C	12×12.0 u	=	144.0 u
22 of H	22×1.0 u	=	22.0 u
11 of O	11×16.0 u	=	176.0 u
	Formula weight	=	342.0 u

EXAMPLE **10.2**

What is the molecular weight of ethyl alcohol, C_2H_5OH? (Answer: 46.0 u)

Percent Composition of Compounds

The formula weight of a compound can provide useful information about the elements making up a compound (Figure 10.3). For example, suppose you want to know how much calcium is provided by a dietary supplement. The label lists the

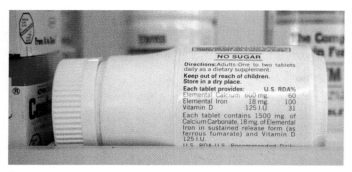
A

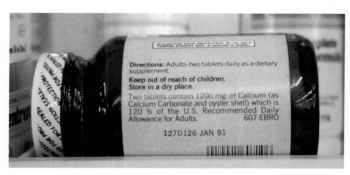
B

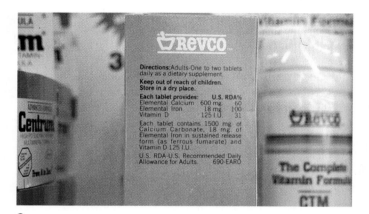
C

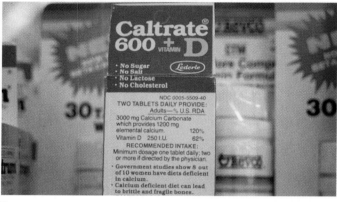
D

FIGURE 10.3

If you know the name of an ingredient, you can write a chemical formula, and the percent composition of a particular substance can be calculated from the formula. This can be useful information for consumer decisions.

main ingredient as calcium carbonate, $CaCO_3$. To find how much calcium is supplied by a pill with a certain mass, you need to find the *mass percentage* of calcium in the compound.

The mass percentage of an element in a compound can be found from

$$\frac{\left(\begin{array}{c}\text{atomic weight}\\\text{of element}\end{array}\right)\left(\begin{array}{c}\text{number of atoms}\\\text{of element}\end{array}\right)}{\text{Formula weight of compound}} \times \begin{array}{c}100\% \text{ of}\\\text{compound}\end{array} = \begin{array}{c}\% \text{ of}\\\text{element}\end{array}$$

equation 10.1

The mass percentage of calcium in $CaCO_3$ can be found in two steps:

Step 1: Determine formula weight:

Atoms	Atomic Weight		Totals
1 of Ca	1×40.1 u	=	40.1 u
1 of C	1×12.0 u	=	12.0 u
3 of O	3×16.0 u	=	48.0 u
	Formula weight	=	100.1 u

Step 2: Determine percentage of Ca:

$$\frac{(40.1 \text{ u CA})(1)}{100.1 \text{u } CaCO_3} \times 100\% \; CaCO_3 = 40.1\% \text{ Ca}$$

Knowing the percentage of the total mass contributed by the calcium, you can multiply this fractional part (as a decimal) by the mass of the supplement pill to find the calcium supplied. The mass percentage of the other elements can also be determined with equation 10.1.

EXAMPLE 10.3

Sodium fluoride is added to water supplies and to some toothpastes for fluoridation. What is the percentage composition of the elements in sodium fluoride?

Solution

Step 1: Write the formula for sodium fluoride, NaF.

Step 2: Determine the formula weight.

Atoms	Atomic Weight		Totals
1 of Na	1×23.0 u	=	23.0 u
1 of F	1×19.0 u	=	19.0 u
	Formula weight	=	42.0 u

Step 3: Determine the percentage of Na and F.

For Na:

$$\frac{(23.0 \text{ u Na})(1)}{42.0 \text{ u NaF}} \times 100\% \text{ NaF} = \boxed{54.7\% \text{ Na}}$$

For F:

$$\frac{(19.0 \text{ u F})(1)}{42.0 \text{ u NaF}} \times 100\% \text{ NaF} = \boxed{45.2\% \text{ F}}$$

The percentage often does not total to exactly 100 percent because of rounding.

EXAMPLE 10.4

Calculate the percentage composition of carbon in table sugar, sucrose, which has a formula of $C_{12}H_{22}O_{11}$. (Answer: 42.1% C)

Concepts Applied

Most for the Money

Chemical fertilizers are added to the soil when it does not contain sufficient elements essential for plant growth (see Figure 10.1). The three critical elements are nitrogen, phosphorus, and potassium, and these are the basic ingredients in most chemical fertilizers. In general, lawns require fertilizers high in nitrogen and gardens require fertilizers high in phosphorus.

Read the labels on commercial packages of chemical fertilizers sold in a garden shop. Find the name of the chemical that supplies each of these critical elements; for example, nitrogen is sometimes supplied by ammonium nitrate NH_4NO_3. Calculate the mass percentage of each critical element supplied according to the label information. Compare these percentages to the grade number of the fertilizer, for example, 10–20–10. Determine which fertilizer brand gives you the most nutrients for the money.

Chemical Equations

Chemical reactions occur when bonds between the outermost parts of atoms are formed or broken. Bonds are formed, for example, when a green plant uses sunlight—a form of energy—to create molecules of sugar, starch, and plant fibers. Bonds are

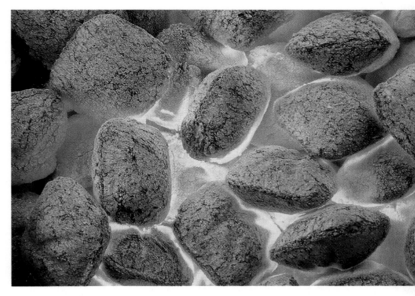

FIGURE 10.4
The charcoal used in a grill is basically carbon. The carbon reacts with oxygen to yield carbon dioxide. The chemical equation for this reaction, $C + O_2 \rightarrow CO_2$, contains the same information as the English sentence but has quantitative meaning as well.

broken and energy is released when you digest the sugars and starches or when plant fibers are burned. Chemical reactions thus involve changes in matter, the creation of new materials with new properties, and energy exchanges. So far you have considered chemical symbols as a concise way to represent elements, and formulas as a concise way to describe what a compound is made of. There is also a concise way to describe a chemical reaction, the **chemical equation.**

Balancing Equations

Word equations are useful in identifying what has happened before and after a chemical reaction. The substances that existed before a reaction are called reactants, and the substances that exist after the reaction are called the products. The equation has a general form of

$$\text{reactants} \longrightarrow \text{products}$$

where the arrow signifies a separation in time; that is, it identifies what existed before the reaction and what exists after the reaction. For example, the charcoal used in a barbecue grill is carbon (Figure 10.4). The carbon reacts with oxygen while burning, and the reaction (1) releases energy and (2) forms carbon dioxide. The reactants and products for this reaction can be described as

$$\text{carbon} + \text{oxygen} \longrightarrow \text{carbon dioxide}$$

The arrow means *yields,* and the word equation is read as, "Carbon reacts with oxygen to yield carbon dioxide." This word equation describes what happens in the reaction but says nothing about the quantities of reactants or products.

Chemical symbols and formulas can be used in the place of words in an equation and the equation will have a whole new meaning. For example, the equation describing carbon reacting with oxygen to yield carbon dioxide becomes

$$C + O_2 \longrightarrow CO_2$$

(balanced)

The new, added meaning is that one atom of carbon (C) reacts with one molecule of oxygen (O_2) to yield one molecule of carbon dioxide (CO_2). Note that the equation also shows one atom of carbon and two atoms of oxygen (recall that oxygen occurs as a diatomic molecule) as reactants on the left side and one atom of carbon and two atoms of oxygen as products on the right side. Since the same number of each kind of atom appears on both sides of the equation, the equation is said to be *balanced.*

You would not want to use a charcoal grill in a closed room because there might not be enough oxygen. An insufficient supply of oxygen produces a completely different product, the poisonous gas carbon monoxide (CO). An equation for this reaction is

$$C + O_2 \longrightarrow CO$$

(not balanced)

As it stands, this equation describes a reaction that violates the **law of conservation of mass,** that matter is neither created nor destroyed in a chemical reaction. From the point of view of an equation, this law states that

mass of reactants = mass of products

Mass of reactants here means all that you start with, including some that might not react. Thus elements are neither created nor destroyed, and this means the elements present and their mass. In any chemical reaction the kind and mass of the reactive elements are identical to the kind and mass of the product elements.

From the point of view of atoms, the law of conservation of mass means that *atoms are neither created nor destroyed in the chemical reaction.* A chemical reaction is the making or breaking of chemical bonds between atoms or groups of atoms. Atoms are not lost or destroyed in the process, nor are they changed to a different kind. The equation for the formation of carbon monoxide has two oxygen atoms in the reactants (O_2) but only one in the product (in CO). An atom of oxygen has disappeared somewhere, and that violates the law of conservation of mass. You cannot fix the equation by changing the CO to a CO_2, because this would change the identity of the compounds. Carbon monoxide is a poisonous gas that is different from carbon dioxide, a relatively harmless product of burning and respiration. *You cannot change the subscript in a formula* because that would change the formula. A different formula means a different composition and thus a different compound.

You cannot change the subscripts of a formula, but you can place a number called a *coefficient* in *front* of the formula. Changing a coefficient changes the *amount* of a substance, not the identity. Thus 2 CO means two molecules of carbon monoxide and 3 CO means three molecules of carbon monoxide. If there is no coefficient, 1 is understood as with subscripts. The meaning of coefficients and subscripts is illustrated in Figure 10.5.

C	means	●	One atom of carbon
O	means	●	One atom of oxygen
O_2	means	 ●●	One molecule of oxygen consisting of two atoms of oxygen
CO	means	 ●●	One molecule of carbon monoxide consisting of one atom of carbon attached to one atom of oxygen
CO_2	means	 ●●●	One molecule of carbon dioxide consisting of one atom of carbon attached to two atoms of oxygen
3 CO_2	means	 ●●● ●●● ●●●	Three molecules of carbon dioxide, each consisting of one atom of carbon attached to two atoms of oxygen

FIGURE 10.5

The meaning of subscripts and coefficients used with a chemical formula. The subscripts tell you how many atoms of a particular element are in a compound. The coefficient tells you about the quantity, or number, of molecules of the compound.

Placing a coefficient of 2 in front of the C and a coefficient of 2 in front of the CO in the equation will result in the same numbers of each kind of atom on both sides:

$$2\,C + O_2 \longrightarrow 2\,CO$$

Reactants: 2 C   Products: 2 C

2 O   2 O

The equation is now balanced.

Suppose your barbecue grill burns natural gas, not charcoal. Natural gas is mostly methane, CH_4. Methane burns by reacting with oxygen (O_2) to produce carbon dioxide (CO_2) and water vapor (H_2O). A balanced chemical equation for this reaction can be written by following a procedure of four steps.

Step 1:   Write the correct formulas for the reactants and products in an unbalanced equation. The reactants and products could have been identified by chemical experiments, or they could have been predicted from what is known about chemical properties. This will be discussed in more detail later. For now, assume that the reactants and products are known and are given in words. For the burning of methane, the unbalanced, but otherwise correct, formula equation would be

$$CH_4 + O_2 \longrightarrow CO_2 + H_2O$$

(not balanced)

Step 2:   Inventory the number of each kind of atom on both sides of the unbalanced equation. In the example there are

Reactants: 1 C   Products: 1 C

4 H   2 H

2 O   3 O

This shows that the H and O are unbalanced.

Step 3: Determine where to place coefficients in front of formulas to balance the equation. It is often best to focus on the simplest thing you can do with whole number ratios. The H and the O are unbalanced, for example, and there are 4 H atoms on the left and 2 H atoms on the right. Placing a coefficient 2 in front of H_2O will balance the H atoms:

$$CH_4 + O_2 \longrightarrow CO_2 + 2\,H_2O$$

(not balanced)

Now take a second inventory:

Reactants:	1 C	Products:	1 C
	4 H		4 H
	2 O		4 O (O_2 + 2 O)

This shows the O atoms are still unbalanced with 2 on the left and 4 on the right. Placing a coefficient of 2 in front of O_2 will balance the O atoms.

$$CH_4 + 2\,O_2 \longrightarrow CO_2 + 2\,H_2O$$

(balanced)

Step 4: Take another inventory to determine if the numbers of atoms on both sides are now equal. If they are, determine if the coefficients are in the lowest possible whole-number ratio. The inventory is now

Reactants:	1 C	Products:	1 C
	4 H		4 H
	4 O		4 O

The number of each kind of atom on each side of the equation is the same, and the ratio of 1:2 → 1:2 is the lowest possible whole-number ratio. The equation is balanced, which is illustrated with sketches of molecules in Figure 10.6.

Reaction:
    Methane reacts with oxygen to yield
    carbon dioxide and water

Balanced equation:
    $CH_4 + 2\,O_2 \longrightarrow CO_2 + 2\,H_2O$

Sketches representing molecules:

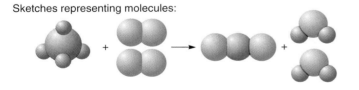

Meaning:

1 molecule of methane	+	2 molecules of oxygen	⟶	1 molecule of carbon dioxide	+	2 molecules of water

FIGURE 10.6

Compare the numbers of each kind of atom in the balanced equation with the numbers of each kind of atom in the sketched representation. Both the equation and the sketch have the same number of atoms in the reactants and in the products.

Balancing chemical equations is mostly a trial-and-error procedure. But with practice, you will find there are a few generalized "role models" that can be useful in balancing equations for many simple reactions. The key to success at balancing equations is to think it out step-by-step while remembering the following:

1. Atoms are neither lost nor gained nor do they change their identity in a chemical reaction. The same kind and number of atoms in the reactants must appear in the products, meaning atoms are conserved.
2. A correct formula of a compound cannot be changed by altering the number or placement of subscripts. Changing subscripts changes the identity of a compound and the meaning of the entire equation.
3. A coefficient in front of a formula multiplies everything in the formula by that number.

There are also a few generalizations that can be helpful for success in balancing equations:

1. Look first to formulas of compounds with the most atoms and try to balance the atoms or compounds they were formed from or decomposed to.
2. Polyatomic ions that appear on both sides of the equation should be treated as independent units with a charge. That is, consider the polyatomic ion as a unit while taking an inventory rather than the individual atoms making up the polyatomic ion. This will save time and simplify the procedure.
3. Both the "crossover technique" and the use of "fractional coefficients" can be useful in finding the least common multiple to balance an equation. All of these generalizations are illustrated in examples 10.5, 10.6, and 10.7.

The physical state of reactants and products in a reaction is often identified by the symbols (g) for gas, (l) for liquid, (s) for solid, and (aq) for an aqueous solution ("aqueous" means water). If a gas escapes, this is identified with an arrow pointing up (↑). A solid formed from a solution is identified with an arrow pointing down (↓). The Greek symbol delta (Δ) is often used under or over the yield sign to indicate a change of temperature or other physical values.

EXAMPLE 10.5

Propane is a liquified petroleum gas (LPG) that is often used as a bottled substitute for natural gas (Figure 10.7). Propane (C_3H_8) reacts with oxygen (O_2) to yield carbon dioxide (CO_2) and water vapor (H_2O). What is the balanced equation for this reaction?

Solution

Step 1: Write the correct formulas of the reactants and products in an unbalanced equation.

$$C_3H_{8(g)} + O_{2(g)} \longrightarrow CO_{2(g)} + H_2O_{(g)}$$

(unbalanced)

FIGURE 10.7

One of two burners is operating at the moment as this hot air balloon ascends. The burners are fueled by propane (C_3H_8), a liquified petroleum gas (LPG). Like other forms of petroleum, propane releases large amounts of heat during the chemical reaction of burning.

Step 2: Inventory the numbers of each kind of atom.

Reactants:	3 C	Products:	1 C
	8 H		2 H
	2 O		3 O

Step 3: Determine where to place coefficients to balance the equation. Looking at the compound with the most atoms (generalization 1), you can see that a propane molecule has 3 C and 8 H. Placing a coefficient of 3 in front of CO_2 and a 4 in front of H_2O will balance these atoms (3 of C and $4 \times 2 = 8$ H atoms on the right has the same number of atoms as C_3H_8 on the left),

$$C_3H_{8(g)} + O_{2(g)} \longrightarrow 3\,CO_{2(g)} + 4\,H_2O_{(g)}$$
(not balanced)

A second inventory shows

Reactants:	3 C
	8 H
	2 O
Products:	3 C
	8 H ($4 \times 2 = 8$)
	10 O [$(3 \times 2) + (4 \times 1) = 10$]

The O atoms are still unbalanced. Place a 5 in front of O_2, and the equation is balanced ($5 \times 2 = 10$). Remember that you cannot

change the subscripts and that oxygen occurs as a diatomic molecule of O_2.

$$C_3H_{8(g)} + 5\,O_{2(g)} \longrightarrow 3\,CO_{2(g)} + 4\,H_2O_{(g)}$$
(balanced)

Step 4: Another inventory shows (a) the number of atoms on both sides are now equal, and (b) the coefficients are 1:5 → 3:4, the lowest possible whole-number ratio. The equation is balanced.

EXAMPLE **10.6**

One type of water hardness is caused by the presence of calcium bicarbonate in solution, $Ca(HCO_3)_2$. One way to remove the troublesome calcium ions from wash water is to add washing soda, which is sodium carbonate, Na_2CO_3. The reaction yields sodium bicarbonate ($NaHCO_3$) and calcium carbonate ($CaCO_3$), which is insoluble. Since $CaCO_3$ is insoluble, the reaction removes the calcium ions from solution. Write a balanced equation for the reaction.

Solution

Step 1: Write the unbalanced equation

$$Ca(HCO_3)_{2(aq)} + Na_2CO_{3(aq)} \longrightarrow NaHCO_{3(aq)} + CaCO_3\downarrow$$
(not balanced)

Step 2: Inventory the numbers of each kind of atom. This reaction has polyatomic ions that appear on both sides, so they should be treated as independent units with a charge (generalization 2). The inventory is

Reactants:	1 Ca	Products:	1 Ca
	2 $(HCO_3)^{1-}$		1 $(HCO_3)^{1-}$
	2 Na		1 Na
	1 $(CO_3)^{2-}$		1 $(CO_3)^{2-}$

Step 3: Placing a coefficient of 2 in front of $NaHCO_3$ will balance the equation,

$$Ca(HCO_3)_{2(aq)} + Na_2CO_{3(aq)} \longrightarrow 2\,NaHCO_{3(aq)} + CaCO_3\downarrow$$
(balanced)

Step 4: An inventory shows

Reactants:	1 Ca	Products:	1 Ca
	2 $(HCO_3)^{1-}$		2 $(HCO_3)^{1-}$
	2 Na		2 Na
	1 $(CO_3)^{2-}$		1 $(CO_3)^{2-}$

The coefficient ratio of 1:1 → 2:1 is the lowest whole-number ratio. The equation is balanced.

EXAMPLE 10.7

Gasoline is a mixture of hydrocarbons, including octane (C_8H_{18}). Combustion of octane produces CO_2 and H_2O, with the release of energy. Write a balanced equation for this reaction.

Solution

Step 1: Write the correct formulas in an unbalanced equation,

$$C_8H_{18(g)} + O_{2(g)} \longrightarrow CO_{2(g)} + H_2O_{(g)}$$
(not balanced)

Step 2: Take an inventory,

Reactants:	8 C	Products:	1 C
	18 H		2 H
	2 O		3 O

(not balanced)

Step 3: Start with the compound with the most atoms (generalization 1) and place coefficients to balance these atoms,

$$C_8H_{18(g)} + O_{2(g)} \longrightarrow 8\,CO_{2(g)} + 9\,H_2O_{(g)}$$
(not balanced)

Redo the inventory,

Reactants:	8 C	Products:	8 C
	18 H		18 H
	2 O		25 O

The O atoms are still unbalanced. There are 2 O atoms in the reactants but 25 O atoms in the products. Since the subscript cannot be changed, it will take 12.5 O_2 to produce 25 oxygen atoms (generalization 3).

$$C_8H_{18(g)} \longrightarrow 12.5\,O_{2(g)} \longrightarrow 8\,CO_{2(g)} + 9\,H_2O_{(g)}$$
(balanced)

Step 4: (a) An inventory will show that the atoms balance,

Reactants:	8 C	Products:	8 C
	18 H		18 H
	25 O		25 O

(b) The coefficients are not in the lowest whole-number ratio (one-half an O_2 does not exist). To make the lowest possible whole-number ratio, all coefficients are multiplied by 2. This results in a correct balanced equation of

$$2\,C_8H_{18(g)} + 25\,O_{2(g)} \longrightarrow 16\,CO_{2(g)} + 18\,H_2O_{(g)}$$
(balanced)

Generalizing Equations

In the previous chapters you learned that the act of classifying, or grouping, something according to some property makes the study of a large body of information less difficult. Generalizing from groups of chemical reactions also makes it possible to predict what will happen in similar reactions. For example, you have studied equations in the previous section describing the combustion of methane (CH_4), propane (C_3H_8), and octane (C_8H_{18}). Each of these reactions involves a *hydrocarbon*, a compound of the elements hydrogen and carbon. Each hydrocarbon reacted with O_2, yielding CO_2 and releasing the energy of combustion. Generalizing from these reactions, you could predict that the combustion of any hydrocarbon would involve the combination of atoms of the hydrocarbon molecule with O_2 to produce CO_2 and H_2O with the release of energy. Such reactions could be analyzed by chemical experiments, and the products could be identified by their physical and chemical properties. You would find your predictions based on similar reactions would be correct, thus justifying predictions from such generalizations. Butane, for example, is a hydrocarbon with the formula C_4H_{10}. The balanced equation for the combustion of butane is

$$2\,C_4H_{10(g)} + 13\,O_{2(g)} \longrightarrow 8\,CO_{2(g)} + 10\,H_2O_{(g)}$$

You could extend the generalization further, noting that the combustion of compounds containing oxygen as well as carbon and hydrogen also produces CO_2 and H_2O (Figure 10.8). These compounds are *carbohydrates*, composed of carbon and water. Glucose, for example, was identified earlier as a compound with

FIGURE 10.8

Hydrocarbons are composed of the elements hydrogen and carbon. Propane (C_3H_8) and gasoline, which contain octane (C_8H_{18}), are examples of hydrocarbons. *Carbohydrates* are composed of the elements hydrogen, carbon, and oxygen. Table sugar, for example, is the carbohydrate $C_{12}H_{22}O_{11}$. Generalizing, all hydrocarbons and carbohydrates react completely with oxygen to yield CO_2 and H_2O.

the formula $C_6H_{12}O_6$. Glucose combines with oxygen to produce CO_2 and H_2O, and the balanced equation is

$$C_6H_{12}O_{6(s)} + 6\ O_{2(g)} \longrightarrow 6\ CO_{2(g)} + 6\ H_2O_{(g)}$$

Note that three molecules of oxygen were not needed from the O_2 reactant because the other reactant, glucose, contains six oxygen atoms per molecule. An inventory of atoms will show that the equation is thus balanced.

Combustion is a rapid reaction with O_2 that releases energy, usually with a flame. A very similar, although much slower reaction takes place in plant and animal respiration. In respiration, carbohydrates combine with O_2 and release energy used for biological activities. This reaction is slow compared to combustion and requires enzymes to proceed at body temperature. Nonetheless, CO_2 and H_2O are the products.

Types of Chemical Reactions

The reactions involving hydrocarbons and carbohydrates with oxygen are examples of an important group of chemical reactions called *oxidation-reduction* reactions. Historically, when the term "oxidation" was first used, it specifically meant reactions involving the combination of oxygen with other atoms. But fluorine, chlorine, and other nonmetals were soon understood to have similar reactions to those of oxygen, so the definition was changed to one concerning the shifts of electrons in the reaction.

An **oxidation-reduction reaction** (or **redox reaction**) is broadly defined as a reaction in which electrons are transferred from one atom to another. As is implied by the name, such a reaction has two parts and each part tells you what happens to the electrons. *Oxidation* is the part of a redox reaction in which there is a loss of electrons by an atom. *Reduction* is the part of a redox reaction in which there is a gain of electrons by an atom. The name also implies that in any reaction in which oxidation occurs reduction must take place, too. One cannot take place without the other.

Substances that take electrons from other substances are called **oxidizing agents.** Oxidizing agents take electrons from the substances being oxidized. Oxygen is the most common oxidizing agent, and several examples have already been given about how it oxidizes foods and fuels. Chlorine is another commonly used oxidizing agent, often for the purposes of bleaching or killing bacteria (Figure 10.9).

A **reducing agent** supplies electrons to the substance being reduced. Hydrogen and carbon are commonly used reducing agents. Carbon is commonly used as a reducing agent to extract metals from their ores. For example, carbon (from coke, which is coal that has been baked) reduces Fe_2O_3, an iron ore, in the reaction

$$2\ Fe_2O_{3(s)} + 3\ C_{(s)} \longrightarrow 4\ Fe_{(s)} + 3\ CO_2\uparrow$$

The Fe in the ore gained electrons from the carbon, the reducing agent in this reaction.

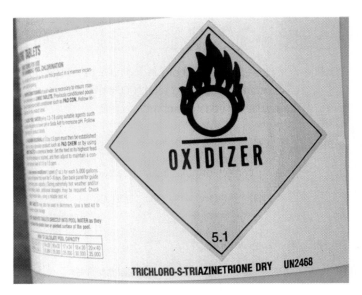

FIGURE 10.9

Oxidizing agents take electrons from other substances that are being oxidized. Oxygen and chlorine are commonly used, strong oxidizing agents.

Concepts Applied

Silver Polish

Silverware and silver-plated objects often become tarnished when the silver is oxidized by sulfur, forming Ag_2S. Commercial silver polishes often act by removing the oxidized layer with an abrasive. The silver can also be polished by reducing the Ag_2S back to metallic silver without removing a layer. Place the tarnished silver in a clean aluminum pan with about 80 g sodium bicarbonate ($NaHCO_3$) and 80 g NaCl dissolved in each liter of near boiling water. A sufficient amount should be prepared to cover the silver object or objects. The salts provide ions to help transfer electrons and facilitate the reaction. The reaction is

$$3\ Ag_2S + 2\ Al + 6\ H_2O \longrightarrow 6\ Ag + 2\ Al(OH)_3 + 3\ H_2S$$

(Note: H_2S has a rotten egg odor.)

Many chemical reactions can be classified as redox or nonredox reactions. Another way to classify chemical reactions is to consider what is happening to the reactants and products. This type of classification scheme leads to four basic categories of chemical reactions, which are (1) *combination,* (2) *decomposition,* (3) *replacement,* and (4) *ion exchange reactions.* The first three categories are subclasses of redox reactions. It is in the ion exchange reactions that you will find the first example of a reaction that is not a redox reaction.

Combination Reactions

A **combination reaction** is a synthesis reaction in which two or more substances combine to form a single compound. The combining substances can be (1) elements, (2) compounds, or (3) combinations of elements and compounds. In generalized form, a combination reaction is

$$X + Y \longrightarrow XY$$

Many redox reactions are combination reactions. For example, metals are oxidized when they burn in air, forming a metal oxide. Consider magnesium, which gives off a bright white light as it burns:

$$2\,Mg_{(s)} + O_{2(g)} \longrightarrow 2\,MgO_{(s)}$$

Note how the magnesium-oxygen reaction follows the generalized form of $x + y \rightarrow xy$.

The rusting of metals is oxidation that takes place at a slower pace than burning, but metals are nonetheless oxidized in the process (Figure 10.10). Again noting the generalized form of a combination reaction, consider the rusting of iron:

$$4\,Fe_{(s)} + 3\,O_{2(g)} \longrightarrow 2\,Fe_2O_{3(s)}$$

Nonmetals are also oxidized by burning in air, for example, when carbon burns with a sufficient supply of O_2:

$$C_{(s)} + O_{2(g)} \longrightarrow CO_{2(g)}$$

Note that all the combination reactions follow the generalized form of $X + Y \rightarrow XY$.

Decomposition Reactions

A **decomposition reaction,** as the term implies, is the opposite of a combination reaction. In decomposition reactions a compound is broken down (1) into the elements that make up the compound, (2) into simpler compounds, or (3) into elements and simpler compounds. Decomposition reactions have a generalized form of

$$XY \longrightarrow X + Y$$

Decomposition reactions generally require some sort of energy, which is usually supplied in the form of heat or electrical energy. An electric current, for example, decomposes water into hydrogen and oxygen:

$$2\,H_2O_{(l)} \xrightarrow{\text{electricity}} 2\,H_{2(g)} + O_{2(g)}$$

Mercury(II) oxide is decomposed by heat, an observation that led to the discovery of oxygen (Figure 10.11):

$$2HgO_{(s)} \xrightarrow{\Delta} 2\,Hg_{(s)} + O_2\uparrow$$

Plaster is a building material made from a mixture of calcium hydroxide, $Ca(OH)_2$, and plaster of Paris, $CaSO_4$. The calcium hydroxide is prepared by adding water to calcium oxide

FIGURE 10.10

Rusting iron is a common example of a combination reaction, where two or more substances combine to form a new compound. Rust is iron(III) oxide formed on these screws from the combination of iron and oxygen under moist conditions.

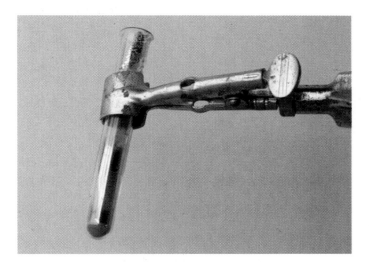

FIGURE 10.11

Mercury(II) oxide is decomposed by heat, leaving the silver-colored element mercury behind as oxygen is driven off. This is an example of a decomposition reaction, $2\,HgO \rightarrow 2\,Hg + O_2\uparrow$. Compare this equation to the general form of a decomposition reaction.

(CaO), which is commonly called quicklime. Calcium oxide is made by heating limestone or chalk ($CaCO_3$), and

$$CaCO_{3(s)} \xrightarrow{\Delta} CaO_{(s)} + CO_2\uparrow$$

Note that all the decomposition reactions follow the generalized form of $XY \rightarrow X + Y$.

Replacement Reactions

In a **replacement reaction,** an atom or polyatomic ion is replaced in a compound by a different atom or polyatomic ion. The replaced part can be either the negative or positive part of the compound. In generalized form, a replacement reaction is

$$XY + Z \longrightarrow XZ + Y$$
(negative part replaced)

or

$$XY + A \longrightarrow AY + X$$
(positive part replaced)

Replacement reactions occur because some elements have a stronger electron-holding ability than other elements. Elements that have the least ability to hold on to their electrons are the most chemically active. Figure 10.12 shows a list of chemical activity of some metals, with the most chemically active at the top. Hydrogen is included because of its role in acids (see chapter 11). Take a few minutes to look over the generalizations listed in Figure 10.12. The generalizations apply to combination, decomposition, and replacement reactions.

Replacement reactions take place as more active metals give up electrons to elements lower on the list with a greater

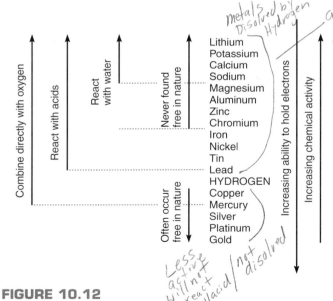

FIGURE 10.12
The activity series for common metals, together with some generalizations about the chemical activities of the metals. The series is used to predict which replacement reactions will take place and which reactions will not occur. (Note that hydrogen is not a metal and is placed in the series for reference to acid reactions.)

FIGURE 10.13
This shows a reaction between metallic aluminum and the blue solution of copper(II) chloride. Aluminum is above copper in the activity series, and aluminum replaces the copper ions from the solution as copper is deposited as a metal. The aluminum loses electrons to the copper and forms aluminum ions in solution.

electron-holding ability. For example, aluminum is higher on the activity series than copper. When aluminum foil is placed in a solution of copper(II) chloride, aluminum is oxidized, losing electrons to the copper. The loss of electrons from metallic aluminum forms aluminum ions in solution, and the copper comes out of solution as a solid metal (Figure 10.13).

$$2\,Al_{(s)} + 3\,CuCl_{2(aq)} \longrightarrow 2\,AlCl_{3(aq)} + 3\,Cu_{(s)}$$

A metal will replace any metal ion in solution that it is above in the activity series. If the metal is listed below the metal ion in solution, no reaction occurs. For example, Ag(s) + $CuCl_2$(aq) → no reaction.

The very active metals (lithium, potassium, calcium, and sodium) react with water to yield metal hydroxides and hydrogen. For example,

$$2\,Na_{(s)} + 2\,H_2O_{(l)} \longrightarrow 2\,NaOH_{(aq)} + H_2\uparrow$$

Acids yield hydrogen ions in solution, and metals above hydrogen in the activity series will replace hydrogen to form a metal salt. For example,

$$Zn_{(s)} + H_2SO_{4(aq)} \longrightarrow ZnSO_{4(aq)} + H_2\uparrow$$

In general, the energy involved in replacement reactions is less than the energy involved in combination or decomposition reactions.

Ion Exchange Reactions

An **ion exchange reaction** is a reaction that takes place when the ions of one compound interact with the ions of another compound, forming (1) a solid that comes out of solution (a precipitate), (2) a gas, or (3) water.

A water solution of dissolved ionic compounds is a solution of ions. For example, solid sodium chloride dissolves in water to become ions in solution,

$$NaCl_{(s)} \longrightarrow Na^+_{(aq)} + Cl^-_{(aq)}$$

If a second ionic compound is dissolved with a solution of another, a mixture of ions results. The formation of a precipitate, a gas, or water, however, removes ions from the solution, and this must occur before you can say that an ionic exchange reaction has taken place. For example, water being treated for domestic use sometimes carries suspended matter that is removed by adding aluminum sulfate and calcium hydroxide to the water. The reaction is

$$3\ Ca(OH)_{2(aq)} + Al_2(SO_4)_{3(aq)} \longrightarrow 3\ CaSO_{4(aq)} + 2\ Al(OH)_3\downarrow$$

The aluminum hydroxide is a jellylike solid, which traps the suspended matter for sand filtration. The formation of the insoluble aluminum hydroxide removed the aluminum and hydroxide ions from the solution, so an ion exchange reaction took place.

In general, an ion exchange reaction has the form

$$AX + BY \longrightarrow AY + BX$$

where one of the products removes ions from the solution. The calcium hydroxide and aluminum sulfate reaction took place as the aluminum and calcium ions traded places. A solubility table such as the one in appendix B will tell you if an ionic exchange reaction has taken place. Aluminum hydroxide is insoluble, according to the table, so the reaction did take place. No ionic exchange reaction occurred if the new products are both soluble.

Another way for an ion exchange reaction to occur is if a gas or water molecule forms to remove ions from the solution. When an acid reacts with a base (an alkaline compound), a salt and water are formed

$$HCl_{(aq)} + NaOH_{(aq)} \longrightarrow NaCl_{(aq)} + H_2O_{(l)}$$

The reactions of acids and bases are discussed in chapter 11.

EXAMPLE **10.8**

Write complete balanced equations for the following, and identify if each reaction is combination, decomposition, replacement, or ion exchange:

(a) silver$_{(s)}$ + sulfur$_{(g)}$ → silver sulfide$_{(s)}$
(b) aluminum$_{(s)}$ + iron(III) oxide$_{(s)}$ → aluminum oxide$_{(s)}$ + iron
(c) sodium chloride$_{(aq)}$ + silver nitrate$_{(aq)}$ → ?
(d) potassium chlorate$_{(s)}$ $\xrightarrow{\Delta}$ potassium chloride$_{(s)}$ + oxygen$_{(g)}$

Solution

(a) The reactants are two elements, and the product is a compound, following the general form $X + Y \rightarrow XY$ of a combination reaction. Table 9.2 gives the charge on silver as

Ag^{1+}, and sulfur (as the other nonmetals in family VIA) is S^{2-}. The balanced equation is

$$2\ Ag_{(s)} + S_{(g)} \longrightarrow Ag_2S_{(s)}$$

Silver sulfide is the tarnish that appears on silverware.

(b) The reactants are an element and a compound that react to form a new compound and an element. The general form is $XY + Z \rightarrow XZ + Y$, which describes a replacement reaction. The balanced equation is

$$2\ Al_{(s)} + Fe_2O_{3(s)} \longrightarrow Al_2O_{3(s)} + 2\ Fe_{(s)}$$

This is known as a "thermite reaction," and in the reaction aluminum reduces the iron oxide to metallic iron with the release of sufficient energy to melt the iron. The thermite reaction is sometimes used to weld large steel pieces, such as railroad rails.

(c) The reactants are water solutions of two compounds with the general form of $AX + BY \rightarrow$, so this must be the reactant part of an ion exchange reaction. Completing the products part of the equation by exchanging parts as shown in the general form and balancing,

$$NaCl_{(aq)} + AgNO_{3(aq)} \longrightarrow NaNO_{3(?)} + AgCl_{(?)}$$

Now consult the solubility chart in appendix B to find out if either of the products is insoluble. $NaNO_3$ is soluble and $AgCl$ is insoluble. Since at least one of the products is insoluble, the reaction did take place, and the equation is rewritten as

$$NaCl_{(aq)} + AgNO_{3(aq)} \longrightarrow NaNO_{3(aq)} + AgCl\downarrow$$

(d) The reactant is a compound, and the products are a simpler compound and an element, following the generalized form of a decomposition reaction, $XY \rightarrow X + Y$. The delta sign (Δ) also means that heat was added, which provides another clue that this is a decomposition reaction. The formula for the chlorate ion is in Table 9.3. The balanced equation is

$$2\ KClO_{3(s)} \xrightarrow{\Delta} 2\ KCl_{(s)} + 3\ O_2\uparrow$$

Information from Chemical Equations

A balanced chemical equation describes what happens in a chemical reaction in a concise, compact way. The balanced equation also carries information about (1) atoms, (2) molecules, and (3) atomic weights. The balanced equation for the combustion of hydrogen, for example, is

$$2\ H_{2(g)} + O_{2(g)} \longrightarrow 2\ H_2O_{(l)}$$

An inventory of each kind of atom in the reactants and products shows

Reactants:	4 hydrogen	Products:	4 hydrogen
	2 oxygen		2 oxygen
Total:	6 atoms	Total:	6 atoms

There are six atoms before the reaction and there are six atoms after the reaction, which is in accord with the law of conservation of mass.

In terms of molecules, the equation says that two diatomic molecules of hydrogen react with one (understood) diatomic molecule of oxygen to yield two molecules of water. The number of coefficients in the equation is the number of molecules involved in the reaction. If you are concerned how two molecules plus one molecule could yield two molecules, remember that *atoms* are conserved in a chemical reaction, not molecules.

Since atoms are conserved in a chemical reaction, their atomic weights should be conserved, too. One hydrogen atom has an atomic weight of 1.0 u, so the formula weight of a diatomic hydrogen molecule must be 2 × 1.0 u, or 2.0 u. The formula weight of O_2 is 2 × 16.0 u, or 32 u. If you consider the equation in terms of atomic weights, then

Equation

$$2\,H_2 + O_2 \longrightarrow 2\,H_2O$$

Formula weights

$$2\,(1.0\,u + 1.0\,u) + (16.0\,u + 16.0\,u) \longrightarrow 2\,(2 \times 1.0\,u + 16.0\,u)$$

$$4\,u + 32\,u \longrightarrow 36\,u$$

$$36\,u \longrightarrow 36\,u$$

The formula weight for H_2O is (1.0 u × 2) + 16 u, or 18 u. The coefficient of 2 in front of H_2O means there are two molecules of H_2O, so the mass of the products is 2 × 18 u, or 36 u. Thus, the reactants had a total mass of 4 u + 32 u, or 36 u, and the products had a total mass of 36 u. Again, this is in accord with the law of conservation of mass.

The equation says that 4 u of hydrogen will combine with 32 u of oxygen. Thus hydrogen and oxygen combine in a mass ratio of 4:32, which reduces to 1:8. So 1 gram of hydrogen will combine with 8 grams of oxygen, and, in fact, they will combine in this ratio no matter what the measurement units are (gram, kilogram, pound, etc.). They always combine in this mass ratio because this is the mass of the individual reactants.

Back in the early 1800s John Dalton attempted to work out a table of atomic weights as he developed his atomic theory. Dalton made two major errors in determining the atomic weights, including (1) measurement errors about mass ratios of combining elements and (2) incorrect assumptions about the formula of the resulting compound. For water, for example, Dalton incorrectly measured that 5.5 g of oxygen combined with 1.0 g of hydrogen. He assumed that one atom of hydrogen combined with one atom of oxygen, resulting in a formula of HO. Thus, Dalton concluded that the atomic mass of oxygen was 5.5 u, and the atomic mass of hydrogen was 1.0 u. Incorrect atomic weights for hydrogen and oxygen led to conflicting formulas for other substances, and no one could show that the atomic theory worked.

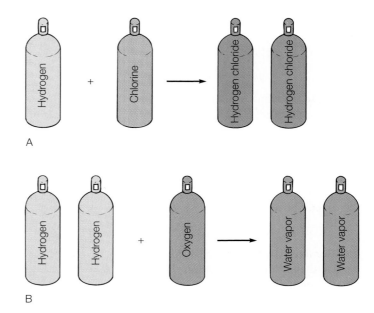

FIGURE 10.14

Reacting gases combine in ratios of small, whole-number volumes when the temperature and pressure are the same for each volume. (A) One volume of hydrogen gas combines with one volume of chlorine gas to yield two volumes of hydrogen chloride gas. (B) Two volumes of hydrogen gas combine with one volume of oxygen gas to yield two volumes of water vapor.

The problem was solved during the first decade of the 1800s through the separate work of a French chemistry professor, Joseph Gay-Lussac, and an Italian physics professor, Amedeo Avogadro. In 1808, Gay-Lussac reported that reacting gases combined in small, whole number *volumes* when the temperature and pressure were constant. Two volumes of hydrogen, for example, combined with one volume of oxygen to form two volumes of water vapor. The term "volume" means any measurement unit, for example, a liter. Other reactions between gases were also observed to combine in small, whole number ratios, and the pattern became known as the *law of combining volumes* (Figure 10.14).

Avogadro proposed an explanation for the law of combining volumes in 1811. He proposed that equal volumes of all gases at the same temperature and pressure *contain the same number of molecules*. Avogadro's hypothesis had two important implications for the example of water. First, since two volumes of hydrogen combine with one volume of oxygen, it means that a molecule of water contains twice as many hydrogen atoms as oxygen atoms. The formula for water must be H_2O, not HO. Second, since *two* volumes of water vapor were produced, each molecule of hydrogen and each molecule of oxygen must be diatomic. Diatomic molecules of hydrogen and oxygen would double the number of hydrogen and oxygen atoms, thus producing twice as much water vapor. These two implications are illustrated in Figure 10.15, along with a balanced equation for the reaction. Note that the coefficients in the equation now have two meanings, (1) the number of molecules of each substance involved in the reaction and (2) the ratios of combining volumes. The coefficient of 2 in

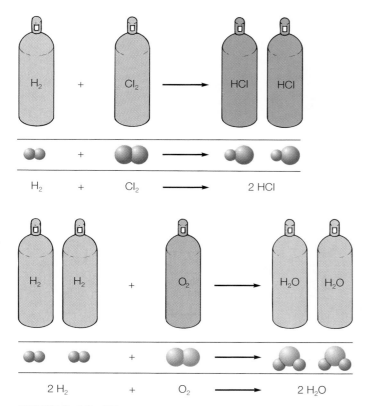

$$H_2 + Cl_2 \longrightarrow 2\ HCl$$

$$2\ H_2 + O_2 \longrightarrow 2\ H_2O$$

FIGURE 10.15

Avogadro's hypothesis of equal volumes of gas having equal numbers of molecules offered an explanation for the law of combining volumes.

front of the H_2, for example, means two molecules of H_2. It also means two volumes of H_2 gas when all volumes are measured at the same temperature and pressure. Recall that equal volumes of any two gases at the same temperature and pressure contain the same number of molecules. Thus, the ratio of coefficients in a balanced equation means a ratio of *any number* of molecules, from 2 of H_2 and 1 of O_2, 20 of H_2 and 10 of O_2, 2,000 of H_2 and 1,000 of O_2, or however many are found in 2 L of H_2 and 1 L of O_2.

EXAMPLE 10.9

Propane is a hydrocarbon with the formula C_3H_8 that is used as a bottled gas. (a) How many liters of oxygen are needed to burn 1 L of propane gas? (b) How many liters of carbon dioxide are produced by the reaction? Assume all volumes to be measured at the same temperature and pressure.

Solution

The balanced equation is

$$C_3H_{8(g)} + 5\ O_{2(g)} \longrightarrow 3\ CO_{2(g)} + 4\ H_2O_{(g)}$$

The coefficients tell you the relative number of molecules involved in the reaction, that 1 molecule of propane reacts with

5 molecules of oxygen to produce 3 molecules of carbon dioxide and 4 molecules of water. Since equal volumes of gases at the same temperature and pressure contain equal numbers of molecules, the coefficients also tell you the relative volumes of gases. Thus 1 L of propane (a) requires 5 L of oxygen and (b) yields 3 L of carbon dioxide (and 4 L of water vapor) when reacted completely.

Units of Measurement Used with Equations

The coefficients in a balanced equation represent a ratio of any *number* of molecules involved in a chemical reaction. The equation has meaning about the atomic *weights* and formula *weights* of reactants and products. The counting of numbers and the use of atomic weights are brought together in a very important measurement unit called a *mole* (from the Latin meaning "a mass"). Here are the important ideas in the mole concept:

1. Recall that the atomic weights of elements are average relative masses of the isotopes of an element. The weights are based on a comparison to carbon-12, with an assigned mass of exactly 12.00 (see chapter 8).
2. The *number* of C-12 atoms in exactly 12.00 g of C-12 has been measured experimentally to be 6.02×10^{23}. This number is called **Avogadro's number,** named after the scientist who reasoned that equal volumes of gases contain equal numbers of molecules.
3. An amount of a substance that contains Avogadro's number of atoms, ions, molecules, or any other chemical unit is defined as a **mole** of the substance. Thus a mole is 6.02×10^{23} atoms, ions, etc., just as a dozen is 12 eggs, apples, etc. The mole is the chemist's measure of atoms, molecules, or other chemical units. A mole of Na^+ ions is 6.02×10^{23} Na^+ ions.
4. A mole of C-12 atoms is defined as having a mass of exactly 12.00 g, a mass that is numerically equal to its atomic mass. So the mass of a mole of C-12 atoms is 12.00 g, or

$$\begin{array}{ccc} \text{mass of one atom} & \times & \text{one mole} & = & \text{mass of a mole of C-12} \\ (12.00\ u) & & (6.02 \times 10^{23}) & = & 12.00\ g \end{array}$$

The masses of all the other isotopes are *based* on a comparison to the C-12 atom. Thus a He-4 atom has one-third the mass of a C-12 atom. An atom of Mg-24 is twice as massive as a C-12 atom. Thus

	1 Atom	×	1 Mole	=	Mass of Mole
C-12:	12.00 u	×	6.02×10^{23}	=	12.00 g
He-4:	4.00 u	×	6.02×10^{23}	=	4.00 g
Mg-24:	24.00 u	×	6.02×10^{23}	=	24.00 g

Therefore, the mass of a mole of any element is numerically equal to its atomic mass. Samples of elements with masses that are the same numerically as their atomic masses are 1 mole measures, and each will contain the same number of atoms (Figure 10.16).

This reasoning can be used to generalize about formula weights, molecular weights, and atomic weights since they are all based on atomic mass units relative to C-12. The

A Each of the following represents one mole of an element:

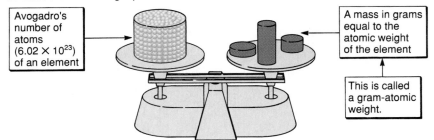

Avogadro's number of atoms (6.02×10^{23}) of an element

A mass in grams equal to the atomic weight of the element

This is called a gram-atomic weight.

B Each of the following represents one mole of a compound:

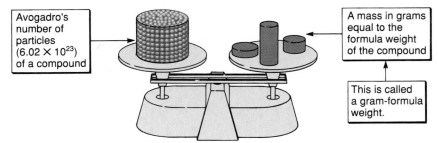

Avogadro's number of particles (6.02×10^{23}) of a compound

A mass in grams equal to the formula weight of the compound

This is called a gram-formula weight.

C Each of the following represents one mole of a molecular substance:

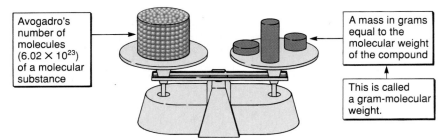

Avogadro's number of molecules (6.02×10^{23}) of a molecular substance

A mass in grams equal to the molecular weight of the compound

This is called a gram-molecular weight.

FIGURE 10.16

The mole concept for (A) elements, (B) compounds, and (C) molecular substances. A mole contains 6.02×10^{23} particles. Since every mole contains the same number of particles, the ratio of the mass of any two moles is the same as the ratio of the masses of individual particles making up the two moles.

gram-atomic weight is the mass in grams of one mole of an element that is numerically equal to its atomic weight. The atomic weight of carbon is 12.01 u; the gram-atomic weight of carbon is 12.01 g. The atomic weight of magnesium is 24.3 u; the gram-atomic weight of magnesium is 24.3 g. Any gram-atomic weight contains Avogadro's number of atoms. Therefore the gram-atomic weights of the elements all contain the same number of atoms.

Similarly, the **gram-formula weight** of a compound is the mass in grams of one mole of the compound that is numerically equal to its formula weight. The **gram-molecular weight** is the gram-formula weight of a molecular compound. Note that one mole of Ne atoms $(6.02 \times 10^{23}$ neon atoms) has a gram-atomic weight of 20.2 g, but one mole of O_2 molecules $(6.02 \times 10^{23}$ oxygen molecules) has a gram-molecular weight of 32.0 g. Stated the other way around, 32.0 g of O_2 and 20.2 g of Ne both contain the same Avogadro's number of particles.

EXAMPLE **10.10**

(a) A 100 percent silver chain has a mass of 107.9 g. How many silver atoms are in the chain? (b) What is the mass of one mole of sodium chloride, NaCl?

Solution

The mole concept and Avogadro's number provide a relationship between numbers and masses. (a) The atomic weight of silver is 107.9 u, so the gram-atomic weight of silver is 107.9 g. A gram-atomic weight is one mole of an element, so the silver chain contains 6.02×10^{23} silver atoms. (b) The formula weight of NaCl is 58.5 u, so the gram-formula weight is 58.5 g. One mole of NaCl has a mass of 58.5 g.

The modern automobile produces two troublesome products in the form of (1) nitrogen monoxide (NO) and (2) hydrocarbons from the incomplete combustion of gasoline. These products from the exhaust enter the air to react in sunlight, eventually producing an irritating haze known as *photochemical smog*. To reduce photochemical smog, modern automobiles are fitted with a catalytic converter as part of the automobile exhaust system (Box Figure 10.1). This reading is about how the catalytic converter combats smog-forming pollutants.

Chemical reactions proceed at a rate that is affected by (1) the concentration of the reactants, (2) the temperature at which a reaction occurs, and (3) the surface area of the reaction. In general, a higher concentration, higher temperatures, and greater surface area mean a faster reaction. The problem with nitrogen monoxide is that it is easily oxidized to nitrogen dioxide (NO_2), a reddish brown, damaging gas that also plays a key role in the formation of photochemical smog. Nitrogen dioxide and the hydrocarbons are oxidized slowly in the air when left to themselves. What is needed is a means to decompose nitrogen monoxide and uncombusted hydrocarbons rapidly before they are released into the air.

The rate at which a chemical reaction proceeds is affected by a *catalyst*, a material that speeds up a chemical reaction without being permanently changed by the reaction. Apparently, molecules require a certain amount of energy to change the chemical bonds that tend to keep them as they are, unreacted. This certain amount of energy is called the *activation energy*, and it represents an energy barrier that must be overcome before a chemical reaction can take place. This explains why chemical reactions proceed at a faster rate at higher temperatures. At higher temperatures, molecules have greater average kinetic energies; thus, they already have part of the minimum energy needed for a reaction to take place.

A catalyst appears to speed a chemical reaction by lowering the activation energy. Molecules become temporarily attached to the surface of the catalyst, which weakens the chemical bonds holding the molecule together. Thus, the weakened molecule is easier to break apart and the activation energy is lowered. Some catalysts do this better with some specific compounds than others, and extensive chemical research programs are devoted to finding new and more effective catalysts.

Automobile catalytic converters use unreactive metals, such as platinum, and transition metal oxides such as copper(II) oxide and chromium(III) oxide. Catalytic reactions that occur in the converter include the following:

$$H_2O + CO \xrightarrow{catalyst} H_2 + CO_2$$
$$2\,NO + 2\,CO \xrightarrow{catalyst} N_2 + 2\,CO_2$$
$$2\,NO + 2\,H_2 \xrightarrow{catalyst} N_2 + 2\,H_2O$$
$$2\,O_2 + CH_4 \xrightarrow{catalyst} CO_2 + 2\,H_2O$$

Thus nitrogen monoxide is reduced to nitrogen gas, and hydrocarbons are oxidized to CO_2 and H_2O.

A catalytic converter can reduce or oxidize about 90 percent of the hydrocarbons, 85 percent of the carbon monoxide, and 40 percent of the nitrogen monoxide from exhaust gases. Other controls, such as exhaust gas recirculation (EGR) are used to further reduce NO formation.

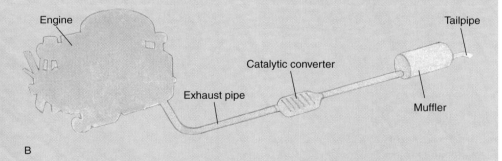

A   B

BOX FIGURE 10.1

Catalytic converters speed up reactions in the exhaust gases from the automobile engine. This results in fewer pollutants being released into the air.

Robert Floyd Curl, Jr. (1933–)

This U.S. chemist, along with colleagues Richard Smalley and Harold Kroto, discovered buckminsterfullerene, a form of carbon (carbon-60) in 1985. They shared the Nobel Prize for Chemistry in 1996 for their discovery.

Born in Alice, Texas, on August 23, 1933, Robert Curl was the son of a Methodist minister. The family lived in various southern Texan towns during his childhood. It was the gift of a chemistry set for Christmas when he was nine that sparked off his interest in chemistry. He received his B.A. from Rice Institute in 1954. Having read about Kenneth Pitzer's discovery of the barriers to internal rotation about single bonds, he decided to go to the University of California at Berkeley to work with him. After receiving his Ph.D. from Berkeley in 1957 he moved on to Harvard and worked with U.S. zoologist E. B. Wilson in microscope spectroscopy for a year, after which he returned to Rice University as an assistant professor. He became full professor in 1967 and was chair of the chemistry department 1992–96.

Curl worked with Smalley and Kroto at Houston, Texas, on the discovery of buckminsterfullerene in 1985. They found a mass-spectrum signal for a molecule of exactly 60 carbon atoms (C-60) in a perfect sphere in which 12 pentagons and 20 hexagons are arranged like the panels on a modern soccer ball (hence the popular name buckyball). The structure also resembles architect Buckminster Fuller's geodesic dome, so Kroto came up with the name buckminsterfullerene, subsequently shortened to fullerene.

Since the discovery, other fullerenes have been identified with 28, 32, 50, 70, and 76 carbon atoms. New molecules have been made based on the buckyball enclosing a metal atom, and buckytubes have been made consisting of cylinders of carbon atoms arranged in hexagons. In 1998 these were proved, in laboratories in Israel and the United States, to be 200 times tougher than any other known fiber. Possible uses for these new molecules are as lubricants, superconductors, and as a starting point for new

drugs. Research in C-60 has continued at Sussex University, England, focusing on the implications of the discovery for several areas of fundamental chemistry including the way it has revolutionized perspectives on carbon-based materials. The research is interdisciplinary involving Curl, Kroto, and Smalley among others, and covers the basic chemistry of fullerenes, fundamental studies of carbon and metal clusters, carbon microparticles and nanotubes, and the study of interstellar and circumstellar molecules and dust. Curl's own research involves the study of the molecules and reactions involved in combustion processes and will enable environmental emissions from vehicles, forest fires, chemical plants, and other sources to be monitored.

Curl is a member of the U.S. Fraternities Phi Beta Kappa, Phi Lambda Upsilon, and Sigma Xi. He received the Clayton Prize of the Institute of Mechanical Engineers in 1957 with Pitzer, and with Smalley and Kroto he received the American Physical Society International Prize for New Materials in 1992.

Quantitative Uses of Equations

A balanced chemical equation can be interpreted in terms of (1) a *molecular ratio* of the reactants and products, (2) a *mole ratio* of the reactants and products, or (3) a *mass ratio* of the reactants and products. Consider, for example, the balanced equation for reacting hydrogen with nitrogen to produce ammonia,

$$3\,H_{2(g)} + N_{2(g)} \longrightarrow 2\,NH_{3(g)}$$

From a *molecular* point of view, the equation says that three molecules of hydrogen combine with one molecule of N_2 to form two molecules of NH_3. The coefficients of $3:1 \rightarrow 2$ thus express a molecular ratio of the reactants and the products.

The molecular ratio leads to the concept of a *mole ratio* since any number of molecules can react as long as they are in the ratio of $3:1 \rightarrow 2$. The number could be Avogadro's number, so $(3) \times (6.02 \times 10^{23})$ molecules of H_2 will combine with $(1) \times$

(6.02×10^{23}) molecules of N_2 to form $(2) \times (6.02 \times 10^{23})$ molecules of ammonia. Since 6.02×10^{23} molecules is the number of particles in a mole, the coefficients therefore represent the *numbers of moles* involved in the reaction. Thus, three moles of H_2 react with one mole of N_2 to produce two moles of NH_3.

The mole ratio of a balanced chemical equation leads to the concept of a *mass ratio* interpretation of a chemical equation. The gram-formula weight of a compound is the mass in grams of *one mole* that is numerically equal to its formula weight. Therefore, the equation also describes the mass ratios of the reactants and the products. The mass ratio can be calculated from the mole relationship described in the equation. The three interpretations are summarized in Table 10.1.

Thus, the coefficients in a balanced equation can be interpreted in terms of molecules, which leads to an interpretation of moles, mass, or any formula unit. The mole concept thus provides the basis for calculations about the quantities of reactants and products in a chemical reaction.

TABLE 10.1

Three interpretations of a chemical equation

Equation: $3 H_2 + N_2 \rightarrow 2 NH_3$

Molecular Ratio:

$$3 \text{ molecules } H_2 + 1 \text{ molecule } N_2 \longrightarrow 2 \text{ molecules } NH_3$$

Mole Ratio:

$$3 \text{ moles } H_2 + 1 \text{ mole } N_2 \longrightarrow 2 \text{ moles } NH_3$$

Mass Ratio:

$$6.0 \text{ g } H_2 + 28.0 \text{ g } N_2 \longrightarrow 34.0 \text{ g } NH_3$$

Concepts Applied

Household Chemistry

Pick a household product that has a list of ingredients with names of covalent compounds or of ions you have met in this chapter. Write the brand name of the product and the type of product (Example: Sani-Flush; toilet-bowl cleaner), then list the ingredients as given on the label, writing them one under the other (column 1). Beside each name put the formula, if you can figure out what it should be (column 2). Also, in a third column, put whatever you know or can guess about the function of that substance in the product. (Example: This is an acid; helps dissolve mineral deposits.)

Summary

A chemical formula is a shorthand way of describing the composition of a compound. An *empirical formula* identifies the simplest whole number ratio of atoms present in a compound. A *molecular formula* identifies the actual number of atoms in a molecule.

The sum of the atomic weights of all the atoms in any formula is called the *formula weight.* The *molecular weight* is the formula weight of a molecular substance. The formula weight of a compound can be used to determine the *mass percentage* of elements making up a compound.

A concise way to describe a chemical reaction is to use formulas in a *chemical equation.* A chemical equation with the same number of each kind of atom on both sides is called a *balanced equation.* A balanced equation is in accord with the *law of conservation of mass,* which states that atoms are neither created nor destroyed in a chemical reaction. To balance a chemical equation, *coefficients* are placed in front of chemical formulas. Subscripts of formulas may not be changed since this would change the formula, meaning a different compound.

One important group of chemical reactions is called *oxidation-reduction reactions,* or *redox* reactions for short. Redox reactions are reactions where shifts of electrons occur. The process of losing electrons

is called *oxidation,* and the substance doing the losing is said to be *oxidized.* The process of gaining electrons is called *reduction,* and the substance doing the gaining is said to be *reduced.* Substances that take electrons from other substances are called *oxidizing agents.* Substances that supply electrons are called *reducing agents.*

Chemical reactions can also be classified as (1) *combination,* (2) *decomposition,* (3) *replacement,* or (4) *ion exchange.* The first three of these are redox reactions, but ion exchange is not.

A balanced chemical equation describes chemical reactions and has quantitative meaning about numbers of atoms, numbers of molecules, and conservation of atomic weights. The coefficients also describe the *volumes* of combining gases. At a constant temperature and pressure gases combine in small, whole number ratios that are given by the coefficients. Each volume at the same temperature and pressure contains the *same number of molecules.*

The number of atoms in exactly 12.00 g of C-12 is called *Avogadro's number,* which has a value of 6.02×10^{23}. Any substance that contains Avogadro's number of atoms, ions, molecules, or any chemical unit is called a *mole* of that substance. The mole is a measure of a number of atoms, molecules, or other chemical units. The mass of a mole of any substance is equal to the atomic mass of that substance.

The mass, number of atoms, and mole concepts are generalized to other units. The *gram-atomic weight* of an element is the mass in grams that is numerically equal to its atomic weight. The *gram-formula weight* of a compound is the mass in grams that is numerically equal to the formula weight of the compound. The *gram-molecular weight* is the gram-formula weight of a molecular compound. The relationships between the mole concept and the mass ratios can be used with a chemical equation for calculations about the quantities of reactants and products in a chemical reaction.

Summary of Equations

10.1

$$\frac{\left(\begin{array}{c}\text{atomic weight}\\\text{of element}\end{array}\right)\left(\begin{array}{c}\text{number of atoms}\\\text{of element}\end{array}\right)}{\text{formula weight of compound}} \times \begin{array}{c}100\% \text{ of}\\\text{compound}\end{array} = \begin{array}{c}\% \text{ of}\\\text{element}\end{array}$$

KEY TERMS

Avogadro's number (p. **267**)

chemical equation (p. **257**)

combination
reaction (p. **263**)

decomposition
reaction (p. **263**)

empirical formula (p. **254**)

formula weight (p. **255**)

gram-atomic weight (p. **268**)

gram-formula
weight (p. **268**)

gram-molecular
weight (p. **268**)

ion exchange reaction (p. **264**)

law of conservation of
mass (p. **258**)

mole (p. **267**)

molecular formula (p. **254**)

molecular weight (p. **255**)

oxidation-reduction
reaction (p. **262**)

oxidizing agents (p. **262**)

redox reaction (p. **262**)

reducing agent (p. **262**)

replacement
reaction (p. **264**)

1. A formula for a compound is given as KCl. This is a (an)
   a. empirical formula.
   b. molecular formula.
   c. structural formula.
   d. formula, but type unknown without further information.

2. A formula for a compound is given as C_8H_{18}. This is a (an)
   a. empirical formula.
   b. molecular formula.
   c. structural formula.
   d. formula, but type unknown without further information.

3. The formula weight of sulfuric acid, H_2SO_4, is
   a. 49 u.
   b. 50 u.
   c. 98 u.
   d. 194 u.

4. A balanced chemical equation has
   a. the same number of molecules on both sides of the equation.
   b. the same kinds of molecules on both sides of the equation.
   c. the same number of each kind of atom on both sides of the equation.
   d. all of the above.

5. The law of conservation of mass means that
   a. atoms are not lost or destroyed in a chemical reaction.
   b. the mass of a newly formed compound cannot be changed.
   c. in burning, part of the mass must be converted into fire in order for mass to be conserved.
   d. molecules cannot be broken apart because this would result in less mass.

6. A chemical equation is balanced by changing (the)
   a. subscripts.
   b. superscripts.
   c. coefficients.
   d. any of the above as necessary to achieve a balance.

7. Since wood is composed of carbohydrates, you should expect what gases to exhaust from a fireplace when complete combustion takes place?
   a. carbon dioxide, carbon monoxide, and pollutants
   b. carbon dioxide and water vapor
   c. carbon monoxide and smoke
   d. It depends on the type of wood being burned.

8. When carbon burns with an insufficient supply of oxygen, carbon monoxide is formed according to the following equation: $2 C + O_2 \rightarrow 2 CO$. What category of chemical reaction is this?
   a. combination
   b. ion exchange
   c. replacement
   d. None of the above, because the reaction is incomplete.

9. According to the activity series for metals, adding metallic iron to a solution of aluminum chloride should result in
   a. a solution of iron chloride and metallic aluminum.
   b. a mixed solution of iron and aluminum chloride.
   c. the formation of iron hydroxide with hydrogen given off.
   d. no metal replacement reaction.

10. In a replacement reaction, elements that have the most ability to hold onto their electrons are
    a. the most chemically active.
    b. the least chemically active.
    c. not generally involved in replacement reactions.
    d. none of the above.

11. Of the following elements, the one with the greatest electron-holding ability is
    a. sodium.
    b. zinc.
    c. copper.
    d. platinum.

12. Of the following elements, the one with the greatest chemical activity is
    a. aluminum.
    b. zinc.
    c. iron.
    d. mercury.

13. You know that an expected ion exchange reaction has taken place if the products include
    a. a precipitate.
    b. a gas.
    c. water.
    d. any of the above.

14. The incomplete equation of $2 KClO_{3(s)} \xrightarrow{\Delta}$ probably represents which type of chemical reaction?
    a. combination
    b. decomposition
    c. replacement
    d. ion exchange

15. In the equation of $2 H_{2(g)} + O_{2(g)} \rightarrow 2 H_2O_{(g)}$,
    a. the total mass of the gaseous reactants is less than the total mass of the liquid product.
    b. the total number of molecules in the reactants is equal to the total number of molecules in the products.
    c. one volume of oxygen combines with two volumes of hydrogen to produce two volumes of water.
    d. All of the above are correct.

16. An amount of a substance that contains Avogadro's number of atoms, ions, or molecules is (a)
    a. mole.
    b. gram-atomic weight.
    c. gram-formula weight.
    d. any of the above.

17. If you have 6.02×10^{23} atoms of metallic iron, you will have how many grams of iron?

   a. 26
   b. 55.8
   c. 334.8
   d. 3.4×10^{25}

Answers

1. a 2. b 3. c 4. c 5. a 6. c 7. b 8. a 9. d 10. b 11. d 12. a 13. d 14. b 15. c 16. d 17. b

QUESTIONS FOR THOUGHT

1. How is an empirical formula like and unlike a molecular formula?
2. Describe the basic parts of a chemical equation. Identify how the physical state of elements and compounds is identified in an equation.
3. What is the law of conservation of mass? How do you know if a chemical equation is in accord with this law?
4. Describe in your own words how a chemical equation is balanced.
5. What is a hydrocarbon? What is a carbohydrate? In general, what are the products of complete combustion of hydrocarbons and carbohydrates?
6. Define and give an example in the form of a balanced equation of (a) a combination reaction, (b) a decomposition reaction, (c) a replacement reaction, and (d) an ion exchange reaction.
7. What must occur in order for an ion exchange reaction to take place? What is the result if this does not happen?
8. Predict the products for the following reactions: (a) The combustion of ethyl alcohol, C_2H_5OH, (b) the rusting of aluminum, and (c) the reaction between iron and sodium chloride.
9. The formula for butane is C_4H_{10}. Is this an empirical formula or a molecular formula? Explain the reason(s) for your answer.
10. How is the activity series for metals used to predict if a replacement reaction will occur or not?
11. What is a gram-formula weight? How is it calculated?
12. What is the meaning and the value of Avogadro's number? What is a mole?

PARALLEL EXERCISES

The exercises in groups A and B cover the same concepts. Solutions to group A exercises are located in appendix D.

Group A

1. Identify the following as empirical formulas or molecular formulas and indicate any uncertainty with (?):
   (a) $CMgCl_2$
   (b) C_2H_2
   (c) BaF_2
   (d) C_8H_{18}
   (e) CH_4
   (f) S_8
2. What is the formula weight for each of the following compounds?
   (a) Copper(II) sulfate
   (b) Carbon disulfide
   (c) Calcium sulfate
   (d) Sodium carbonate

3. What is the mass percentage composition of the elements in the following compounds?
   (a) Fool's gold, FeS_2
   (b) Boric acid, H_3BO_3
   (c) Baking soda, $NaHCO_3$
   (d) Aspirin, $C_9H_8O_4$

Group B

1. Identify the following as empirical formulas or molecular formulas and indicate any uncertainty with (?):
   (a) CH_2O
   (b) $C_6H_{12}O_6$
   (c) NaCl
   (d) CH_4
   (e) F_6
   (f) CaF_2
2. Calculate the formula weight for each of the following compounds:
   (a) Dinitrogen monoxide
   (b) Lead(II) sulfide
   (c) Magnesium sulfate
   (d) Mercury(II) chloride

3. What is the mass percentage composition of the elements in the following compounds?
   (a) Potash, K_2CO_3
   (b) Gypsum, $CaSO_4$
   (c) Saltpeter, KNO_3
   (d) Caffeine, $C_8H_{10}N_4O_2$

4. Write balanced chemical equations for each of the following unbalanced reactions:
(a) $SO_2 + O_2 \rightarrow SO_3$
(b) $P + O_2 \rightarrow P_2O_5$
(c) $Al + HCl \rightarrow AlCl_3 + H_2$
(d) $NaOH + H_2SO_4 \rightarrow Na_2SO_4 + H_2O$
(e) $Fe_2O_3 + CO \rightarrow Fe + CO_2$
(f) $Mg(OH)_2 + H_3PO_4 \rightarrow Mg_3(PO_4)_2 + H_2O$

5. Identify the following as combination, decomposition, replacement, or ion exchange reactions:
(a) $NaCl_{(aq)} + AgNO_{3(aq)} \rightarrow NaNO_{3(aq)} + AgCl\downarrow$
(b) $H_2O_{(l)} + CO_{2(g)} \rightarrow H_2CO_{3(l)}$
(c) $2\,NaHCO_{3(s)} \rightarrow Na_2CO_{3(s)} + H_2O_{(g)} + CO_{2(g)}$
(d) $2\,Na_{(s)} + Cl_{2(g)} \rightarrow 2\,NaCl_{(s)}$
(e) $Cu_{(s)} + 2\,AgNO_{3(aq)} \rightarrow Cu(NO_3)_{2(aq)} + 2\,Ag_{(s)}$
(f) $CaO_{(s)} + H_2O_{(l)} \rightarrow Ca(OH)_{2(aq)}$

6. Write complete, balanced equations for each of the following reactions:
(a) $C_5H_{12(g)} + O_{2(g)} \rightarrow$
(b) $HCl_{(aq)} + NaOH_{(aq)} \rightarrow$
(c) $Al_{(s)} + Fe_2O_{3(s)} \rightarrow$
(d) $Fe_{(s)} + CuSO_{4(aq)} \rightarrow$
(e) $MgCl_{(aq)} + Fe(NO_3)_{2(aq)} \rightarrow$
(f) $C_6H_{10}O_{5(s)} + O_{2(g)} \rightarrow$

7. Write complete, balanced equations for each of the following decomposition reactions. Include symbols for physical states, heating, and others as needed:
(a) Solid potassium chloride and oxygen gas are formed when solid potassium chlorate is heated.
(b) Upon electrolysis, molten bauxite (aluminum oxide) yields solid aluminum metal and oxygen gas.
(c) Upon heating, solid calcium carbonate yields solid calcium oxide and carbon dioxide gas.

8. Write complete, balanced equations for each of the following replacement reactions. If no reaction is predicted, write "no reaction" as the product:
(a) $Na_{(s)} + H_2O_{(l)} \rightarrow$
(b) $Au_{(s)} + HCl_{(aq)} \rightarrow$
(c) $Al_{(s)} + FeCl_{2(aq)} \rightarrow$
(d) $Zn_{(s)} + CuCl_{2(aq)} \rightarrow$

9. Write complete, balanced equations for each of the following ion exchange reactions. If no reaction is predicted, write "no reaction" as the product:
(a) $NaOH_{(aq)} + HNO_{3(aq)} \rightarrow$
(b) $CaCl_{2(aq)} + KNO_{3(aq)} \rightarrow$
(c) $Ba(NO_3)_{2(aq)} + Na_3PO_{4(aq)} \rightarrow$
(d) $KOH_{(aq)} + ZnSO_{4(aq)} \rightarrow$

10. The gas welding torch is fueled by two tanks, one containing acetylene (C_2H_2) and the other pure oxygen (O_2). The very hot flame of the torch is produced as acetylene burns,

$$2\,C_2H_{2(g)} + O_{2(g)} \rightarrow 4\,CO_{2(g)} + H_2O_{(g)}$$

According to this equation, how many liters of oxygen are required to burn 1 liter of acetylene?

4. Write balanced chemical equations for each of the following unbalanced reactions:
(a) $NO + O_2 \rightarrow NO_2$
(b) $KClO_3 \rightarrow KCl + O_2$
(c) $NH_4Cl + Ca(OH)_2 \rightarrow CaCl_2 + NH_3 + H_2O$
(d) $NaNO_3 + H_2SO_4 \rightarrow Na_2SO_4 + HNO_3$
(e) $PbS + H_2O_2 \rightarrow PbSO_4 + H_2O$
(f) $Al_2(SO_4)_3 + BaCl_2 \rightarrow AlCl_3 + BaSO_4$

5. Identify the following as combination, decomposition, replacement, or ion exchange reactions:
(a) $ZnCO_{3(s)} \rightarrow ZnO_{(s)} + CO_2\uparrow$
(b) $2\,NaBr_{(aq)} + Cl_{2(g)} \rightarrow 2\,NaCl_{(aq)} + Br_{2(g)}$
(c) $2\,Al_{(s)} + 3\,Cl_{2(g)} \rightarrow 2\,AlCl_{3(s)}$
(d) $Ca(OH)_{2(aq)} + H_2SO_{4(aq)} \rightarrow CaSO_{4(aq)} + 2\,H_2O_{(l)}$
(e) $Pb(NO_3)_{2(aq)} + H_2S_{(g)} \rightarrow 2\,HNO_{3(aq)} + PbS\downarrow$
(f) $C_{(s)} + ZnO_{(s)} \rightarrow Zn_{(s)} + CO\uparrow$

6. Write complete, balanced equations for each of the following reactions:
(a) $C_3H_{6(g)} + O_{2(g)} \rightarrow$
(b) $H_2SO_{4(aq)} + KOH_{(aq)} \rightarrow$
(c) $C_6H_{12}O_{6(s)} + O_{2(g)} \rightarrow$
(d) $Na_3PO_{4(aq)} + AgNO_{3(aq)} \rightarrow$
(e) $NaOH_{(aq)} + Al(NO_3)_{3(aq)} \rightarrow$
(f) $Mg(OH)_{2(aq)} + H_3PO_{4(aq)} \rightarrow$

7. Write complete, balanced equations for each of the following decomposition reactions. Include symbols for physical states, heating, and others as needed:
(a) When solid zinc carbonate is heated, solid zinc oxide and carbon dioxide gas are formed.
(b) Liquid hydrogen peroxide decomposes to liquid water and oxygen gas.
(c) Solid ammonium nitrite decomposes to liquid water and nitrogen gas.

8. Write complete, balanced equations for each of the following replacement reactions. If no reaction is predicted, write "no reaction" as the product:

(a) $Zn_{(s)} + FeCl_{2(aq)} \rightarrow$
(b) $Zn_{(s)} + AlCl_{3(aq)} \rightarrow$
(c) $Cu_{(s)} + HgCl_{2(aq)} \rightarrow$
(d) $Al_{(s)} + HCl_{(aq)} \rightarrow$

9. Write complete, balanced equations for each of the following ion exchange reactions. If no reaction is predicted, write "no reaction" as the product:
(a) $Ca(OH)_{2(aq)} + H_2SO_{4(aq)} \rightarrow$
(b) $NaCl_{(aq)} + AgNO_{3(aq)} \rightarrow$
(c) $NH_4NO_{3(aq)} + Mg_3(PO_4)_{2(aq)} \rightarrow$
(d) $Na_3PO_{4(aq)} + AgNO_{3(aq)} \rightarrow$

10. Iron(III) oxide, or hematite, is one mineral used as an iron ore. Other iron ores are magnetite (Fe_3O_4) and siderite ($FeCO_3$). Assume that you have pure samples of all three ores that will be reduced by reaction with carbon monoxide. Which of the three ores will have the highest yield of metallic iron?

Water is often referred to as the *universal solvent* because it makes so many different kinds of solutions. Eventually, moving water can dissolve solid rock, carrying it away in solution.

CHAPTER

11

Water and Solutions

What do you think about when you see a stream (Figure 11.1)? Do you wonder about the water quality and what might be dissolved in the water? Do you wonder where the stream comes from and if it will ever run out of water?

Many people can look at a stream, but they might think about different things. A farmer might think about how the water could be diverted and used for his crops. A city planner might wonder if the water is safe for domestic use, and if not, what it would cost to treat the water. Others might wonder if the stream has large fish they could catch. Many large streams can provide water for crops, domestic use, and recreation, and still meet the requirements for a number of other uses.

It is the specific properties of water that make it important for agriculture, domestic use, and recreation. Living things evolved in a watery environment, so water and its properties are essential to life on the earth. Some properties of water, such as the ability to dissolve almost anything, also make water very easy to pollute. This chapter is concerned with some of the unique properties of water, water solutions, and the household use of water.

Household Water

Water is an essential resource, not only because it is required for life processes but also because of its role in a modern society. Water is used in the home for drinking and cooking (2%), cleaning dishes (6%), laundry (11%), bathing (23%), toilets (29%), and for maintaining lawns and gardens (29%).

The water supply is obtained from streams, lakes, and reservoirs on the surface, or from groundwater pumped from below the surface. Surface water contains more sediments, bacteria, and possible pollutants than water from a well because it is exposed to the atmosphere and water runs off the land into streams and rivers. Surface water requires filtering to remove suspended particles, treatment to kill bacteria, and sometimes processing to remove pollution. Well water is generally cleaner but still might require treatment to kill bacteria and remove pollution that has seeped through the ground from waste dumps, agricultural activities, or industrial sites.

Most pollutants are usually too dilute to be considered a significant health hazard, but there are exceptions. There are five types of contamination found in U.S. drinking water that are responsible for the most widespread danger, and these are listed in Table 11.1. In spite of these general concerns and other occasional local problems, the U.S. water supply is considered to be among the cleanest in the world.

The demand for domestic water sometimes exceeds the immediate supply in some metropolitan areas. This is most common during the summer, when water demand is high and rainfall is often low. Communities in these areas often have public education campaigns designed to help reduce the demand for water. For example, did you know that taking a tub bath can use up to 135 liters (about 36 gal) of water compared to only 95 liters (about 25 gal) for a regular shower? Even more water is saved by a shower that does not run continuously—wetting down, soaping up, and rinsing off uses only 15 liters (about 4 gal) of water. You can also save about 35 liters (about 9 gal) of water by not letting the water run continuously while brushing your teeth.

It is often difficult to convince people to conserve water when it is viewed as an inexpensive, limitless resource. However, efforts to conserve water increase dramatically as the cost to the household consumer increases.

The issues involved in maintaining a safe water supply are better understood by considering some of the properties of water and water solutions. These are the topics of the following sections.

Properties of Water

Water is essential for life since living organisms are made up of cells filled with water and a great variety of dissolved substances. Foods are mostly water, with fruits and vegetables containing up to 95 percent water and meat consisting of about 50 percent water. Your body is over 70 percent water by weight. Since water is such a large component of living things, understanding the properties of water is important to understanding life. One important property is water's unusual ability to act as a solvent. Water is called a "universal solvent" because of its ability to dissolve most molecules. In living things these dissolved molecules can be transported from one place to another by diffusion or by some kind of a circulatory system.

The usefulness of water does not end with its unique abilities as a solvent and transporter; it has many more properties that are useful, although unusual. For example, unlike other liquids, water in its liquid phase has a greater density than solid water (ice). This important property enables solid ice to float on the surface of liquid water, insulating the water below and permitting fish and other water organisms to survive the winter. If ice were denser than water, it would sink, freezing all lakes and rivers from the bottom up. Fish and most organisms that live in water would not be able to survive in a lake or river of solid ice.

As described in chapter 4, water is also unusual because it has a high specific heat. The same amount of sunlight falling on equal masses of soil and water will warm the soil 5°C for each

FIGURE 11.1
A freshwater stream has many potential uses.

TABLE 11.1
Possible pollution problems in the U.S. water supply

Pollutant	Source	Risk
Lead	Lead pipes in older homes; solder in copper pipes; brass fixtures	Nerve damage, miscarriage, birth defects, high blood pressure, hearing problems
Chlorinated solvents	Industrial pollution	Cancer
Trihalomethanes	Chlorine disinfectant reacting with other pollutants	Liver damage, kidney damage, possible cancer
PCBs	Industrial waste, older transformers	Liver damage, possible cancer
Bacteria and viruses	Septic tanks, outhouses, overflowing sewer lines	Gastrointestinal problems, serious disease

1°C increase in water temperature. Thus, it will take five times more sunlight to increase the temperature of the water as much as the soil temperature change. This enables large bodies of water to moderate the temperature, making it more even.

A high latent heat of vaporization is yet another unusual property of water. This property enables people to dissipate large amounts of heat by evaporating a small amount of water. Since people carry this evaporative cooling system with them, they can survive some very warm desert temperatures, for example.

Finally, other properties of water are not crucial for life, but are interesting nonetheless. For example, why do all snowflakes have six sides? Is it true that no two snowflakes are alike? The unique structure of the water molecule will explain water's unique solvent abilities, why solid water is less dense than liquid water, its high specific heat, its high latent heat of vaporization, and perhaps why no two snowflakes seem to be alike.

Structure of Water Molecules

In chapter 9 you learned that atoms combine in two ways. Atoms from opposite sides of the periodic table form ionic bonds after transferring one or more electrons. Atoms from the right side of the periodic table form covalent bonds by sharing one or more pairs of electrons. This distinction is clear-cut in many compounds, but not in water. The way atoms share electrons in a water molecule is not exactly covalent, but it is not ionic, either. As you learned in chapter 9, the bond that is not exactly covalent or ionic is called a *polar covalent bond*.

In a water molecule an oxygen atom shares a pair of electrons with each of two hydrogen atoms with polar covalent bonds. Oxygen has six outer electrons and needs two more to satisfy the octet rule, achieving the noble gas structure of eight. Each hydrogen atom needs one more electron to fill its outer orbital with two. Therefore, one oxygen atom bonds with two hydrogen atoms, forming H_2O. Both oxygen and hydrogen are more stable with the outer orbital configuration of the noble gases (neon and helium in this case).

Electrons are shared in a water molecule, but not equally. Oxygen, with its eight positive protons, has a greater attraction for the shared electrons than do either of the hydrogens with a single proton. Therefore, the shared electrons spend more time around the oxygen part of the molecule than they do around the hydrogen part. This results in the oxygen end of the molecule being more negative than the hydrogen end. When electrons in a covalent bond are not equally shared the molecule is said to be polar. A **polar molecule** has a *dipole* ("di" = two; "pole" = side or end), meaning it has a positive end and a negative end.

A water molecule has a negative center at the oxygen end and a positive center at the hydrogen end. The positive charges on the hydrogen end are separated, giving the molecule a bent

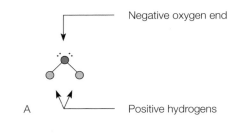

Negative oxygen end

Positive hydrogens

A

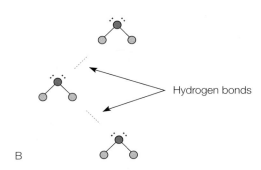

Hydrogen bonds

B

FIGURE 11.2

(A) The water molecule is polar, with centers of positive and negative charges. (B) Attractions between these positive and negative centers establish hydrogen bonds between adjacent molecules.

rather than straight-line arrangement. Figure 11.2A shows a model of a water molecule showing its polar nature.

It is the polar structure of the water molecule that is responsible for many of the unique properties of water. Polar molecules of any substances have attractions between the positive end of a molecule and the negative end of another molecule. When the polar molecule has hydrogen at one end and fluorine, oxygen, or nitrogen on the other, the attractions are strong enough to make a type of bonding called **hydrogen bonding.** Hydrogen bonding is a strong bond that occurs between the hydrogen end of a molecule and the fluorine, oxygen, or nitrogen end of similar molecules. A better name for this would be a hydrogen-fluorine bond, a hydrogen-oxygen bond, or a hydrogen-nitrogen bond. However, for brevity the second part of the bond is not named and all the hydrogen-something bonds are simply known as "hydrogen" bonds. The dotted line between the hydrogen and oxygen molecules in Figure 11.2B represents a hydrogen bond. A dotted line is used to represent a bond that is not as strong as the bond represented by the solid line of a covalent compound.

Hydrogen bonding accounts for the physical properties of water, including its unusual density changes with changes in temperature. Figure 11.3 shows the hydrogen-bonded structure of ice. Water molecules form a six-sided hexagonal structure that extends out for billions of molecules. The large channels, or holes, in the structure result in ice being less dense than water. The shape of the hexagonal arrangement also suggests why snowflakes always have six sides. Why does it seem like no two snowflakes are alike? Perhaps the answer can be found in the

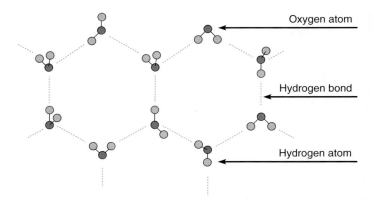

Oxygen atom

Hydrogen bond

Hydrogen atom

FIGURE 11.3

The hexagonal structure of ice. Hydrogen bonding between the oxygen atom and two hydrogen atoms of other water molecules results in a tetrahedral arrangement, which forms the open, hexagonal structure of ice.

almost infinite variety of shapes that can be built from billions and billions of tiny hexagons of ice crystals.

When ice is warmed, the increased vibrations of the molecules begin to expand and stretch the hydrogen bond structure. When ice melts, about 15 percent of the hydrogen bonds break and the open structure collapses into the more compact arrangement of liquid water. As the liquid water is warmed from 0°C still more hydrogen bonds break down, and the density of the water steadily increases. At 4°C the expansion of water from the increased molecular vibrations begins to predominate and the density decreases steadily with further warming (Figure 11.4). Thus, water has its greatest density at a temperature of 4°C.

The heat of fusion, specific heat, and heat of vaporization of water are unusually high when compared to other, but chemically

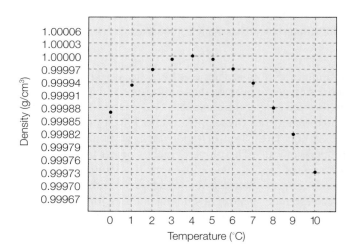

FIGURE 11.4

The density of water from 0°C to 10°C. The density of water is at a maximum at 4°C, becoming less dense as it is cooled or warmed from this temperature. Hydrogen bonding explains this unusual behavior.

similar, substances. These high values are accounted for by the additional energy needed to break hydrogen bonds.

The Dissolving Process

A **solution** is a homogeneous mixture of ions or molecules of two or more substances. *Dissolving* is the process of making a solution. During dissolving, the different components that make up the solution become mixed. For example, when sugar dissolves in water the molecules of sugar become uniformly dispersed throughout the molecules of water. The uniform taste of sweetness of any part of the sugar solution is a result of this uniform mixing.

The general terms *solvent* and *solute* identify the components of a solution. The solvent is the component present in the larger amount. The solute is the component that dissolves in the solvent. Atmospheric air, for example, is about 78 percent nitrogen, so nitrogen is considered the solvent. Oxygen (about 21 percent), argon (about 0.9 percent), and other gases make up the solutes. If one of the components of a solution is a liquid, it is usually identified as the solvent. An *aqueous solution* is a solution of a solid, a liquid, or a gas in water.

A solution is formed when the molecules or ions of two or more substances become homogeneously mixed. But the process of dissolving must be more complicated than the simple mixing together of particles because (1) solutions become saturated, meaning there is a limit on solubility, and (2) some substances are *insoluble,* not dissolving at all or at least not noticeably. In general, the forces of attraction between molecules or ions of the solvent and solute determine if something will dissolve and if there are any limits on the solubility. These forces of attraction and their role in the dissolving process will be considered in the following examples.

First, consider the dissolving process in gaseous and liquid solutions. In a gas, the intermolecular forces are small, so gases can mix in any proportion. Fluids that can mix in any proportion without separating into phases are called **miscible fluids.** Fluids that do not mix are called *immiscible fluids.* Air is a mixture of gases, so gases (including vapors) are miscible.

Liquid solutions can dissolve a gas, another liquid, or a solid. Gases are miscible in liquids, and a carbonated beverage (your favorite cola) is the common example, consisting of carbon dioxide dissolved in water. Whether or not two given liquids form solutions depends on some similarities in their molecular structures. The water molecule, for example, is a polar molecule with a negative end and a positive end. On the other hand, carbon tetrachloride (CCl_4) is a molecule with covalent bonds that are symmetrically arranged. Because of the symmetry, CCl_4 has no negative or positive ends, so it is nonpolar. Thus some liquids have polar molecules, and some have nonpolar molecules. The general rule for forming solutions is *like dissolves like.* A nonpolar compound, such as carbon tetrachloride, will dissolve oils and greases because they are nonpolar compounds. Water, a polar compound, will not dissolve the nonpolar oils and greases. Carbon tetrachloride was at one time used as a cleaning solvent because of its oil and grease dissolving abilities. Its use is no longer recommended because it causes liver damage.

Some molecules, such as soap, have a part of the molecule that is polar and a part that is nonpolar. Washing with water alone will not dissolve oils because water and oil are immiscible. When soap is added to the water, however, the polar end of the soap molecule is attracted to the polar water molecules, and the nonpolar end is absorbed into the oil. A particle (larger than a molecule) is formed, and the oil is washed away with the water (Figure 11.5).

Concepts Applied

How to Mix

Obtain a small, clear bottle water container with a screw-on cap. Fill the bottle halfway with water, then add some food coloring and swirl to mix. Now add enough mineral or cooking oil to almost fill the bottle. Seal the bottle tightly with the cap.

Describe the oil and water in the bottle. Shake the bottle vigorously for about 30 seconds, then observe what happens when you stop shaking. Does any of the oil and water mix?

Try mixing the oil and water again, this time after adding a squirt of liquid dishwashing soap. Describe what happens before and after adding the soap. What does this tell you about the structure of the oil and water molecules? How did soap overcome these differences?

The "like dissolves like" rule applies to solids and liquid solvents as well as liquids and liquid solvents. Polar solids, such as salt, will readily dissolve in water, which has polar molecules, but do not dissolve readily in oil, grease, or other nonpolar solvents. Polar water readily dissolves salt because the charged polar water molecules are able to exert an attraction on the ions, pulling them away from the crystal structure. Thus, ionic compounds dissolve in water.

Ionic compounds vary in their solubilities in water. This difference is explained by the existence of two different forces involved in an ongoing "tug of war." One force is the attraction between an ion on the surface of the crystal and a water molecule, an *ion-polar molecule force.* When solid sodium chloride and water are mixed together, the negative ends of the water molecules (the oxygen ends) become oriented toward the positive sodium ions on the crystal. Likewise, the positive ends of water molecules (the hydrogen ends) become oriented toward the negative chlorine ions. The attraction of water molecules for ions is called *hydration.* If the force of hydration is greater than the attraction between the ions in the solid, they are pulled away from the solid, and dissolving

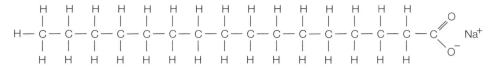
Structural formula of a soap molecule

Hydrocarbon end
(soluble in oil)

Ionic end
(soluble in water)

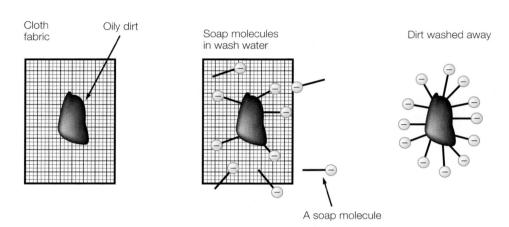

Cloth fabric Oily dirt

Soap molecules
in wash water

Dirt washed away

A soap molecule

FIGURE 11.5

Soap cleans oil and grease because one end of the soap molecule is soluble in water, and the other end is soluble in oil and grease. Thus, the soap molecule provides a link between two substances that would otherwise be immiscible.

occurs (Figure 11.6). Not considering the role of water in this dissolving process, the equation is

$$\text{Na}^+\text{Cl}^-_{(s)} \longrightarrow \text{Na}^+_{(aq)} + \text{Cl}^-_{(aq)}$$

which shows that the ions were separated from the solid to become a solution of ions. In other compounds the attraction between the ions in the solid might be greater than the energy of hydration. In this case, the ions of the solid would win the "tug-of-war," and the ionic solid is insoluble.

The saturation of soluble compounds is explained in terms of hydration eventually occupying a large number of the polar water molecules. Fewer available water molecules means less attraction on the ionic solid, with more solute ions being pulled back to the surface of the solid. The tug-of-war continues back and forth as an equilibrium condition is established.

Concentration of Solutions

The relative amounts of solute and solvent are described by the **concentration** of a solution. In general, a solution with a large amount of solute is *concentrated,* and a solution with much less solute is *dilute.* The terms "dilute" and "concentrated" are somewhat arbitrary, and it is sometimes difficult to know the

difference between a solution that is "weakly concentrated" and one that is "not very diluted." More meaningful information is provided by measurement of the *amount of solute in a solution.* There are different ways to express concentration measurements, each lending itself to a particular kind of solution or to how the information will be used. For example, you read about concentrations of parts per million in an article about pollution, but most of the concentrations of solutions sold in stores are reported in percent by volume or percent by weight (Figure 11.7). Each of these concentrations is concerned with the amount of *solute* in the *solution.*

Concentration ratios that describe small concentrations of solute are sometimes reported as a ratio of *parts per million* (ppm) or *parts per billion* (ppb). This ratio could mean ppm by volume or ppm by weight, depending on whether the solution is a gas or a liquid. For example, a drinking water sample with 1 ppm Na^+ by weight has 1 weight measure of solute, sodium ions, *in* every 1,000,000 weight measures of the total solution. By way of analogy, 1 ppm expressed in money means 1 cent in every $10,000 (which is one million cents). A concentration of 1 ppb means 1 cent in $10,000,000. Thus, the concentrations of very dilute solutions, such as certain salts in seawater, minerals in drinking water, and pollutants in water or in the atmosphere are often reported in ppm or ppb.

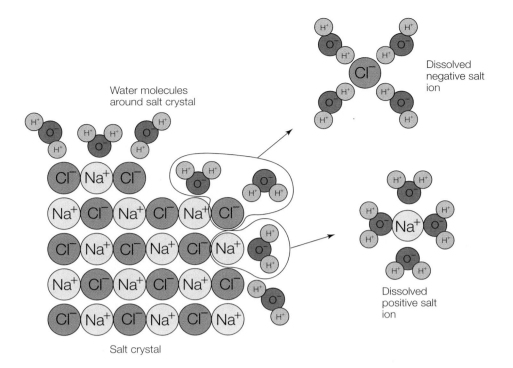

FIGURE 11.6

An ionic solid dissolves in water because the number of water molecules around the surface is greater than the number of other ions of the solid. The attraction between polar water molecules and a charged ion enables the water molecules to pull ions away from the crystal, so the salt crystals dissolve in the water.

Concepts Applied

ppm or ppb to Percent

Sometimes it is useful to know the conversion factors between ppm or ppb and the more familiar percent concentration by weight. These factors are ppm ÷ (1×10^4) = percent concentration and ppb ÷ (1×10^7) = percent concentration. For example, very hard water (water containing Ca^{2+} or Mg^{2+} ions), by definition, contains more than 300 ppm of the ions. This is a percent concentration of 300 ÷ 1 × 10^4, or 0.03 percent. To be suitable for agricultural purposes, irrigation water must not contain more than 700 ppm of total dissolved salts, which means a concentration no greater than 0.07 percent salts.

The concentration term of *percent by volume* is defined as the *volume of solute in 100 volumes of solution*. This concentration term is just like any other percentage ratio, that is, "part" divided by the "whole" times 100 percent. The distinction is that the part and the whole are concerned with a volume of solute and a volume of solution. Knowing the meaning of percent by volume can be useful in consumer decisions. Rubbing alcohol, for example, can be purchased at a wide range of prices. The various brands range from a concentration, according to the labels, of "12% by volume" to "70% by volume." If the volume unit is mL, a "12% by volume" concentration contains 12 mL of pure isopropyl (rubbing) alcohol in every 100 mL of solution. The "70% by volume" contains 70 mL of isopropyl alcohol in every 100 mL of solution. The relationship for % by volume is

$$\frac{\text{volume solute}}{\text{volume solution}} \times 100\% \text{ solution} = \% \text{ solute}$$

or

$$\frac{V_{\text{solute}}}{V_{\text{solution}}} \times 100\% \text{ solution} = \% \text{ solute}$$

equation 11.1

The concentration term of *percent by weight* is defined as the *weight of solute in 100 weight units of solution*. This concentration term is just like any other percentage composition, the difference being that it is concerned with the weight of solute (the part) in a weight of solution (the whole). Hydrogen peroxide, for example, is usually sold in a concentration of "3% by weight." This means that 3 oz (or other weight units) of pure hydrogen peroxide are in 100 oz of solution. Since weight is proportional to mass in a given location, mass units such as grams are sometimes used to calculate a percent by weight. The relationship for percent by weight (using mass units) is

A Solution strength by parts

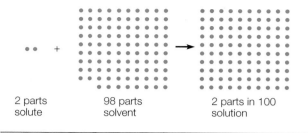

2 parts   98 parts   2 parts in 100
solute   solvent   solution

B Solution strength by percent (volume)

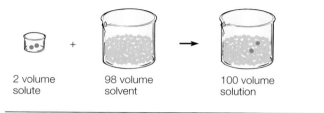

2 volume   98 volume   100 volume
solute   solvent   solution

C Solution strength by percent (weight)

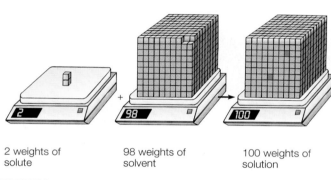

2 weights of   98 weights of   100 weights of
solute   solvent   solution

FIGURE 11.7

Three ways to express the amount of solute in a solution: (A) as parts (e.g., parts per million), this is 2 parts per 100; (B) as a percent by volume, this is 2 percent by volume; (C) as percent by weight, this is 2 percent by weight.

$$\frac{\text{mass of solute}}{\text{mass of solution}} \times 100\% \text{ solution} = \% \text{ solute}$$

or

$$\frac{m_{solute}}{m_{solution}} \times 100\% \text{ solution} = \% \text{ solute}$$

equation 11.2

EXAMPLE **11.1**

Vinegar that is prepared for table use is a mixture of acetic acid in water, usually 5.00% by weight. How many grams of pure acetic acid are in 25.0 g of vinegar?

Solution

The percent by weight is given (5.00%), the mass of the solution is given (25.0 g), and the mass of the solute ($H_2C_2H_3O_2$) is the unknown. The relationship between these quantities is found in equation 11.2, which can be solved for the mass of the solute:

$$\% \text{ solute} = 5.00\%$$
$$m_{solution} = 25.0 \text{ g}$$
$$m_{solute} = ?$$

$$\frac{m_{solute}}{m_{solution}} \times 100\% \text{ solution} = \% \text{ solute}$$

$$\therefore$$

$$m_{solute} = \frac{(m_{solution})(\% \text{ solute})}{100\% \text{ solution}}$$

$$= \frac{(m_{solution})(\% \text{ solute})}{100\% \text{ solution}}$$

$$= \frac{(25.0 \text{ g})(5.00)}{100} \text{ solute}$$

$$= \boxed{1.25 \text{ g solute}}$$

EXAMPLE **11.2**

A solution used to clean contact lenses contains 0.002% by volume of thimerosal as a preservative. How many mL of this preservative are needed to make 100,000 L of the cleaning solution? (Answer: 2.0 mL)

Both percent by volume and percent by weight are defined as the volume or weight per 100 units of solution because percent *means* parts per hundred. The measure of dissolved salts in seawater is called *salinity*. **Salinity** is defined as the mass of salts dissolved in 1,000 g of solution. As illustrated in Figure 11.8, evaporation of 965 g of water from 1,000 g of seawater will leave

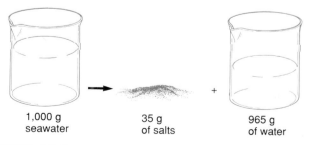

1,000 g   35 g   965 g
seawater   of salts   of water

FIGURE 11.8

Salinity is a measure of the amount of salts dissolved in 1 kg of solution. If 1,000 g of seawater were evaporated, 35.0 g of salts would remain as 965.0 g of water leave.

an average of 35 g salts. Thus, the average salinity of the seawater is 35‰. Note the ‰, which means parts per thousand just as % means parts per hundred. The equivalent percent measure for salinity is 3.5%, which equals 35‰.

Recall from chapter 10 that a *mole* is a measure of amount used in chemistry. One mole is defined as the amount of a substance that contains the same number of elementary units as there are atoms in exactly 12 grams of the carbon-12 isotope. The number of units in this case is called *Avogadro's number,* which is 6.02×10^{23}—a very large number. This measure can be compared with identifying amounts in the grocery store by the dozen. You know that a dozen is twelve of something. Now you know that a mole is 6.02×10^{23} of whatever you are measuring.

Chemists use a measure of concentration that is convenient for considering chemical reactions of solutions. The measure is based on moles of solute since a mole is a known number of particles (atoms, molecules, or ions). The concentration term of **molarity (M)** is defined as the number of moles of solute dissolved in one liter of solution. Thus,

$$\text{Molarity (M)} = \frac{\text{moles of solute}}{\text{liters of solution}}$$

equation 11.3

An aqueous solution of NaCl that has a molarity of 1.0 contains 1.0 mole NaCl per liter of solution. To make such a solution you would place 58.5 g (1.0 mole) NaCl in a beaker, then add water to make 1 liter of solution.

Solubility

There is a limit to how much solid can be dissolved in a liquid. You may have noticed that a cup of hot tea will dissolve several teaspoons of sugar, but the limit of solubility is reached quickly in a glass of iced tea. The limit of how much sugar will dissolve seems to depend on the temperature of the tea. More sugar added to the cold tea after the limit is reached will not dissolve, and solid sugar granules begin to accumulate at the bottom of the glass. At this limit the sugar and tea solution is said to be *saturated*. Dissolving does not actually stop when a solution becomes saturated, and undissolved sugar continues to enter the solution. However, dissolved sugar is now returning to the undissolved state at the same rate as it is dissolving. The overall equilibrium condition of sugar dissolving as sugar is coming out of solution is called a *saturated solution*. A saturated solution is a *state of equilibrium that exists between dissolving solute and solute coming out of solution*. You actually cannot see the dissolving and coming out of solution that occurs in a saturated solution because the exchanges are taking place with particles the size of molecules or ions.

Not all compounds dissolve as sugar does, and more or less of a given compound may be required to produce a saturated solution at a particular temperature. In general, the difficulty of dissolving a given compound is referred to as *solubility*. More specifically, the **solubility** of a solute is defined as the *concentration that is reached in a saturated solution at a*

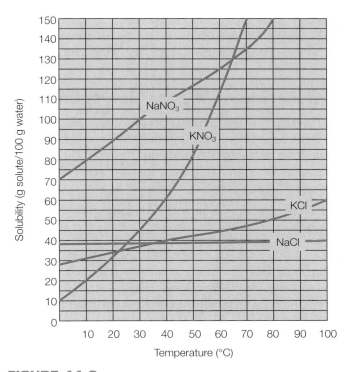

FIGURE 11.9

Approximate solubility curves for sodium nitrate, potassium nitrate, potassium chloride, and sodium chloride.

particular temperature. Solubility varies with the temperature as the sodium and potassium salt examples show in Figure 11.9. These solubility curves describe the amount of solute required to reach the saturation equilibrium at a particular temperature. In general, the solubilities of most ionic solids increase with temperature, but there are exceptions. In addition, some salts release heat when dissolved in water, and other salts absorb heat when dissolved. The "instant cold pack" used for first aid is a bag of water containing a second bag of ammonium nitrate (NH_4NO_3). When the bag of ammonium nitrate is broken, the compound dissolves and absorbs heat.

You can usually dissolve more of a solid, such as salt or sugar, as the temperature of the water is increased. Contrary to what you might expect, gases usually become *less* soluble in water as the temperature increases. As a glass of water warms, small bubbles collect on the sides of the glass as dissolved air comes out of solution. The first bubbles that appear when warming a pot of water to boiling are also bubbles of dissolved air coming out of solution. This is why water that has been boiled usually tastes "flat." The dissolved air has been removed by the heating. The "normal" taste of water can be restored by pouring the boiled water back and forth between two glasses. The water dissolves more air during this process, restoring the usual taste.

Changes in pressure have no effect on the solubility of solids in liquids but greatly affect the solubility of gases. The release of bubbles (fizzing) when a bottle or can of soda is opened occurs because pressure is reduced on the beverage and dissolved carbon dioxide comes out of solution. In general, *gas solubility decreases with temperature and increases with pressure.* As usual, there are exceptions to this generalization.

Properties of Water Solutions

Pure solvents have characteristic physical and chemical properties that are changed by the presence of the solute. Following are some of the more interesting changes.

Electrolytes

Water solutions of ionic substances will conduct an electric current, so they are called **electrolytes.** Ions must be present and free to move in a solution to carry the charge, so electrolytes are solutions containing ions. Pure water will not conduct an electric current because it is a covalent compound, which ionizes only very slightly. Water solutions of sugar, alcohol, and most other covalent compounds are nonconductors, so they are called *nonelectrolytes.* Nonelectrolytes are covalent compounds that form molecular solutions, so they cannot conduct an electric current (Figure 11.10).

Some covalent compounds are nonelectrolytes as pure liquids but become electrolytes when dissolved in water. Pure hydrogen chloride (HCl), for example, does not conduct an electric current, so you can assume that it is a molecular substance. When dissolved in water, hydrogen chloride does conduct a current, so it must now contain ions. Evidently, the hydrogen chloride has become *ionized* by the water. The process of forming ions from molecules is called *ionization.* Hydrogen

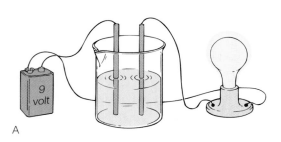

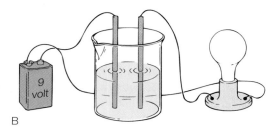

FIGURE 11.10

(*A*) Water solutions that conduct an electric current are called *electrolytes.* (*B*) Water solutions that do not conduct electricity are called *nonelectrolytes.*

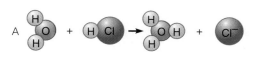

B   H:Ö: + H:Cl: → H:Ö:H⁺ + :Cl:⁻
       Ḧ   Ḧ

C H_2O + HCl → H_3O^+ + Cl^-

FIGURE 11.11

Three representations of water and hydrogen chloride in an ionizing reaction. (*A*) Sketches of molecules involved in the reaction. (*B*) Electron dot equation of the reaction. (*C*) The chemical equation for the reaction. Each of these representations shows the hydrogen being pulled away from the chlorine atom to form H_3O^+, the hydronium ion.

chloride, like water, has polar molecules. The positive hydrogen atom on the HCl molecule is attracted to the negative oxygen end of a water molecule, and the force of attraction is strong enough to break the hydrogen-chlorine bond, forming charged particles (Figure 11.11). The reaction is

$$HCl_{(l)} + H_2O_{(l)} \longrightarrow H_3O^+_{(aq)} + Cl^-_{(aq)}$$

The H_3O^+ ion is called a **hydronium ion.** A hydronium ion is basically a molecule of water with an attached hydrogen ion. The presence of the hydronium ion gives the solution new chemical properties; the solution is no longer hydrogen chloride but is *hydrochloric acid.* Hydrochloric acid, and other acids, will be discussed shortly.

Boiling Point

Boiling occurs when the pressure of the vapor escaping from a liquid is equal to the atmospheric pressure on the liquid. The *normal* boiling point is defined as the temperature at which the vapor pressure is equal to the average atmospheric pressure at sea level. For pure water, this temperature is 100°C (212°F). It is important to remember that boiling is a purely physical process. No bonds within water molecules are broken during boiling.

The vapor pressure over a solution is *less* than the vapor pressure over the pure solvent at the same temperature. Molecules of a liquid can escape into the air only at the surface of the liquid, and the presence of molecules of a solute means that fewer solvent molecules can be at the surface to escape. Thus, the vapor pressure over a solution is less than the vapor pressure over a pure solvent (Figure 11.12).

Because the vapor pressure over a solution is less than that over the pure solvent, the solution boils at a higher temperature. A higher temperature is required to increase the vapor pressure to that of the atmospheric pressure. Some cooks have been observed to add a "pinch" of salt to a pot of water before boiling.

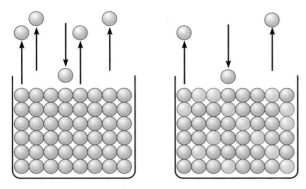

FIGURE 11.12

The rate of evaporation, and thus the vapor pressure, is less for a solution than for a solvent in the pure state. The greater the solute concentration, the less the vapor pressure.

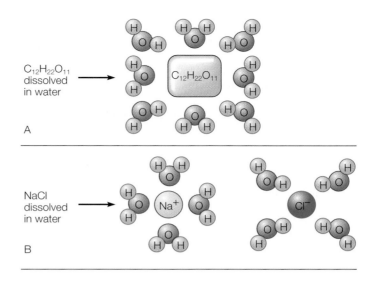

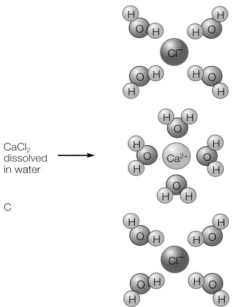

FIGURE 11.13

Since ionic compounds dissolve by the separation of ions, they provide more particles in solution than molecular compounds. (A) A mole of sugar provides Avogadro's number of particles. (B) A mole of NaCl provides two times Avogadro's number of particles. (C) A mole of $CaCl_2$ provides three times Avogadro's number of particles.

Is this to increase the boiling point, and therefore cook the food more quickly? How much does a pinch of salt increase the boiling temperature? The answers are found in the relationship between the concentration of a solute and the boiling point of the solution.

It is the number of solute particles (ions or molecules) at the surface of a solution that increases the boiling point. Recall that a mole is a measure that can be defined as a number of particles called Avogadro's number. Since the number of particles at the surface is proportional to the ratio of particles in the solution, the concentration of the solute will directly influence the increase in the boiling point. In other words, the boiling point of any dilute solution is increased proportional to the concentration of the solute. For water, the boiling point is increased 0.521°C for every mole of solute dissolved in 1,000 g of water. Thus, any water solution will boil at a higher temperature than pure water. Since it boils at a higher temperature, it also takes a longer time to reach the boiling point.

It makes no difference what substance is dissolved in the water; one mole of solute in 1,000 g of water will elevate the boiling point by 0.521°C. A mole contains Avogadro's number of particles, so a mole of any solute will lower the vapor pressure by the same amount. Sucrose, or table sugar, for example, is $C_{12}H_{22}O_{11}$ and has a gram-formula weight of 342 g. Thus 342 g of sugar in 1,000 g of water (about a liter) will increase the boiling point by 0.521°C. Therefore, if you measure the boiling point of a sugar solution you can determine the concentration of sugar in the solution. For example, pancake syrup that boils at 100.261°C (sea-level pressure) must contain 171 g of sugar dissolved in 1,000 g of water. You know this because the increase of 0.261°C over 100°C is one-half of 0.521°C. If the boiling point were increased by 0.521°C over 100°C, the syrup would have the full gram-formula weight (342 g) dissolved in a kg of water.

Since it is the number of particles of solute in a specific sample of water that elevates the boiling point, different effects are observed in dissolved covalent and dissolved ionic compounds (Figure 11.13). Sugar is a covalent compound, and the solute is molecules of sugar moving between the water molecules. Sodium chloride, on the other hand, is an ionic compound and dissolves by the separation of ions, or

$$Na^+Cl^-{}_{(s)} \longrightarrow Na^+{}_{(aq)} + Cl^-{}_{(aq)}$$

This equation tells you that one mole of NaCl separates into one mole of sodium ions and one mole of chlorine ions for a total of *two* moles of solute. The boiling point elevation of a solution made from one mole of NaCl (58.5 g) is therefore multiplied by two, or $2 \times 0.521°C = 1.04°C$. The boiling point of a solution made by adding 58.5 g of NaCl to 1,000 g of water is therefore 101.04°C at normal sea-level pressure.

Now back to the question of how much a pinch of salt increases the boiling point of a pot of water. Assuming the pot contains about a liter of water (about a quart), and assuming that a "pinch" of salt has a mass of about 0.2 gram, the boiling point will be increased by 0.0037°C. Thus, there must be some reason other than increasing the boiling point that a cook adds a pinch of salt to a pot of boiling water. Perhaps the salt is for seasoning?

Freezing Point

Freezing occurs when the kinetic energy of molecules has been reduced sufficiently so the molecules can come together, forming the crystal structure of the solid. Reduced kinetic energy of the molecules, that is, reduced temperature, results in a specific freezing point for each pure liquid. The *normal* freezing point for pure water, for example, is 0°C (32°F) under normal pressure. The presence of solute particles in a solution interferes with the water molecules as they attempt to form the six-sided hexagonal structure. The water molecules cannot get by the solute particles until the kinetic energy of the solute particles is reduced, that is, until the temperature is below the normal freezing point. Thus, the presence of solute particles lowers the freezing point, and solutions freeze at a lower temperature than the pure solvent.

The freezing-point depression of a solution has a number of interesting implications for solutions such as seawater. When seawater freezes, the water molecules must work their way around the salt particles as was described earlier. Thus, the solute particles are *not* normally included in the hexagonal structure of ice. Ice formed in seawater is practically pure water. Since the solute was *excluded* when the ice formed, the freezing of seawater increases the salinity. Increased salinity means increased concentration, so the freezing point of seawater is further depressed and more ice forms only at a lower temperature. When this additional ice forms more pure water is removed, and the process goes on. Thus, seawater does not have a fixed freezing point but has a lower and lower freezing point as more and more ice freezes.

The depression of the freezing point by a solute has a number of interesting applications in colder climates. Salt, for example, is spread on icy roads to lower the freezing point (and thus the melting point) of the ice. Calcium chloride, $CaCl_2$, is a salt that is often used for this purpose. Water in a car radiator would also freeze in colder climates if a solute, called antifreeze, were not added to the radiator water. Methyl alcohol has been used as an antifreeze because it is soluble in water and does not damage the cooling system. Methyl alcohol, however, has a low boiling point and tends to boil away. Ethylene glycol has a higher boiling point, so it is called a "permanent" antifreeze. Like other solutes, ethylene glycol also raises the boiling point, which is an added benefit for summer driving.

Acids, Bases, and Salts

The electrolytes known as *acids, bases,* and *salts* are evident in environmental quality, foods, and everyday living. Environmental quality includes the hardness of water, which is determined by the presence of certain salts, the acidity of soils, which determines how well plants grow, and acid rain, which is a by-product of industry and automobiles. Many concerns about air and water pollution are often related to the chemistry concepts of acids, bases, and salts. These concepts, and uses of acids, bases, and salts, will be considered in this section.

Properties of Acids and Bases

Acids and bases are classes of chemical compounds that have certain characteristic properties. These properties can be used to identify if a substance is an acid or a base (Tables 11.2 and 11.3). The following are the properties of *acids* dissolved in water:

1. Acids have a sour taste such as the taste of citrus fruits.
2. Acids change the color of certain substances; for example, litmus changes from blue to red when placed in an acid solution (Figure 11.14A).
3. Acids react with active metals, such as magnesium or zinc, releasing hydrogen gas.
4. Acids *neutralize* bases, forming water and salts from the reaction.

TABLE 11.2

Some common acids

Name	Formula	Comment
Acetic acid	CH_3COOH	A weak acid found in vinegar
Boric acid	H_3BO_3	A weak acid used in eyedrops
Carbonic acid	H_2CO_3	The weak acid of carbonated beverages
Formic acid	HCOOH	Makes the sting of insects and certain plants
Hydrochloric acid	HCl	Also called muriatic acid; used in swimming pools, soil acidifiers, and stain removers
Lactic acid	$CH_3CHOHCOOH$	Found in sour milk, sauerkraut, and pickles; gives tart taste to yogurt
Nitric acid	HNO_3	A strong acid
Phosphoric acid	H_3PO_4	Used in cleaning solutions; added to carbonated beverages for tartness
Sulfuric acid	H_2SO_4	Also called oil of vitriol; used as battery acid and in swimming pools

TABLE 11.3
Some common bases

Name	Formula	Comment
Sodium hydroxide	NaOH	Also called lye or caustic soda; a strong base used in oven cleaners and drain cleaners
Potassium hydroxide	KOH	Also called caustic potash; a strong base used in drain cleaners
Ammonia	NH_3	A weak base used in household cleaning solutions
Calcium hydroxide	$Ca(OH)_2$	Also called slaked lime; used to make brick mortar
Magnesium hydroxide	$Mg(OH)_2$	Solution is called milk of magnesia; used as antacid and laxative

Likewise, *bases* have their own characteristic properties. Bases are also called alkaline substances, and the following are the properties of bases dissolved in water:

1. Bases have a bitter taste, for example, the taste of caffeine.
2. Bases reverse the color changes that were caused by acids. Red litmus is changed back to blue when placed in a solution containing a base (Figure 11.14B).
3. Basic solutions feel slippery on the skin. They have a *caustic* action on plant and animal tissue, converting tissue into soluble materials. A strong base, for example, reacts with fat to make soap and glycerine. This accounts for the slippery feeling on the skin.
4. Bases *neutralize* acids, forming water and salts from the reaction.

Tasting an acid or base to see if it is sour or bitter can be hazardous, since some are highly corrosive or caustic. Many organic acids are not as corrosive and occur naturally in foods. Citrus fruit, for example, contains citric acid, vinegar is a solution of acetic acid, and sour milk contains lactic acid. The stings or bites of some insects (bees, wasps, and ants) and some plants (stinging nettles) are painful because an organic acid, formic acid, is injected by the insect or plant. Your stomach contains a solution of hydrochloric acid. In terms of relative strength, the hydrochloric acid in your stomach is about ten times stronger than the carbonic acid (H_2CO_3) of carbonated beverages.

Examples of bases include solutions of sodium hydroxide (NaOH), which has a common name of lye or caustic soda, and potassium hydroxide (KOH), which has a common name of caustic potash. These two bases are used in products known as drain cleaners. They open plugged drains because of their caustic action, turning grease, hair, and other organic "plugs" into soap and other soluble substances that are washed away. A weaker base is a solution of ammonia (NH_3), which is often used as a household cleaner. A solution of magnesium hydroxide, $Mg(OH)_2$, has a common name of milk of magnesia and is sold as an antacid and laxative.

Many natural substances change color when mixed with acids or bases. You may have noticed that tea changes color slightly, becoming lighter, when lemon juice (which contains citric acid) is added. Some plants have flowers of one color when grown in acidic soil and flowers of another color when grown in basic soil. A vegetable dye that changes color in the presence of acids or bases can be used as an **acid-base indicator.** An indicator is simply a vegetable dye that is used to distinguish between acid and base solutions by a color change. Litmus, for example, is an acid-base indicator made from a dye extracted from certain species of lichens. The dye is applied to paper strips, which turn red in acidic solutions and blue in basic solutions.

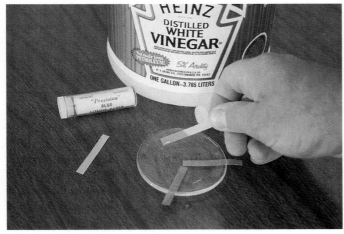

A

B

FIGURE 11.14
(*A*) Acid solutions will change the color of blue litmus to red. (*B*) Solutions of bases will change the color of red litmus to blue.

Explaining Acid-Base Properties

Comparing the lists in Tables 11.2 and 11.3, you can see that acids and bases appear to be chemical opposites. Notice in Table 11.2 that the acids all have an H, or hydrogen atom, in their formulas. In Table 11.3, most of the bases have a hydroxide ion, OH^-, in their formulas. Could this be the key to acid-base properties?

The modern concept of an acid considers the properties of acids in terms of the hydronium ion, H_3O^+. As was mentioned earlier, the hydronium ion is a water molecule to which an H^+ ion is attached. Since a hydrogen ion is a hydrogen atom without its single electron, it could be considered as an ion consisting of a single proton. Thus the H^+ ion can be called a *proton*. An **acid** is defined as any substance that is a *proton donor* when dissolved in water, increasing the hydronium ion concentration. For example, hydrogen chloride dissolved in water has the following reaction:

$$\overset{\frown}{(H)Cl_{(aq)}} \quad + \quad H_2O_{(l)} \quad \longrightarrow \quad H_3O^+_{(aq)} \quad + \quad Cl^-_{(aq)}$$

The dotted circle and arrow were added to show that the hydrogen chloride donated a proton to a water molecule. The resulting solution contains H_3O^+ ions and has acid properties, so the solution is called hydrochloric acid. It is the H_3O^+ ion that is responsible for the properties of an acid.

The bases listed in Table 11.3 all appear to have a hydroxide ion, OH^-. Water solutions of these bases do contain OH^- ions, but the definition of a base is much broader. A **base** is defined as any substance that is a *proton acceptor* when dissolved in water, increasing the hydroxide ion concentration. For example, ammonia dissolved in water has the following reaction:

$$NH_{3(g)} \quad + \quad \overset{\frown}{(H_2)O_{(l)}} \quad \longrightarrow \quad (NH_4)^+ \quad + \quad OH^-$$

The dotted circle and arrow show that the ammonia molecule accepted a proton from a water molecule, providing a hydroxide ion. The resulting solution contains OH^- ions and has basic properties, so a solution of ammonium hydroxide is a base.

Carbonates, such as sodium carbonate (Na_2CO_3), form basic solutions because the carbonate ion reacts with water to produce hydroxide ions.

$$(CO_3)^{2-}_{(aq)} + H_2O_{(l)} \longrightarrow (HCO_3)^-_{(aq)} + OH^-_{(aq)}$$

Thus, sodium carbonate produces a basic solution.

Acids could be thought of as simply solutions of hydronium ions in water, and bases could be considered solutions of hydroxide ions in water. The proton donor and proton acceptor definition is much broader, and it does include the definition of acids and bases as hydronium and hydroxide compounds. The broader, more general definition covers a wider variety of reactions and is therefore more useful.

The modern concept of acids and bases explains why the properties of acids and bases are **neutralized,** or lost, when acids and bases are mixed together. For example, consider the hydronium ion produced in the hydrochloric acid solution and the hydroxide ion produced in the ammonia solution. When these solutions are mixed together, the hydronium ion reacts with the hydroxide ion, and

$$H_3O^+_{(aq)} + OH^+_{(aq)} \longrightarrow H_2O_{(l)} + H_2O_{(l)}$$

Thus, a proton is transferred from the hydronium ion (an acid), and the proton is accepted by the hydroxide ion (a base). Water is produced, and both the acid and base properties disappear or are neutralized.

Strong and Weak Acids and Bases

Acids and bases are classified according to their degree of ionization when placed in water. *Strong acids* ionize completely in water, with all molecules dissociating into ions. Nitric acid, for example, reacts completely in the following equation:

$$HNO_{3(aq)} + H_2O_{(l)} \longrightarrow H_3O^+_{(aq)} + (NO_3)^-_{(aq)}$$

Nitric acid, hydrochloric acid (Figure 11.15), and sulfuric acid are common strong acids.

Acids that ionize only partially and produce fewer hydronium ions are weaker acids. *Weak acids* are only partially ionized. Vinegar, for example, contains acetic acid that reacts with water in the following reaction:

$$HC_2H_3O_2 + H_2O \longrightarrow H_3O^+ + (C_2H_3O_2)^-$$

Only about 1 percent or less of the acetic acid molecules ionize, depending on the concentration.

Bases are also classified as strong or weak. A *strong base* is completely ionic in solution and has hydroxide ions. Sodium hydroxide, or lye, is the most common example of a strong base. It dissolves in water to form a solution of sodium and hydroxide ions:

$$Na^+OH^-_{(s)} \longrightarrow Na^+_{(aq)} + OH^-_{(aq)}$$

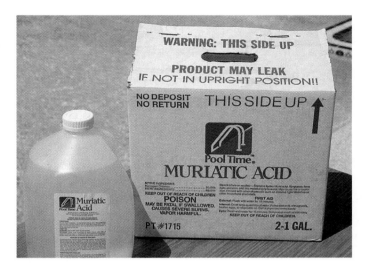

FIGURE 11.15

Hydrochloric acid (HCl) has the common name of *muriatic* acid. Hydrochloric acid is a strong acid used in swimming pools, soil acidifiers, and stain removers.

A *weak base* is only partially ionized. Ammonia, magnesium hydroxide, and calcium hydroxide are examples of weak bases. Magnesium and calcium hydroxide are only slightly soluble in water, and this reduces the *concentration* of hydroxide ions in a solution. It would appear that $Ca(OH)_2$ would produce two moles of hydroxide ions. It would, if it were completely soluble and reacted completely. It is the concentration of hydroxide ions in solution that determines if a base is weak or strong, not the number of ions per mole.

The pH Scale

The strength of an acid or a base is usually expressed in terms of a range of values called a **pH scale.** The pH scale is based on the concentration of the hydronium ion (in moles/L) in an acidic or a basic solution. To understand how the scale is able to express both acid and base strength in terms of the hydronium ion, first note that pure water is very slightly ionized in the reaction:

$$H_2O_{(l)} + H_2O_{(l)} \longrightarrow H_3O^+_{(aq)} + OH^-_{(aq)}$$

The amount of self-ionization by water has been determined through measurements. In pure water at 25°C or any neutral water solution at that temperature, the H_3O^+ concentration is 1×10^{-7} moles/L, and the OH^- concentration is also 1×10^{-7} moles/L. Since both ions are produced in equal numbers, then the H_3O^+ concentration equals the OH^- concentration, and pure water is neutral, neither acidic nor basic.

In general, adding an acid substance to pure water increases the H_3O^+ concentration. Adding a base substance to pure water increases the OH^- concentration. Adding a base also *reduces* the H_3O^+ concentration as the additional OH^- ions are able to combine with more of the hydronium ions to produce un-ionized water. Thus, at a given temperature, an increase in OH^- concentration is matched by a *decrease* in H_3O^+ concentration.

TABLE 11.4

The pH and hydronium ion concentration (moles/L)

Hydronium Ion Concentration (moles/L)	Reciprocal of Hydronium Ion Concentration	pH	Meaning
10^0	10^0	0	
10^{-1}	10^1	①	
10^{-2}	10^2	②	Increasing acidity
10^{-3}	10^3	3	
10^{-4}	10^4	4	
10^{-5}	10^5	5	
10^{-6}	10^6	6	
10^{-7}	10^7	7	neutral
10^{-8}	10^8	8	
10^{-9}	10^9	9	
10^{-10}	10^{10}	10	Increasing basicity
10^{-11}	10^{11}	11	
10^{-12}	10^{12}	12	
10^{-13}	10^{13}	13	
10^{-14}	10^{14}	14	

The concentration of the hydronium ion can be used as a measure of acidic, neutral, and basic solutions. In general, (1) acidic solutions have H_3O^+ concentrations above 1×10^{-7} moles/L, (2) neutral solutions have H_3O^+ concentrations equal to 1×10^{-7} moles/L, and (3) basic solutions have H_3O^+ concentrations less than 1×10^{-7} moles/L. These three statements lead directly to the pH scale, which is named from the French *pouvoir hydrogene,* meaning "hydrogen power." Power refers to the exponent of the hydronium ion concentration, and the pH is a *power of ten notation that expresses the H_3O^+ concentration* (Table 11.4).

A neutral solution has a pH of 7.0. Acidic solutions have pH values below 7, and smaller numbers mean greater acidic properties. Increasing the OH^- concentration decreases the H_3O^+ concentration, so the strength of a base is indicated on the same scale with values greater than 7. Note that the pH scale is logarithmic, so a pH of 2 is ten times as acidic as a pH of 3. Likewise, a pH of 10 is one hundred times as basic as a pH of 8. Table 11.5 compares the pH of some common substances (Figure 11.16).

Concepts Applied

Acid or Base?

Pick some household product that probably has an acid or base character (Example: pH increaser for aquariums). On a separate paper write the listed ingredients and identify any you believe would be distinctly acidic or basic in a water solution. Tell whether you expect the product to be an acid or a base. Describe your findings from a litmus paper test.

TABLE 11.5

The approximate pH of some common substances

Substance	pH (or pH Range)
Hydrochloric acid (4%)	0
Gastric (stomach) solution	1.6–1.8
Lemon juice	2.2–2.4
Vinegar	2.4–3.4
Carbonated soft drinks	2.0–4.0
Grapefruit	3.0–3.2
Oranges	3.2–3.6
Acid rain	4.0–5.5
Tomatoes	4.2–4.4
Potatoes	5.7–5.8
Natural rainwater	5.6–6.2
Milk	6.3–6.7
Pure water	7.0
Seawater	7.0–8.3
Blood	7.4
Sodium bicarbonate solution	8.4
Milk of magnesia	10.5
Ammonia cleaning solution	11.9
Sodium hydroxide solution	13.0

FIGURE 11.16

The pH increases as the acidic strength of these substances decreases from left to right. Did you know that lemon juice is more acidic than vinegar? That a soft drink is more acidic than orange juice or grapefruit juice?

Properties of Salts

Salt is produced by a neutralization reaction between an acid and a base. A **salt** is defined as any ionic compound except those with hydroxide or oxide ions. Table salt, NaCl, is but one example of this large group of ionic compounds. As an example of a salt produced by a neutralization reaction, consider the reaction of HCl (an acid in solution) with $Ca(OH)_2$ (a base in solution). The reaction is

$$2\,HCl_{(aq)} + Ca(OH)_{2(aq)} \longrightarrow CaCl_{2(aq)} + 2\,H_2O_{(l)}$$

This is an ionic exchange reaction that forms molecular water, leaving Ca^{2+} and Cl^- in solution. As the water is evaporated, these ions begin forming ionic crystal structures as the solution concentration increases. When the water is all evaporated, the white crystalline salt of $CaCl_2$ remains.

If sodium hydroxide had been used as the base instead of calcium hydroxide, a different salt would have been produced:

$$HCl_{(aq)} + NaOH_{(aq)} \longrightarrow NaCl_{(aq)} + H_2O_{(l)}$$

Salts are also produced when elements combine directly, when an acid reacts with a metal, and by other reactions.

Salts are essential in the diet both as electrolytes and as a source of certain elements, usually called *minerals* in this context. Plants must have certain elements that are derived from water-soluble salts. Potassium, nitrates, and phosphate salts are often used to supply the needed elements. There is no scientific evidence that plants prefer to obtain these elements from natural sources, as compost, or from chemical fertilizers. After all, a nitrate ion is a nitrate ion, no matter what its source. Table 11.6 lists some common salts and their uses.

Hard and Soft Water

Salts vary in their solubility in water, and a solubility chart appears in appendix B. Table 11.7 lists some generalizations concerning the various common salts. Some of the salts are dissolved by water that will eventually be used for domestic supply. When the salts are soluble calcium or magnesium compounds, the water will contain calcium or magnesium ions in solution. A solution of Ca^{2+} of Mg^{2+} ions is said to be *hard water* because it is hard to make soap lather in the water. "Soft" water, on the other hand, makes a soap lather easily. The difficulty occurs because soap is a sodium or potassium compound that is soluble in water. The calcium or magnesium ions, when present, replace the sodium or potassium ions in the soap compound, forming an insoluble compound. It is this insoluble compound that forms a "bathtub ring" and also collects on clothes being washed, preventing cleansing.

The key to "softening" hard water is to remove the troublesome calcium and magnesium ions (Figure 11.17). If the hardness is caused by magnesium or calcium *bicarbonates*, the removal is accomplished by simply heating the water. Upon heating, they decompose, forming an insoluble compound that effectively removes the ions from solution. The decomposition reaction for calcium bicarbonate is

$$Ca^{2+}(HCO_3)_{2(aq)} \longrightarrow CaCO_{3(s)} + H_2O_{(l)} + CO_2 \uparrow$$

The reaction is the same for magnesium bicarbonate. As the solubility chart in appendix B shows, magnesium and calcium carbonates are insoluble, so the ions are removed from solution in the solid that is formed. Perhaps you have noticed such a white

TABLE 11.6

Some common salts and their uses

Common Name	Formula	Use
Alum	$KAl(SO_4)_2$	Medicine, canning, baking powder
Baking soda	$NaHCO_3$	Fire extinguisher, antacid, deodorizer, baking powder
Bleaching powder (chlorine tablets)	$CaOCl_2$	Bleaching, deodorizer, disinfectant in swimming pools
Borax	$Na_2B_4O_7$	Water softener
Chalk	$CaCO_3$	Antacid tablets, scouring powder
Cobalt chloride	$CoCl_2$	Hygrometer (pink in damp weather, blue in dry weather)
Chile saltpeter	$NaNO_3$	Fertilizer
Epsom salt	$MgSO_4 \cdot 7\,H_2O$	Laxative
Fluorspar	CaF_2	Metallurgy flux
Gypsum	$CaSO_4 \cdot 2\,H_2O$	Plaster of Paris, soil conditioner
Lunar caustic	$AgNO_3$	Germicide and cauterizing agent
Niter (or saltpeter)	KNO_3	Meat preservative, makes black gunpowder (75 parts KNO_3, 15 of carbon, 10 of sulfur)
Potash	K_2CO_3	Makes soap, glass
Rochelle salt	$KNaC_4H_4O_6$	Baking powder ingredient
TSP	Na_3PO_4	Water softener, fertilizer

TABLE 11.7

Generalizations about salt solubilities

Salts	Solubility	Exceptions
Sodium Potassium Ammonium	Soluble	None
Nitrate Acetate Chlorate	Soluble	None
Chlorides	Soluble	Ag and Hg (I) are insoluble
Sulfates	Soluble	Ba, Sr, and Pb are insoluble
Carbonates Phosphates Silicates	Insoluble	Na, K, and NH_4 are soluble
Sulfides	Insoluble	Na, K, and NH_4 are soluble; Mg, Ca, Sr, and Ba decompose

compound forming around faucets if you live where bicarbonates are a problem. Commercial products to remove such deposits usually contain an acid, which reacts with the carbonate to make a new, soluble salt that can be washed away.

Water hardness is also caused by magnesium or calcium *sulfate,* which requires a different removal method. Certain chemicals such as sodium carbonate (washing soda), trisodium phosphate (TSP), and borax will react with the troublesome ions, forming an insoluble solid that removes them from solution. For example, washing soda and calcium sulfate react as follows:

$$Na_2CO_{3(aq)} + CaSO_{4(aq)} \longrightarrow Na_2SO_{4(aq)} + CaCO_3 \downarrow$$

Calcium carbonate is insoluble; thus the calcium ions are removed from solution before they can react with the soap. Many laundry detergents have Na_2CO_3, TSP, or borax $(Na_2B_4O_7)$ added to soften the water. TSP causes other problems, however, because the additional phosphates in the waste water can act as a fertilizer, stimulating the growth of algae to such an extent that other organisms in the water die.

A water softener unit is an ion exchanger. The unit contains a mineral that exchanges sodium ions for calcium and magnesium ions as water is run through it. The softener is regenerated

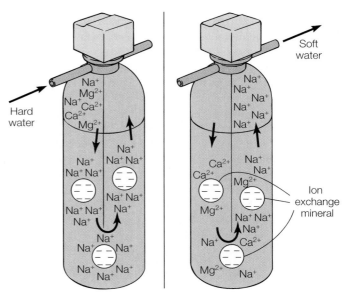

FIGURE 11.17

A water softener exchanges sodium ions for the calcium and magnesium ions of hard water. Thus, the water is now soft, but it contains the same number of ions as before.

periodically by flushing with a concentrated sodium chloride solution. The sodium ions replace the calcium and magnesium ions, which are carried away in the rinse water. The softener is then ready for use again. The frequency of renewal cycles depends on the water hardness, and each cycle can consume from four to twenty pounds of sodium chloride per renewal cycle. In general, water with less than 75 ppm calcium and magnesium ions is called soft water; with greater concentrations, it is called hard water. The greater the concentration above 75 ppm, the harder the water.

Acid rain is a general term used to describe any acidic substances, wet or dry, that fall from the atmosphere. Wet acidic deposition could be in the form of rain, but snow, sleet, and fog could also be involved. Dry acidic deposition could include gases, dust, or any solid particles that settle out of the atmosphere to produce an acid condition.

Pure, unpolluted rain is naturally acidic. Carbon dioxide in the atmosphere is absorbed by rainfall, forming carbonic acid (H_2CO_3). Carbonic acid lowers the pH of pure rainfall to a range of 5.6 to 6.2. Decaying vegetation in local areas can provide more CO_2, making the pH even lower. A pH range of 4.5 to 5.0, for example, has been measured in remote areas of the Amazon jungle. Human-produced exhaust emissions of sulfur and nitrogen oxides can lower the pH of rainfall even more, to a 4.0 to 5.5 range. This is the pH range of acid rain.

The sulfur and nitrogen oxides that produce acid rain come from exhaust emissions of industries and electric utilities that burn coal and from the exhaust of cars, trucks, and buses (Box Figure 11.1). The emissions are sometimes called "SO_x" and "NO_x," which is read "socks" and "knox." The x subscript implies the variable presence of any or all of the oxides, for example, nitrogen monoxide (NO), nitrogen dioxide (NO_2), and dinitrogen tetroxide (N_2O_4) for NO_x.

SO_x and NO_x are the raw materials of acid rain and are not themselves acidic. They react with other atmospheric chemicals to form sulfates and nitrates, which combine with water vapor to form sulfuric acid (H_2SO_4) and nitric acid (HNO_3). These are the chemicals of concern in acid rain.

Many variables influence how much and how far SO_x and NO_x are carried in the atmosphere and if they are converted to acid rain or simply return to the surface as a dry gas or particles. During the 1960s and 1970s, concerns about local levels of pollution led to the replacement of short smokestacks of about 60 m (about 200 ft) with taller smokestacks of about 200 m (about 650 ft). This did reduce the local levels of pollution by dumping the exhaust higher in the atmosphere where winds could carry it away. It also set the stage for longer-range transport of SO_x and NO_x and their eventual conversion into acids.

There are two main reaction pathways by which SO_x and NO_x are converted to acids: (1) reactions in the gas phase and (2) reactions in the liquid phase, such as in water droplets in clouds and fog. In the gas phase, SO_x and NO_x are oxidized to acids, mainly by hydroxyl ions and ozone, and the acid is absorbed by cloud droplets and precipitated as rain or snow. Most of the nitric acid in acid rain and about one-fourth of the sulfuric acid is formed in gas-phase reactions. Most of the liquid-phase reactions that produce sulfuric acid involve the absorbed SO_x and hydrogen peroxide (H_2O_2), ozone, oxygen, and particles of carbon, iron oxide, and manganese oxide particles. These particles also come from the exhaust of fossil fuel combustion.

Acid rain falls on the land, bodies of water, forests, crops, buildings, and people. The concerns about acid rain center on its environmental impact on lakes, forests, crops, materials, and human health. Lakes in different parts of the world, for example, have been increasing in acidity over the past fifty years. Lakes in northern New England, the Adirondacks, and parts of Canada now have a pH of less than 5.0, and correlations have been established between lake acidity and decreased fish populations. Trees, mostly conifers, are dying at unusually rapid rates in the northeastern United States. Red spruce in Vermont's Green Mountains and the mountains of New York and New Hampshire have been affected by acid rain as have pines in New Jersey's Pine Barrens. It is believed that acid rain leaches essential nutrients, such as calcium, from the soil and also mobilizes

BOX FIGURE 11.1

Natural rainwater has a pH of 5.6 to 6.2. Exhaust emissions of sulfur and nitrogen oxides can lower the pH of rainfall to a range of 4.0 to 5.5. The exhaust emissions come from industries, electric utilities, and automobiles. Not all emissions are as visible as those pictured in this illustration.

aluminum ions. The aluminum ions disrupt the water equilibrium of fine root hairs, and when the root hairs die, so do the trees.

Human-produced emissions of sulfur and nitrogen oxides from burning fossil fuels are the cause of acid rain. The heavily industrialized northeastern part of the United States, from the Midwest through New England, release sulfur and nitrogen emissions that result in a precipitation pH of 4.0 to 4.5. This region is the geographic center of the nation's acid rain problem. The solution to the problem is found in (1) using fuels other than fossil fuels and (2) reducing the thousands of tons of SO_x and NO_x that are dumped into the atmosphere per day when fossil fuels are used.

Johannes Nicolaus Brönsted (1879–1947)

Johannes Brönsted was a Danish physical chemist whose work in solution chemistry, particularly electrolytes, resulted in a new theory of acids and bases.

Brönsted was born on February 22, 1879, in Varde, Jutland, the son of a civil engineer. He was educated at local schools before going to study chemical engineering at the Technical Institute of the University of Copenhagen in 1897. He graduated two years later and then turned to chemistry, in which he qualified in 1902. After a short time in industry, he was appointed an assistant in the university's chemical laboratory in 1905, becoming Professor of Physical and Inorganic Chemistry in 1908. In his later years he turned to politics, being elected to the Danish parliament in 1947. He died on December 17 in that year, before he could take his seat.

Brönsted's early work was wide ranging, particularly in the fields of electrochemistry, the measurement of hydrogen ion concentrations, amphoteric electrolytes, and the behavior of indicators. He discovered a method of eliminating potentials in the measurement of hydrogen ion concentrations, and devised a simple equation that connects the activity and osmotic coefficients of an electrolyte, and another that relates activity coefficients to reaction velocities. From the absorption spectra of chromic—chromium (III)—salts he concluded that strong electrolytes are completely dissociated, and that the changes of molecular

conductivity and freezing point that accompany changes in concentration are caused by the electrical forces between ions in solution. He related the stages of ionization of polybasic acids to their molecular structure and the specific heat capacities of steam and carbon dioxide to their band spectra. In 1912 he published work with Herman Nernst on the specific heat capacities of steam and carbon dioxide at high temperatures. Two years later he laid the foundations of the theory of the infrared spectra of polyatomic molecules by introducing the so-called valence force-field. Brönsted also applied the newly developed quantum theory of specific heat capacities to gases and published papers about the factors that determine the pH and fertility of soils.

In 1887 Svante Arrhenius had proposed a theory of acidity that explained its nature on an atomic level. He defined an acid as a compound that could generate hydrogen ions in aqueous solution, and an alkali as a compound that could generate hydroxyl ions. A strong acid is completely ionized (dissociated) and produces many hydrogen ions, whereas a weak acid is only partly dissociated and produces few hydrogen ions. Conductivity measurements confirm the theory, as long as the solutions are not too concentrated.

In 1923 Brönsted published (simultaneously with Thomas Lowry in Britain) a new theory of acidity, which has certain important advantages over that of Arrhe-

nius. Brönsted defined an acid as a proton donor and a base as a proton acceptor. The definition applies to all solvents, not just water. It also explains the different behavior of pure acids in solution. Pure dry liquid sulfuric acid or acetic (ethanoic) acid does not change the color of indicators nor react with carbonates or metals. But as soon as water is added, all of these reactions occur.

In Brönsted's scheme, every acid is related to a conjugate base, and every base to a conjugate acid. When hydrogen chloride dissolves in water, for example, a reaction takes place and an equilibrium is established:

$$HCl + H_2O \leftrightarrow H_3O^+ + Cl^-$$
$$\text{Acid 1} \quad \text{Base 2} \quad \text{Acid 2} \quad \text{Base 1}$$

HCl is an acid for the forward reaction, but the hydronium ion (H_3O^+) is an acid in the reverse reaction; it is the conjugate acid (acid 2) of water (base 2). Similarly, the chloride ion (Cl^-, base 1) accepts protons in the reverse reaction to form its conjugate acid, (HCl, acid 1). In this theory acids are not confined to neutral species or positive ions. For example, the negatively charged hydrogen sulfate ion can behave as an acid.:

$$HSO_4^-{}_{(aq)} + H_2O_{(1)} \leftrightarrow H_3O^+ + SO_4^{2-}{}_{(aq)}$$

It donates a proton to form the hydronium ion.

Source: From the Hutchinson *Dictionary of Scientific Biography*. © Research Machines plc [2003] All Rights Reserved. Helicon Publishing is a division of Research Machines.

Summary

A water molecule consists of two hydrogen atoms and an oxygen atom with covalent bonding. Oxygen has more positive protons than either of the hydrogens, so electrons spend more time around the oxygen, producing a *polar molecule*, with centers of negative and positive charge. Polar water molecules interact with an attractive force between the negative center of one molecule and the positive center of another. This force is called a *hydrogen bond*. The hydrogen bond accounts for the decreased density of ice, the high heat of fusion, and the high heat

of vaporization of water. The hydrogen bond is also involved in the *dissolving* process.

A *solution* is a homogeneous mixture of ions or molecules of two or more substances. The substance present in the large amount is the *solvent*, and the *solute* is dissolved in the solvent. If one of the components is a liquid, however, it is called the solvent. Fluids that mix in any proportion are called *miscible fluids*, and *immiscible fluids* do not mix. Polar substances dissolve in polar solvents, but not nonpolar solvents, and the general rule is *like dissolves like*. Thus oil, a nonpolar substance, is immiscible in water, a polar substance.

The relative amount of solute in a solvent is called the *concentration* of a solution. Concentrations are measured (1) in *parts per million* (ppm) or *parts per billion* (ppb), (2) *percent by volume,* the volume of a solute per 100 volumes of solution, (3) *percent by weight,* the weight of solute per 100 weight units of solution, and (4) *salinity,* the mass of salts in 1 kg of solution.

A limit to dissolving solids in a liquid occurs when the solution is *saturated.* A *saturated solution* is one with equilibrium between solute dissolving and solute coming out of solution. The *solubility* of a solid is the concentration of a saturated solution at a particular temperature.

Water solutions that carry an electric current are called *electrolytes,* and nonconductors are called *nonelectrolytes.* In general, ionic substances make electrolyte solutions, and molecular substances make nonelectrolyte solutions. Polar molecular substances may be *ionized* by polar water molecules, however, making an electrolyte from a molecular solution.

The *boiling point of a solution* is greater than the boiling point of the pure solvent, and the increase depends only on the concentration of the solute (at a constant pressure). For water, the boiling point is increased 0.521°C for each mole of solute in each kg of water. The *freezing point of a solution* is lower than the freezing point of the pure solvent, and the depression also depends on the concentration of the solute.

Acids, bases, and salts are chemicals that form ionic solutions in water, and each can be identified by simple properties. These properties are accounted for by the modern concepts of each. *Acids* are *proton donors* that form *hydronium ions* (H_3O^+) in water solutions. *Bases* are *proton acceptors* that form *hydroxide ions* (OH^-) in water solutions. *Strong acids* and *strong bases* ionize completely in water, and *weak acids* and *weak bases* are only partially ionized. The strength of an acid or base is measured on the *pH scale,* a power of ten notation of the hydronium ion concentration. On the scale, numbers from 0 up to 7 are acids, 7 is neutral, and numbers above 7 and up to 14 are bases. Each unit represents a tenfold increase or decrease in acid or base properties.

A *salt* is any ionic compound except those with hydroxide or oxide ions. Salts provide plants and animals with essential elements. The solubility of salts varies with the ions that make up the compound. Solutions of magnesium or calcium produce *hard water,* water in which it is hard to make soap lather. Hard water is softened by removing the magnesium and calcium ions.

Summary of Equations

11.1

Percent by volume

$$\frac{V_{solute}}{V_{solution}} \times 100\% \text{ solution} = \% \text{ solute}$$

11.2

Percent by weight (mass)

$$\frac{m_{solute}}{m_{solution}} \times 100\% \text{ solution} = \% \text{ solute}$$

11.3

$$\text{Molarity (M)} = \frac{\text{moles of solute}}{\text{liters of solution}}$$

KEY TERMS

acid (p. **288**)
acid-base indicator (p. **287**)
base (p. **288**)
concentration (p. **280**)
electrolytes (p. **284**)
hydrogen bonding (p. **278**)
hydronium ion (p. **284**)
miscible fluids (p. **279**)

molarity (M) (p. **283**)
neutralized (p. **288**)
pH scale (p. **289**)
polar molecule (p. **277**)
salinity (p. **282**)
salt (p. **290**)
solubility (p. **283**)
solution (p. **279**)

APPLYING THE CONCEPTS

1. Which of the following is *not* a solution?
   a. seawater
   b. carbonated water
   c. sand
   d. brass

2. Atmospheric air is a homogeneous mixture of gases that is mostly nitrogen gas. The nitrogen is therefore (the)
   a. solvent.
   b. solution.
   c. solute.
   d. none of the above.

3. A homogeneous mixture is made up of 95 percent alcohol and 5 percent water. In this case the water is (the)
   a. solvent.
   b. solution.
   c. solute.
   d. none of the above.

4. The solution concentration terms of parts per million, percent by volume, and percent by weight are concerned with the amount of
   a. solvent in the solution.
   b. solute in the solution.
   c. solute compared to solvent.
   d. solvent compared to solute.

5. A concentration of 500 ppm is reported in a news article. This is the same concentration as
   a. 0.005%.
   b. 0.05%.
   c. 5%.
   d. 50%.

6. According to the label, a bottle of vodka has a 40% by volume concentration. This means the vodka contains 40 mL of pure alcohol
   a. in each 140 mL of vodka.
   b. to every 100 mL of water.
   c. to every 60 mL of vodka.
   d. mixed with water to make 100 mL vodka.

7. A bottle of vinegar is 4% by weight, so you know that the solution contains 4 weight units of pure vinegar with
   a. 96 weight units of water.
   b. 100 weight units of water.
   c. 104 weight units of water.

8. If a salt solution has a salinity of 40‰, what is the equivalent percentage measure?
   a. 400%
   b. 40%
   c. 4%
   d. 0.4%

9. A salt solution has solid salt on the bottom of the container and salt is dissolving at the same rate that it is coming out of solution. You know the solution is
   a. an electrolyte.
   b. a nonelectrolyte.
   c. a buffered solution.
   d. a saturated solution.

10. As the temperature of water *decreases,* the solubility of carbon dioxide gas in the water
    a. increases.
    b. decreases.
    c. remains the same.
    d. increases or decreases, depending on the specific temperature.

11. Water has the greatest density at what temperature?
    a. 100°C
    b. 20°C
    c. 4°C
    d. 0°C

12. An example of a hydrogen bond is a weak-to-moderate bond between
    a. any two hydrogen atoms.
    b. a hydrogen of one polar molecule and an oxygen of another polar molecule.
    c. two hydrogen atoms on two nonpolar molecules.
    d. a hydrogen atom and any nonmetal atom.

13. Whether two given liquids form solutions or not depends on some similarities in their
    a. electronegativities.
    b. polarities.
    c. molecular structures.
    d. hydrogen bonds.

14. A solid salt is insoluble in water so the strongest force must be the
    a. ion-water molecule force.
    b. ion-ion force.
    c. force of hydration.
    d. polar molecule force.

15. Which of the following will conduct an electric current?
    a. pure water
    b. a water solution of a covalent compound
    c. a water solution of an ionic compound
    d. All of the above are correct.

16. Ionization occurs upon solution of
    a. ionic compounds.
    b. some polar molecules.
    c. nonpolar molecules.
    d. none of the above.

17. Adding sodium chloride to water raises the boiling point of water because
    a. sodium chloride has a higher boiling point.
    b. sodium chloride ions occupy space at the water surface.
    c. sodium chloride ions have stronger ion-ion bonds than water.
    d. the energy of hydration is higher.

18. The ice that forms in freezing seawater is
    a. pure water.
    b. the same salinity as liquid seawater.
    c. more salty than liquid seawater.
    d. more dense than liquid seawater.

19. Salt solutions freeze at a lower temperature than pure water because
    a. more ionic bonds are present.
    b. salt solutions have a higher vapor pressure.
    c. ions get in the way of water molecules trying to form ice.
    d. salt naturally has a lower freezing point than water.

20. Which of the following would have a pH of *less* than 7?
    a. a solution of ammonia
    b. a solution of sodium chloride
    c. pure water
    d. carbonic acid

21. Which of the following would have a pH of *more* than 7?
    a. a solution of ammonia
    b. a solution of sodium chloride
    c. pure water
    d. carbonic acid

22. The condition of two opposing reactions happening at the same time and at the same rate is called
    a. neutralization.
    b. chemical equilibrium.
    c. a buffering reaction.
    d. cancellation.

23. Solutions of acids, bases, and salts have what in common? All have
    a. proton acceptors.
    b. proton donors.
    c. ions.
    d. polar molecules.

24. When a solution of an acid and a base are mixed together,
    a. a salt and water are formed.
    b. they lose their acid and base properties.
    c. both are neutralized.
    d. All of the above are correct.

25. A substance that ionizes completely into hydronium ions is known as a
    a. strong acid.
    b. weak acid.
    c. strong base.
    d. weak base.
26. A scale of values that expresses the hydronium ion concentration of a solution is known as
    a. an acid-base indicator.
    b. the pH scale.
    c. the solubility scale.
    d. the electrolyte scale.
27. Substance A has a pH of 2 and substance B has a pH of 3. This means that
    a. substance A has more basic properties than substance B.
    b. substance B has more acidic properties than substance A.
    c. substance A is ten times more acidic than substance B.
    d. substance B is ten times more acidic than substance A.

Answers

1. c 2. a 3. c 4. b 5. b 6. d 7. a 8. c 9. d 10. a 11. c 12. b 13. c 14. b 15. c
16. b 17. b 18. a 19. c 20. d 21. a 22. b 23. c 24. d 25. a 26. b 27. c

QUESTIONS FOR THOUGHT

1. How is a solution different from other mixtures?
2. Explain why some ionic compounds are soluble while others are insoluble in water.
3. Explain why adding salt to water increases the boiling point.
4. A deep lake in Minnesota is covered with ice. What is the water temperature at the bottom of the lake? Explain your reasoning.
5. Explain why water has a greater density at 4°C than at 0°C.
6. What is hard water? How is it softened?
7. According to the definition of an acid and the definition of a base, would the pH increase, decrease, or remain the same when NaCl is added to pure water? Explain.
8. What is a hydrogen bond? Explain how a hydrogen bond forms.
9. What feature of a soap molecule gives it cleaning ability?
10. What ion is responsible for (a) acidic properties? (b) for basic properties?
11. Explain why a pH of 7 indicates a neutral solution—why not some other number?

PARALLEL EXERCISES

The exercises in groups A and B cover the same concepts. Solutions to group A exercises are located in appendix D.

Group A

1. A 50.0 g sample of a saline solution contains 1.75 g NaCl. What is the percentage by weight concentration?
2. A student attempts to prepare a 3.50 percent by weight saline solution by dissolving 3.50 g NaCl in 100 g of water. Since equation 11.2 calls for 100 g of solution, the correct amount of solvent should have been 96.5 g water ($100 - 3.5 = 96.5$). What percent by weight solution did the student actually prepare?
3. Seawater contains 30,113 ppm by weight dissolved sodium and chlorine ions. What is the percent by weight concentration of sodium chloride in seawater?
4. What is the mass of hydrogen peroxide, H_2O_2, in 250 grams of a 3.0% by weight solution?
5. How many mL of pure alcohol are in a 200 mL glass of wine that is 12 percent alcohol by volume?
6. How many mL of pure alcohol are in a single cocktail made with 50 mL of 40% vodka? (Note: "Proof" is twice the percent, so 80 proof is 40%.)
7. If fish in a certain lake are reported to contain 5 ppm by weight DDT, (a) what percentage of the fish meat is DDT? (b) How much of this fish would have to be consumed to reach a poisoning accumulation of 17.0 grams of DDT?

Group B

1. What is the percent by weight of a solution containing 2.19 g NaCl in 75 g of the solution?
2. What is the percent by weight of a solution prepared by dissolving 10 g of NaCl in 100 g of H_2O?
3. A concentration of 0.5 ppm by volume SO_2 in air is harmful to plant life. What is the percent by volume of this concentration?
4. What is the volume of water in a 500 mL bottle of rubbing alcohol that has a concentration of 70% by volume?
5. If a definition of intoxication is an alcohol concentration of 0.05 percent by volume in blood, how much alcohol would be present in the average (155 lb) person's 6,300 mL of blood if that person was intoxicated?
6. How much pure alcohol is in a 355 mL bottle of a "wine cooler" that is 5.0% alcohol by volume?
7. In the 1970s, when lead was widely used in "ethyl" gasoline, the blood level of the average American contained 0.25 ppm lead. The danger level of lead poisoning is 0.80 ppm. (a) What percent of the average person was lead? (b) How much lead would be in an average 80 kg person? (c) How much more lead would the average person need to accumulate to reach the danger level?

8. For each of the following reactants, draw a circle around the proton donor and a box around the proton acceptor. Label which acts as an acid and which acts as a base.

(a) $HC_2H_3O_{2(aq)} + H_2O_{(l)} \rightarrow H_3O^+_{(aq)} + C_2H_3O_2^-_{(aq)}$

(b) $C_6H_6NH_{2(l)} + H_2O_{(l)} \rightarrow C_6H_6NH_3^+_{(aq)} + OH^-_{(aq)}$

acid (c) $\widehat{HClO_{4(aq)}} + \boxed{HC_2H_3O_{2(aq)}} \rightarrow H_2C_2H_3O_2^+_{(aq)} + ClO_4^-_{(aq)}$

(d) $H_2O_{(l)} + H_2O_{(l)} \rightarrow H_3O^+_{(aq)} + OH^-_{(aq)}$

base

8. Draw a circle around the proton donor and a box around the proton acceptor for each of the reactants and label which acts as an acid and which acts as a base.

(a) $H_3PO_{4(aq)} + H_2O_{(l)} \rightarrow H_3O^+_{(aq)} + H_2PO_4^-_{(aq)}$

(b) $N_2H_{4(l)} + H_2O_{(l)} \rightarrow N_2H_5^+_{(aq)} + OH^-_{(aq)}$

(c) $HNO_{3(aq)} + HC_2H_3O_{2(aq)} \rightarrow H_2C_2H_3O_2^+_{(aq)} + NO_3^-_{(aq)}$

(d) $2\,NH_4^+_{(aq)} + Mg_{(s)} \rightarrow Mg^{2+}_{(aq)} + 2\,NH_3^+_{(aq)} + H_{2(g)}$

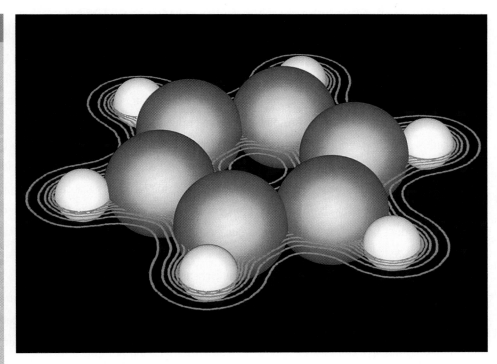

This is a computer-generated model of a benzene molecule, showing six carbon atoms (gold) and six hydrogen atoms (white). Benzene is a hydrocarbon, an organic compound made up of the elements carbon and hydrogen.

CHAPTER

12

Organic Chemistry

The impact of ancient Aristotelian ideas on the development of understandings of motion, elements, and matter was discussed in earlier chapters. Historians also trace the "vitalist theory" back to Aristotle. According to Aristotle's idea, all living organisms are composed of the four elements (earth, air, fire, and water) and have in addition an *actuating force,* the life or soul that makes the organism different from nonliving things made of the same four elements. Plants, as well as animals, were considered to have this actuating, or vital, force in the Aristotelian scheme of things.

There were strong proponents of the vitalist theory as recently as the early 1800s. Their basic argument was that organic matter, the materials and chemical compounds recognized as being associated with life, could not be produced in the laboratory. Organic matter could only be produced in a living organism, they argued, because the organism had a vital force that is not present in laboratory chemicals. Then, in 1828, a German chemist named Friedrich Wöhler decomposed a chemical that was *not organic* to produce urea (N_2H_4CO), a known *organic* compound that occurs in urine. Wöhler's production of an organic compound was soon followed by the production of other organic substances by other chemists. The vitalist theory gradually disappeared with each new reaction, and a new field of study, organic chemistry, emerged.

This chapter is an introductory survey of the field of organic chemistry, which is concerned with compounds and reactions of compounds that contain carbon. You will find this an interesting, informative introduction, particularly if you have ever wondered about synthetic materials, natural foods and food products, or any of the thousands of carbon-based chemicals you use every day. The survey begins with the simplest of organic compounds, those consisting of only carbon and hydrogen atoms, compounds known as hydrocarbons. Hydrocarbons are the compounds of crude oil, which is the source of hundreds of petroleum products (Figure 12.1).

Most common organic compounds can be considered derivatives of the hydrocarbons, such as alcohols, ethers, fatty acids, and esters. Some of these are the organic compounds that give flavors to foods, and others are used to make hundreds of commercial products, from face cream to margarine. The main groups, or classes, of derivatives will be briefly introduced, along with some interesting examples of each group. Some of the important organic compounds of life, including proteins, carbohydrates, and fats, are discussed next. The chapter concludes with an introduction to synthetic polymers, what they are, and how they are related to the fossil fuel supply.

Organic Compounds

Today, **organic chemistry** is defined as the study of compounds in which carbon is the principal element, whether the compound was formed by living things or not. The study of compounds that do not contain carbon as a central element is called **inorganic chemistry.** An *organic compound* is thus a compound that contains carbon as the principal element, and an *inorganic compound* is any other compound.

Organic compounds, by definition, must contain carbon while all the inorganic compounds can contain all the other elements. Yet, the majority of known compounds are organic. Several million organic compounds are known and thousands of new ones are discovered every year. You use organic compounds every day, including gasoline, plastics, grain alcohol, foods, flavorings, and many others.

It is the unique properties of carbon that allow it to form so many different compounds. A carbon atom has a valence of four and can combine with one, two, three, or four *other carbon atoms* in addition to a wide range of other kinds of atoms (Figure 12.2). The number of possible molecular combinations is almost limitless, which explains why there are so many organic compounds. Fortunately, there are patterns of groups of carbon atoms and groups of other atoms that lead to similar chemical characteristics, making the study of organic chemistry less difficult. The key to success in studying organic chemistry is to recognize patterns and to understand the code and meaning of organic chemical names.

Hydrocarbons

A **hydrocarbon** is an organic compound consisting of only two elements. As the name implies, these elements are hydrogen and carbon. The simplest hydrocarbon has one carbon atom and four hydrogen atoms (Figure 12.3), but since carbon atoms can

FIGURE 12.1

Refinery and tank storage facilities, like this one in Texas, are needed to change the hydrocarbons of crude oil to many different petroleum products. The classes and properties of hydrocarbons form one topic of study in organic chemistry.

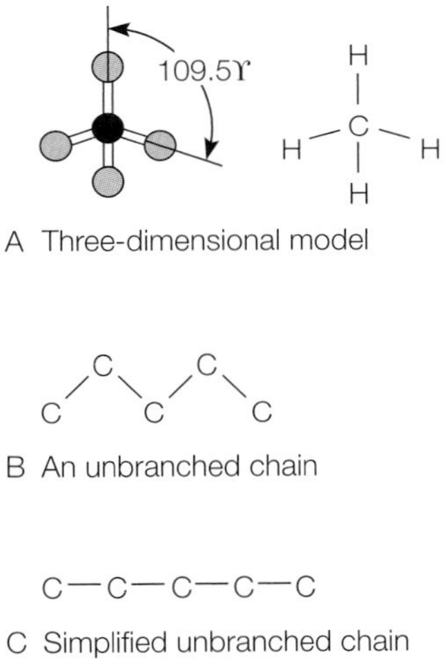

A Three-dimensional model

B An unbranched chain

$C—C—C—C—C$

C Simplified unbranched chain

FIGURE 12.2

(A) The carbon atom forms bonds in a tetrahedral structure with a bond angle of 109.5°. (B) Carbon-to-carbon bond angles are 109.5°, so a chain of carbon atoms makes a zigzag pattern. (C) The unbranched chain of carbon atoms is usually simplified in a way that looks like a straight chain, but it is actually a zigzag, as shown in (B).

combine with one another, there are thousands of possible structures and arrangements. The carbon-to-carbon bonds are covalent and can be single, double, or triple (Figure 12.4). Recall that the dash in a structural formula means one shared electron pair, a covalent bond. To satisfy the octet rule, this means that each carbon atom must have a total of four dashes around it, no more and no less. Note that when the carbon atom has double

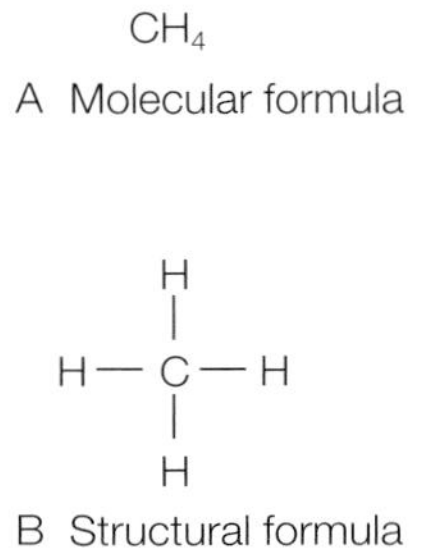

CH_4

A Molecular formula

B Structural formula

FIGURE 12.3

A molecular formula (A) describes the numbers of different kinds of atoms in a molecule, and a structural formula (B) represents a two-dimensional model of how the atoms are bonded to each other. Each dash represents a bonding pair of electrons.

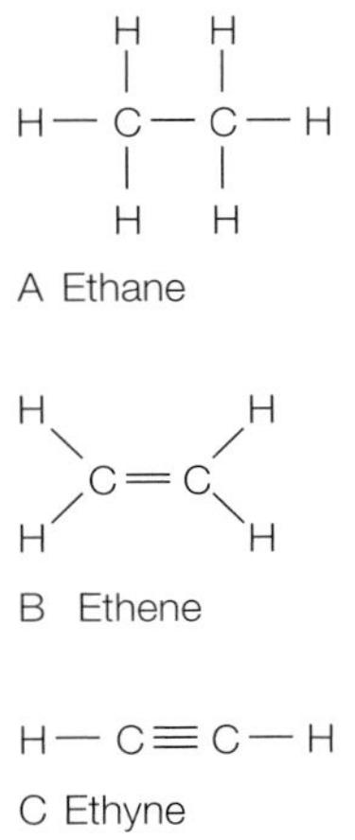

A Ethane

B Ethene

$H—C{\equiv}C—H$

C Ethyne

FIGURE 12.4

Carbon-to-carbon bonds can be single (A), double (B), or triple (C). Note that in each example, each carbon atom has four dashes, which represent four bonding pairs of electrons, satisfying the octet rule.

or triple bonds, fewer hydrogen atoms can be attached as the octet rule is satisfied. There are four groups of hydrocarbons that are classified according to how the carbon atoms are put together, the (1) *alkanes*, (2) *alkenes*, (3) *alkynes*, and (4) *aromatic hydrocarbons*.

The **alkanes** are *hydrocarbons with single covalent bonds* between the carbon atoms. Alkanes that are large enough to form chains of carbon atoms occur with a straight structure, a branched structure, or a ring structure as shown in Figure 12.5. (The "straight" structure is actually a zigzag as shown in Figure 12.2.) You are familiar with many alkanes, for they make up the bulk of petroleum and petroleum products, which will be discussed shortly. The clues and codes in the names of the alkanes will be considered first.

The alkanes are also called the *paraffin series*. The alkanes are not as chemically reactive as the other hydrocarbons, and the term *paraffin* means "little affinity." They are called a series because *each higher molecular weight alkane has an additional CH_2*.

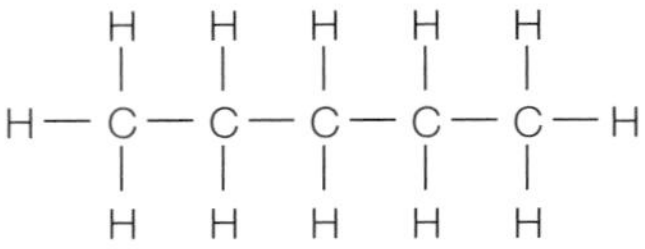

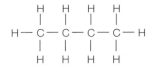

A Straight chain for C$_5$H$_{12}$

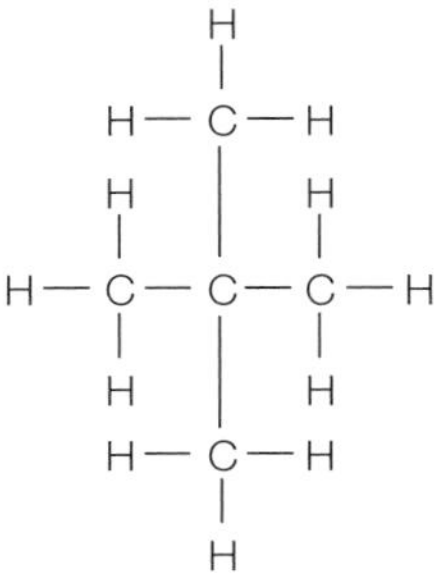

B Branched chain for C$_5$H$_{12}$

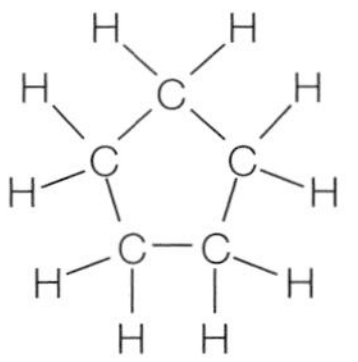

C Ring chain for C$_5$H$_{10}$

FIGURE 12.5

Carbon-to-carbon chains can be (A) straight, (B) branched, or (C) in a closed ring. (Some carbon bonds are drawn longer, but are actually the same length.)

A n-butane, C$_4$H$_{10}$

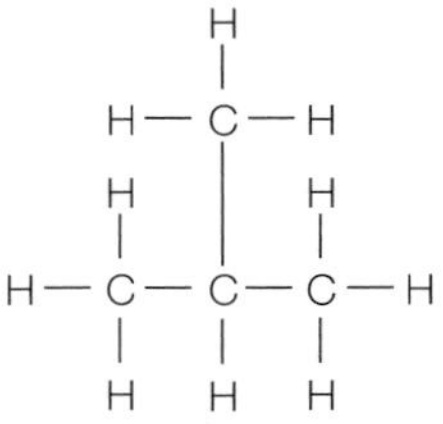

B Isobutane (2-methylpropane), C$_4$H$_{10}$

FIGURE 12.6

(A) A straight-chain alkane is identified by the prefix n- for "normal" in the common naming system. (B) A branched-chain alkane isomer is identified by the prefix iso- for "isomer" in the common naming system. In the IUPAC name, isobutane is 2-methylpropane. (Carbon bonds are actually the same length.)

The simplest alkane is methane, CH$_4$, and the next highest molecular weight alkane is ethane, C$_2$H$_6$. As you can see, C$_2$H$_6$ is CH$_4$ with an additional CH$_2$. If you compare the first ten alkanes in Table 12.1, you will find that each successive compound in the series always has an additional CH$_2$.

Note the names of the alkanes listed in Table 12.1. After pentane the names have a consistent prefix and suffix pattern. The prefix and suffix pattern is a code that provides a clue about the compound. The Greek prefix tells you the *number of carbon atoms* in the molecule, for example, "oct-" means eight, so *octane* has eight carbon atoms. The suffix "-ane" tells you this hydrocarbon is a member of the alk*ane* series, so it has single bonds only. With the general alkane formula of C$_n$H$_{2n+2}$, you can now write the formula when you hear the name. Octane has eight carbon atoms with single bonds and $n = 8$. Two times 8 plus 2 $(2n + 2)$ is 18, so the formula for octane is C$_8$H$_{18}$. Most organic chemical names provide clues like this.

The alkanes in Table 12.1 all have straight chains. A straight, continuous chain is identified with the term *normal,* which is abbreviated *n.* Figure 12.6A shows *n*-butane with a straight chain and a molecular formula of C$_4$H$_{10}$. Figure 12.6B shows a different branched structural formula that has the

same C$_4$H$_{10}$ molecular formula. Compounds with the same molecular formulas with different structures are called **isomers.** Since the straight-chained isomer is called *n*-butane, the branched isomer is called *isobutane.* The isomers of a particular alkane, such as butane, have different physical and chemical properties because they have different structures. Isobutane, for example, has a boiling point of −10°C. The boiling point of *n*-butane, on the other hand, is −0.5°C. In the next section you will learn that the various isomers of the octane hydrocarbon perform differently in automobile engines, requiring the "reforming" of *n*-octane to *iso-octane* before it can be used.

Methane, ethane, and propane can have only one structure each, and butane has two isomers. The number of possible isomers for a particular molecular formula increases rapidly as the number of carbon atoms increases. After butane, hexane has five isomers, octane eighteen isomers, and decane seventy-five isomers. Because they have different structures, each isomer has different physical properties. A different naming system is needed because there are just too many isomers to keep track of. The system of naming the branched-chain alkanes is described by rules agreed upon by a professional organization, the International Union of Pure and Applied Chemistry, or IUPAC. Here are the steps in naming the alkane isomers.

Step 1: The longest continuous chain of carbon atoms determines the *base name* of the molecule. The longest continuous chain is not necessarily straight and can take any number of right-angle turns as long as the continuity is not broken. The base name corresponds to the number of carbon atoms in this chain as in Table 12.1. For example, the structure has six carbon atoms in the longest chain, so the base name is *hexane.*

TABLE 12.1
The first ten straight-chain alkanes

Name	Molecular Formula	Structural Formula		Name	Molecular Formula	Structural Formula
Methane	CH_4	H–C–H (with H above and below)		Hexane	C_6H_{14}	6-carbon chain
Ethane	C_2H_6	H–C–C–H (2 carbons)		Heptane	C_7H_{16}	7-carbon chain
Propane	C_3H_8	3-carbon chain		Octane	C_8H_{18}	8-carbon chain
Butane	C_4H_{10}	4-carbon chain		Nonane	C_9H_{20}	9-carbon chain
Pentane	C_5H_{12}	5-carbon chain		Decane	$C_{10}H_{22}$	10-carbon chain

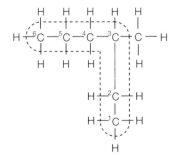

Step 2: The locations of other groups of atoms attached to the base chain are identified by counting carbon atoms from either the left or from the right. The direction selected is the one that results in the *smallest* numbers for attachment locations. For example, the hexane

chain has a CH_3 attached to the third or the fourth carbon atom, depending on which way you count. The third atom direction is chosen since it results in a smaller number.

Step 3: The hydrocarbon groups attached to the base chain are named from the number of carbons in the group by changing the alkane suffix "-ane" to "-yl." Thus, a hydrocarbon group attached to a base chain that has one carbon atom is called meth*yl*. Note that the "-yl" hydrocarbon groups have one less hydrogen than the corresponding alkane. Therefore, methane is CH_4, and a *methyl group* is CH_3. The first ten alkanes and their corresponding hydrocarbon group names are listed in Table 12.2. In the example, a methyl group is attached to the third carbon atom of the base hexane chain. The name and address of this hydrocarbon group is 3-methyl. The compound is named 3-methylhexane.

Step 4: The prefixes "di-," "tri-," and so on are used to indicate if a particular hydrocarbon group appears on the main chain more than once. For example,

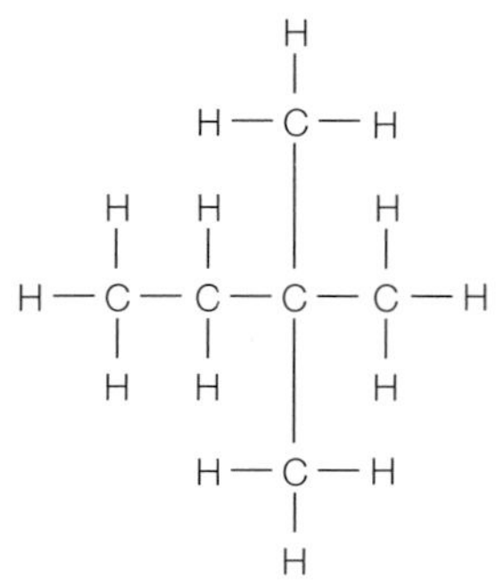

(or)

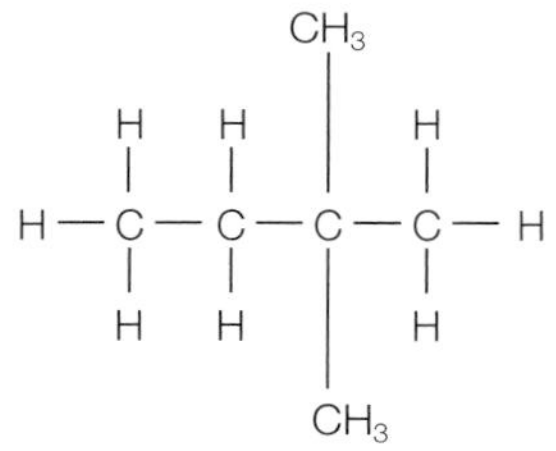

is 2,2-dimethylbutane and

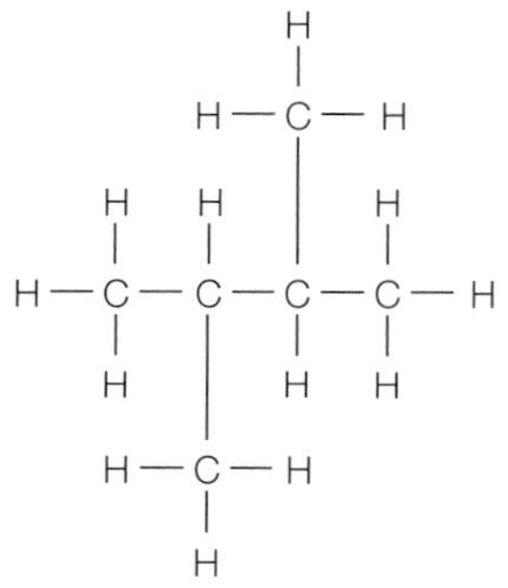

(or)

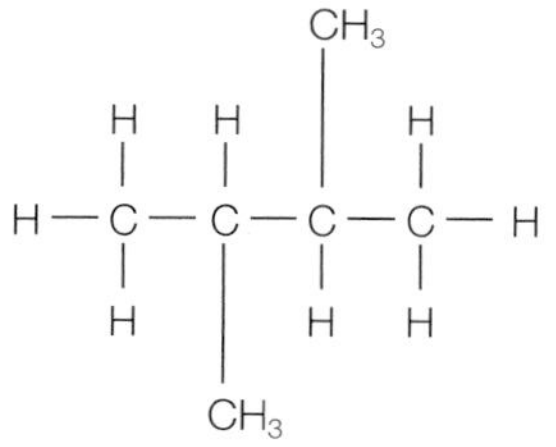

is 2,3-dimethylbutane.

TABLE 12.2
Alkane hydrocarbons and corresponding hydrocarbon groups

Alkane Name	Molecular Formula	Hydrocarbon Group	Molecular Formula
Methane	CH_4	Methyl	$-CH_3$
Ethane	C_2H_6	Ethyl	$-C_2H_5$
Propane	C_3H_8	Propyl	$-C_3H_7$
Butane	C_4H_{10}	Butyl	$-C_4H_9$
Pentane	C_5H_{12}	Amyl	$-C_5H_{11}$
Hexane	C_6H_{14}	Hexyl	$-C_6H_{13}$
Heptane	C_7H_{16}	Heptyl	$-C_7H_{15}$
Octane	C_8H_{18}	Octyl	$-C_8H_{17}$
Nonane	C_9H_{20}	Nonyl	$-C_9H_{19}$
Decane	$C_{10}H_{22}$	Decyl	$-C_{10}H_{21}$

Note: $-CH_3$ means

where * denotes unattached. The attachment takes place on a base chain or functional group.

If hydrocarbon groups with different numbers of carbon atoms are on a main chain, they are listed in alphabetic order. For example,

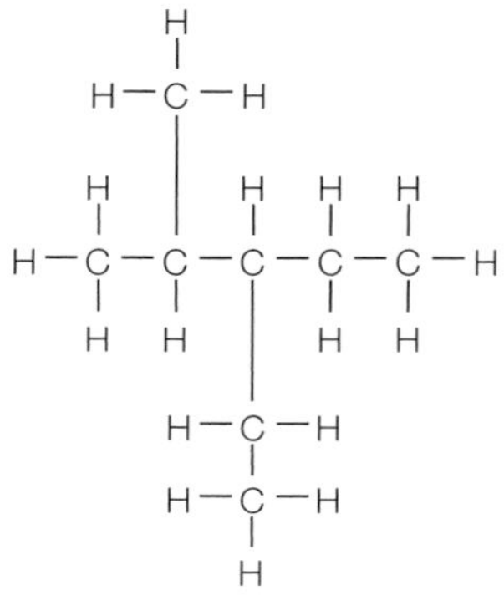

(or)

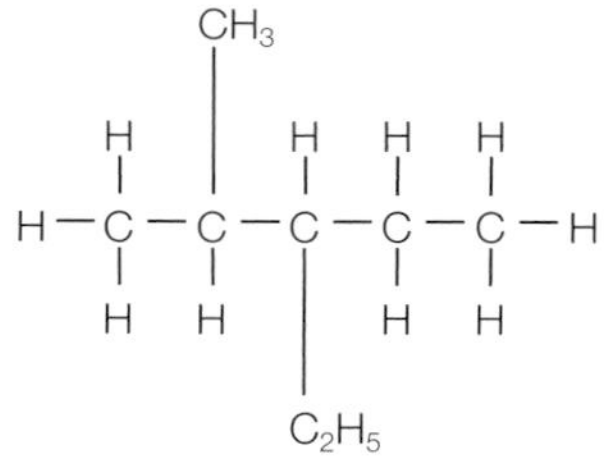

is named 3-ethyl-2-methylpentane. Note how numbers are separated from names by hyphens.

EXAMPLE 12.1

What is the name of an alkane with the following formula?

$$H-\overset{\overset{\displaystyle H}{|}}{\underset{\underset{\displaystyle H}{|}}{C}}-H$$

(structural formula as drawn)

TABLE 12.3

The general molecular formulas and molecular structures of the alkanes, alkenes, and alkynes

Group	General Molecular Formula	Example Compound	Molecular Structure
Alkanes	C_nH_{2n+2}	Ethane	(structure)
Alkenes	C_nH_{2n}	Ethene	(structure)
Alkynes	C_nH_{2n-2}	Ethyne	$H-C\equiv C-H$

Solution

The longest continuous chain has seven carbon atoms, so the base name is heptane. The smallest numbers are obtained by counting from right to left and counting the carbons on this chain; there is a methyl group in carbon atom 2, a second methyl group on atom 4, and an ethyl group on atom 5. There are two methyl groups, so the prefix "di-" is needed, and the "e" of the ethyl group comes first in the alphabet so ethyl is listed first. The name of the compound is 5-ethyl-2,4-dimethylheptane.

EXAMPLE 12.2

Write the structural formula for 2,2-dichloro-3-methyloctane.
Answer:

(structural formula as drawn)

Alkenes and Alkynes

The **alkenes** are *hydrocarbons with a double covalent carbon-to-carbon bond.* To denote the presence of a double bond the "-ane" suffix of the alkanes is changed to "-ene" as in alk*ene* (Table 12.3). Figure 12.4 shows the structural formula for (A) ethane, C_2H_6, and (B) ethene, C_2H_4. Alkenes have room for two

fewer hydrogen atoms because of the double bond, so the general alkene formula is C_nH_{2n}. Note the simplest alkene is called ethene, but is commonly known as ethylene.

Ethylene is an important raw material in the chemical industry. Obtained from the processing of petroleum, about half of the commercial ethylene is used to produce the familiar polyethylene plastic. It is also produced by plants to ripen fruit, which explains why unripe fruit enclosed in a sealed plastic bag with ripe fruit will ripen more quickly (Figure 12.7). The ethylene produced by the ripe fruit acts on the unripe fruit. Commercial fruit packers sometimes use small quantities of ethylene gas to quickly ripen fruit that was picked while green.

Perhaps you have heard the terms "saturated" and "unsaturated" in advertisements for cooking oil and margarine. An organic molecule, such as a hydrocarbon, that does not contain the maximum number of hydrogen atoms is an *unsaturated*

FIGURE 12.7

Ethylene is the gas that ripens fruit, and a ripe fruit emits the gas, which will act on unripe fruit. Thus, a ripe tomato placed in a sealed bag with green tomatoes will help ripen them.

hydrocarbon. For example, ethylene can add more hydrogen atoms by reacting with hydrogen gas to form ethane:

$$H_2C=CH_2 + H_2 \longrightarrow H_3C-CH_3$$

Ethylene + Hydrogen ⟶ Ethane

The ethane molecule has all the hydrogen atoms possible, so ethane is a *saturated* hydrocarbon. Unsaturated molecules are less stable, which means that they are more chemically reactive than saturated molecules.

Alkenes are named as the alkanes are, except (1) the longest chain of carbon atoms must contain the double bond, (2) the base name now ends in "-ene," (3) the carbon atoms are numbered from the end nearest the double bond, and (4) the base name is given a number of its own, which identifies the address of the double bond. For example,

is named 4-methyl-1-pentene. The 1-pentene tells you there is a double bond (-ene), and the 1 tells you the double bond is after the first carbon atom in the longest chain containing the double bond. The methyl group is on the fourth carbon atom in this chain.

An **alkyne** is a *hydrocarbon with a carbon-to-carbon triple bond* and the general formula of C_nH_{2n-2}. The alkynes are highly reactive, and the simplest one, ethyne, has a common name of acetylene. Acetylene is commonly burned with oxygen gas in a welding torch because the flame reaches a temperature of about 3,000°C. Acetylene is also an important raw material in the production of plastics. The alkynes are named as the alkenes are, except the longest chain must contain the triple bond, and the base name suffix is changed to "-yne."

Cycloalkanes and Aromatic Hydrocarbons

The hydrocarbons discussed up until now have been straight or branched open-ended chains of carbon atoms. Carbon atoms can also bond to each other to form a ring, or cyclic, structure. Figure 12.8 shows the structural formulas for some of these cyclic structures. Note that the cycloalkanes have the same molecular formulas as the alkenes, and thus they are isomers of the alkenes. They are, of course, very different compounds, with different physical and chemical properties. This shows the importance of structural, rather than simply molecular formulas in referring to organic compounds.

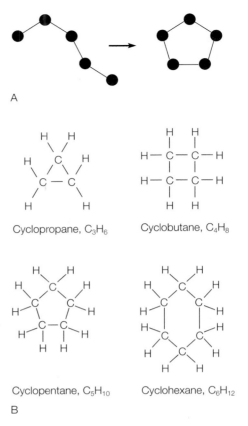

A

Cyclopropane, C_3H_6 Cyclobutane, C_4H_8

Cyclopentane, C_5H_{10} Cyclohexane, C_6H_{12}

B

FIGURE 12.8

(*A*) The "straight" chain has carbon atoms that are able to rotate freely around their single bonds, sometimes linking up in a closed ring. (*B*) Ring compounds of the first four cycloalkanes.

The six-carbon ring structure shown in Figure 12.9A has three double bonds that do not behave like the double bonds in the alkenes. In this six-carbon ring the double bonds are not localized in one place but are spread over the whole molecule. Instead of alternating single and double bonds, all the bonds are something in between. This gives the C_6H_6 molecule increased stability. As a result, the molecule does not behave like other unsaturated compounds, that is, it does not readily react in order to add hydrogen to the ring. The C_6H_6 molecule is the organic compound named *benzene.* Organic compounds that are based on the benzene ring structure are called **aromatic hydrocarbons.** To denote the six-carbon ring with delocalized electrons, benzene is represented by the symbol shown in Figure 12.9B.

The circle in the six-sided benzene symbol represents the delocalized electrons. Figure 12.9B illustrates how this benzene ring symbol is used to show the structural formula of some aromatic hydrocarbons. You may have noticed some of the names on labels of paints, paint thinners, and lacquers. Toluene and the xylenes are commonly used in these products as solvents. A benzene ring attached to another molecule or functional group is given the name *phenyl.*

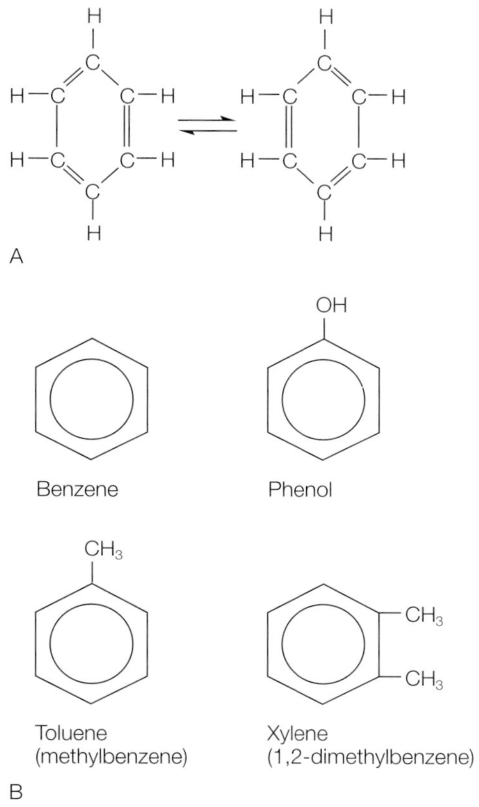

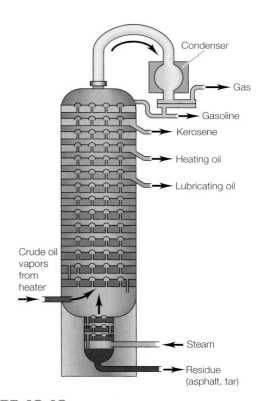

A

Benzene

Phenol

Toluene
(methylbenzene)

Xylene
(1,2-dimethylbenzene)

B

FIGURE 12.9

(*A*) The bonds in C_6H_6 are something between single and double, which gives it different chemical properties than double-bonded hydrocarbons. (*B*) The six-sided symbol with a circle represents the benzene ring. Organic compounds based on the benzene ring are called *aromatic hydrocarbons* because of their aromatic character.

Petroleum

Petroleum is a mixture of alkanes, cycloalkanes, and some aromatic hydrocarbons. The origin of petroleum is uncertain, but it is believed to have formed from the slow decomposition of buried marine life, primarily microscopic plankton and algae in the absence of oxygen (i.e., anaerobic). Time, temperature, pressure, and perhaps bacteria are considered important in the formation of petroleum. As the petroleum formed, it was forced through porous rock until it reached a rock type or rock structure that stopped it. Here, it accumulated to saturate the porous rock, forming an accumulation called an *oil field*. The composition of petroleum varies from one oil field to the next. The oil from a given field might be dark or light in color, and it might have an asphalt base or a paraffin base. Some oil fields contain oil with a high quantity of sulfur, referred to as "sour crude." Because of such variations, some fields have oil with more desirable qualities than oil from other fields.

Early settlers found oil seeps in the eastern United States and collected the oil for medicinal purposes. One enterprising oil peddler tried to improve the taste by running the petroleum

through a whiskey still. He obtained a clear liquid by distilling the petroleum and, by accident, found that the liquid made an excellent lamp oil. This was fortunate timing, for the lamp oil used at that time was whale oil, and whale oil production was declining. This clear liquid obtained by distilling petroleum is today known as *kerosene.*

Wells were drilled, and crude oil refineries were built to produce the newly discovered lamp oil. Gasoline was a by-product of the distillation process and was used primarily as a spot remover. With Henry Ford's automobile production and Edison's electric light invention, the demand for gasoline increased, and the demand for kerosene decreased. The refineries were converted to produce gasoline, and the petroleum industry grew to become one of the world's largest industries.

Crude oil is petroleum that is pumped from the ground, a complex and variable mixture of hydrocarbons with an upper limit of about fifty carbon atoms. This thick, smelly black mixture is not usable until it is refined, that is, separated into usable groups of hydrocarbons called petroleum products. Petroleum products are separated from crude oil by distillation, and any particular product has a boiling point range, or "cut" of the distilled vapors (Figure 12.10). Thus, each product, such as gasoline, heating oil, and so forth is made up of hydrocarbons within

FIGURE 12.10

Fractional distillation is used to separate petroleum into many products. This simplified illustration shows how the 30-foot tower is used to separate the different "fractions" by differences in their boiling points.

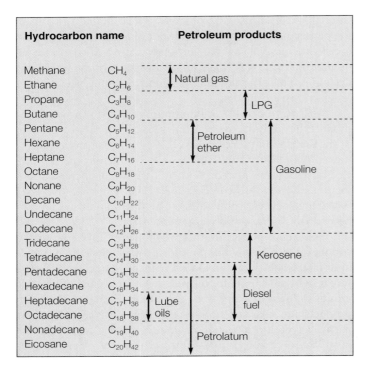

Hydrocarbon name		Petroleum products
Methane	CH_4	Natural gas
Ethane	C_2H_6	
Propane	C_3H_8	LPG
Butane	C_4H_{10}	
Pentane	C_5H_{12}	Petroleum ether
Hexane	C_6H_{14}	
Heptane	C_7H_{16}	
Octane	C_8H_{18}	Gasoline
Nonane	C_9H_{20}	
Decane	$C_{10}H_{22}$	
Undecane	$C_{11}H_{24}$	
Dodecane	$C_{12}H_{26}$	
Tridecane	$C_{13}H_{28}$	
Tetradecane	$C_{14}H_{30}$	Kerosene
Pentadecane	$C_{15}H_{32}$	
Hexadecane	$C_{16}H_{34}$	
Heptadecane	$C_{17}H_{36}$	Lube oils / Diesel fuel
Octadecane	$C_{18}H_{38}$	
Nonadecane	$C_{19}H_{40}$	Petrolatum
Eicosane	$C_{20}H_{42}$	

FIGURE 12.11

Petroleum products and the ranges of hydrocarbons in each product.

a range of carbon atoms per molecule (Figure 12.11). The products, their boiling ranges, and ranges of carbon atoms per molecule are listed in Table 12.4.

The hydrocarbons that have one to four carbon atoms (CH_4 to C_4H_{10}) are gases at room temperature. They can be pumped from certain wells as a gas, but they also occur dissolved in crude oil. *Natural gas* is a mixture of hydrocarbon gases, but it is about 95 percent methane (CH_4). Propane (C_3H_8) and butane (C_4H_{10}) are liquified by compression and cooling and are sold as liquified petroleum gas, or *LPG*. LPG is used where natural gas is not available for cooking or heat-

ing and is widely used as a fuel in barbecue grills and camp stoves.

Gasoline is a mixture of hydrocarbons that may have five to twelve carbon atoms per molecule. Gasoline distilled from crude oil consists mostly of straight-chain molecules not suitable for use as an automotive fuel. Straight-chain molecules burn too rapidly in an automobile engine, producing more of an explosion than a smooth burn. You hear these explosions as a knocking or pinging in the engine, and they indicate poor efficiency and could damage the engine. On the other hand, branched-chain molecules burn slower by comparison, without the pinging or knocking explosions. The burning rate of gasoline is described by the *octane number* scale. The scale is based on pure *n*-heptane, straight-chain molecules that are assigned an octane number of 0, and a multiple branched isomer of octane, 2,2,4-trimethylpentane, which is assigned an octane number of 100 (Figure 12.12). Most unleaded gasolines have an octane rating of 87, which could be obtained with a mixture that is 87 percent 2,2,4-trimethylpentane and 13 percent *n*-heptane. Gasoline, however, is a much more complex mixture.

It is expensive to produce unleaded gasoline because some of the straight-chain hydrocarbon molecules must be converted into branched molecules. The process is one of "cracking and reforming" some of the straight-chain molecules. First, the gasoline is passed through metal tubes heated to 500°C to 800°C (932°F to 1,470°F). At this high temperature, and in the absence of oxygen, the hydrocarbon molecules decompose by breaking into smaller carbon-chain units. These smaller hydrocarbons are then passed through tubes containing a catalyst, which causes them to reform into branched-chain molecules. Unleaded gasoline is produced by the process. Without the reforming

A *n*-heptane, C_7H_{16}

B 2,2,4-trimethylpentane (or iso-octane), C_8H_{18}

FIGURE 12.12

The octane rating scale is a description of how rapidly gasoline burns. It is based on (A) *n*-heptane, with an assigned octane number of 0, and (B) 2,2,4-trimethylpentane, with an assigned number of 100.

TABLE 12.4

Petroleum products

Name	Boiling Range (°C)	Carbon Atoms per Molecule
Natural gas	Less than 0	C_1 to C_4
Petroleum ether	35–100	C_5 to C_7
Gasoline	35–215	C_5 to C_{12}
Kerosene	35–300	C_{12} to C_{15}
Diesel fuel	300–400	C_{15} to C_{18}
Motor oil, grease	350–400	C_{16} to C_{18}
Paraffin	Solid, melts at about 55	C_{20}
Asphalt	Boiler residue	C_{40} or more

that produces unleaded gasoline, low-numbered hydrocarbons (such as ethylene) can be produced. Ethylene is used as a raw material for many plastic materials, antifreeze, and other products. Cracking is also used to convert higher-numbered hydrocarbons, such as heating oil, into gasoline.

Kerosene is a mixture of hydrocarbons that have from twelve to fifteen carbon atoms. The petroleum product called kerosene is also known by other names, depending on its use. Some of these names are lamp oil (with coloring and odorants added), jet fuel (with a flash flame retardant added), heating oil, #1 fuel oil, and in some parts of the country, "coal oil."

Diesel fuel is a mixture of a group of hydrocarbons that have from fifteen to eighteen carbon atoms per molecule. Diesel fuel also goes by other names, again depending on its use, for example, diesel fuel, distillate fuel oil, heating oil, or #2 fuel oil. During the summer season there is a greater demand for gasoline than for heating oil, so some of the supply is converted to gasoline by the cracking process.

Motor oil and *lubricating oils* have sixteen to eighteen carbon atoms per molecule. Lubricating grease is heavy oil that is thickened with soap. *Petroleum jelly,* also called petrolatum (or Vaseline), is a mixture of hydrocarbons with sixteen to thirty-two carbon atoms per molecule. *Mineral oil* is a light lubricating oil that has been decolorized and purified.

Depending on the source of the crude oil, varying amounts of *paraffin* wax (C_{20} or greater) or *asphalt* (C_{36} or more) may be present. Paraffin is used for candles, waxed paper, and home canning. Asphalt is mixed with gravel and used to surface roads. It is also mixed with refinery residues and lighter oils to make a fuel called #6 fuel oil or residual fuel oil. Industries and utilities often use this semisolid material that must be heated before it will pour. Number 6 fuel oil is used as a boiler fuel, costing about half as much as #2 fuel oil.

Hydrocarbon Derivatives

The hydrocarbons account for only about 5 percent of the known organic compounds, but the other 95 percent can be considered hydrocarbon derivatives. **Hydrocarbon derivatives** are formed when *one or more hydrogen atoms on a hydrocarbon have been replaced by some element or group of elements other than hydrogen.* For example, the halogens (F_2, Cl_2, Br_2) react with an alkane in sunlight or when heated, replacing a hydrogen:

In this particular *substitution reaction* a hydrogen atom on methane is replaced by a chlorine atom to form methyl chloride. Replacement of any number of hydrogen atoms is possible, and a few *organic halides* are illustrated in Figure 12.13.

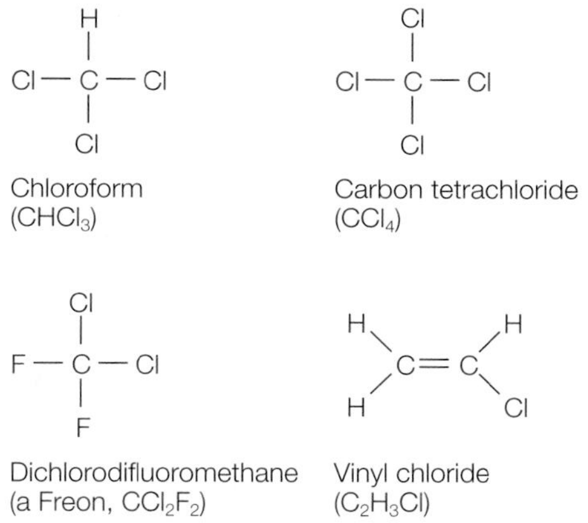

FIGURE 12.13
Common examples of organic halides.

If a hydrocarbon molecule is unsaturated (has a multiple bond), a hydrocarbon derivative can be formed by an *addition reaction:*

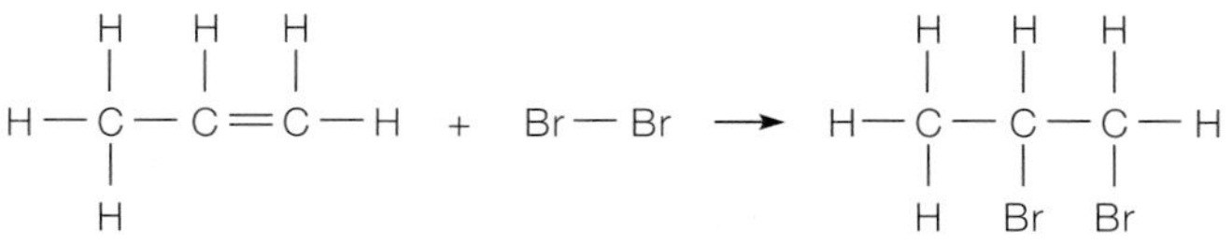

The bromine atoms add to the double bond on propene, forming 1,2-dibromopropane.

Alkene molecules can also add to each other in an addition reaction to form a very long chain consisting of hundreds of molecules. A long chain of repeating units is called a **polymer** (poly- = many; -mer = segment), and the reaction is called *addition polymerization.* Ethylene, for example, is heated under pressure with a catalyst to form *polyethylene.* Heating breaks the double bond,

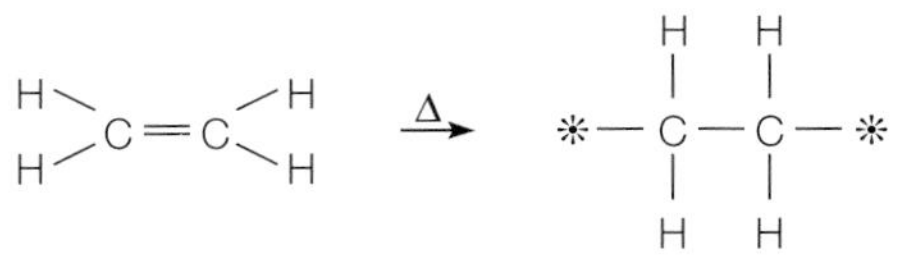

which provides sites for single covalent bonds to join the ethylene units together,

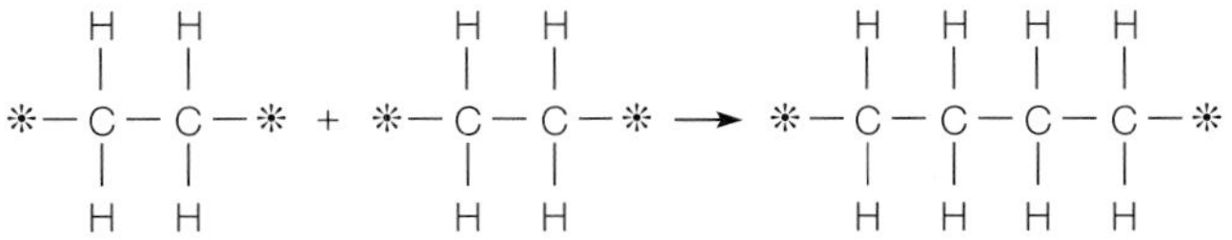

which continues the addition polymerization until the chain is hundreds of units long. Synthetic polymers such as polyethylene are discussed in a later section.

The addition reaction and the addition polymerization reaction can take place because of the double bond of the alkenes, and, in fact, the double bond is the site of most alkene reactions. The atom or group of atoms in an organic molecule that is the

TABLE 12.5
Selected organic functional groups

Name of Functional Group	General Formula	General Structure
Organic Halide	RCl	R—C̈l:
Alcohol	ROH	R—Ö—H
Ether	ROR′	R—Ö—R′
Aldehyde	RCHO	R—C—H ‖ :O:
Ketone	RCOR′	R—C—R′ ‖ :O:
Organic Acid	RCOOH	R—C—Ö—H ‖ :O:
Ester	RCOOR′	R—C—Ö—R′ ‖ :O:
Amine	RNH₂	R—N̈—H | H

The name of the hydrocarbon group (Table 12.2) determines the name of the alcohol. If the hydrocarbon group in ROH is methyl, for example, the alcohol is called *methyl alcohol.* Using the IUPAC naming rules, the name of an alcohol has the suffix "-ol." Thus, the IUPAC name of methyl alcohol is *methanol.*

All alcohols have the hydroxyl functional group, and all are chemically similar (Figure 12.14). Alcohols are toxic to humans, except that ethanol can be consumed in limited quantities. Consumption of other alcohols such as 2-propanol (isopropyl alcohol, or "rubbing alcohol") can result in serious gastric distress. Consumption of methanol can result in blindness and death. Ethanol, C_2H_5OH, is produced by the action of yeast or by a chemical reaction of ethylene derived from petroleum refining. Yeast acts on sugars to produce ethanol and CO_2. When beer, wine, and other such beverages are the desired products, the CO_2 escapes during fermentation, and the alcohol remains in solution. In baking, the same reaction utilizes the CO_2 to make the dough rise, and the alcohol is evaporated during baking. Most alcoholic beverages are produced by the yeast

site of a chemical reaction is identified as a **functional group.** *It is the functional group that is responsible for the chemical properties of an organic compound.* Functional groups usually have (1) multiple bonds or (2) lone pairs of electrons that cause them to be sites of reactions. Table 12.5 lists some of the common hydrocarbon functional groups. Look over this list, comparing the structure of the functional group with the group name. Some of the more interesting examples from a few of these groups will be considered next. Note that the R and R′ (pronounced, "R prime") stand for one or more of the hydrocarbon groups from Table 12.2. For example, in the reaction between methane and chlorine, the product is methyl chloride. In this case the R in RCl stands for methyl, but it could represent any hydrocarbon group.

Alcohols

An *alcohol* is an organic compound formed by replacing one or more hydrogens on an alkane with a hydroxyl functional group (−OH). The hydroxyl group should not be confused with the hydroxide ion, OH^-. The hydroxyl group is attached to an organic compound and does not form ions in solution as the hydroxide ion does. It remains attached to a hydrocarbon group (R), giving the compound its set of properties that are associated with alcohols.

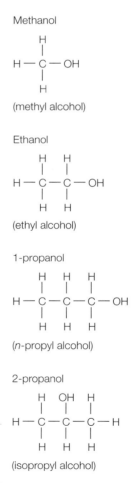

FIGURE 12.14

Four different alcohols. The IUPAC name is given above each structural formula, and the common name is given below.

FIGURE 12.15

Gasoline is a mixture of hydrocarbons (C_8H_{18} for example) that contain no atoms of oxygen. Gasohol contains ethyl alcohol, C_2H_5OH, which does contain oxygen. The addition of alcohol to gasoline, therefore, adds oxygen to the fuel. Since carbon monoxide forms when there is an insufficient supply of oxygen, the addition of alcohol to gasoline helps cut down on carbon monoxide emissions. An atmospheric inversion, with increased air pollution, is likely during the dates shown on the pump, so that is when the ethanol is added.

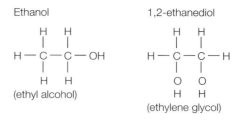

FIGURE 12.16

Common examples of alcohols with one, two, and three hydroxyl groups per molecule. The IUPAC name is given above each structural formula, and the common name is given below.

fermentation reaction, but some are made from ethanol derived from petroleum refining.

Alcohols with six or fewer carbon atoms per molecule are soluble in both alkanes and water. A solution of ethanol and gasoline is called *gasohol* (Figure 12.15). Alcoholic beverages are a solution of ethanol and water. The *proof* of such a beverage is double the ethanol concentration by volume. Therefore, a solution of 40 percent ethanol by volume in water is 80 proof, and wine that is 12 percent alcohol by volume is 24 proof. Distillation alone will produce a 190 proof concentration, but other techniques are necessary to obtain 200 proof absolute alcohol. *Denatured alcohol* is ethanol with acetone, formaldehyde, and other chemicals in solution that are difficult to separate by distillation. Since these denaturants make consumption impossible, denatured alcohol is sold without the consumption tax.

Methanol, ethanol, and isopropyl alcohol all have one hydroxyl group per molecule. An alcohol with two hydroxyl groups per molecule is called a *glycol*. Ethylene glycol is perhaps the best-known glycol since it is used as an antifreeze. An alcohol with three hydroxyl groups per molecule is called *glycerol* (or glycerin). Glycerol is a by-product in the making of soap. It is added to toothpastes, lotions, and some candies to retain moisture and softness. Ethanol, ethylene glycol, and glycerol are compared in Figure 12.16.

Glycerol reacts with nitric acid in the presence of sulfuric acid to produce glyceryl trinitrate, commonly known as *nitro-glycerine*. Nitroglycerine is a clear oil that is violently explosive, and when warmed, it is extremely unstable. In 1867, Alfred Nobel discovered that a mixture of nitroglycerine and siliceous earth was more stable than pure nitroglycerine but was nonetheless explosive. The mixture is packed in a tube and is called *dynamite*. Old dynamite tubes, however, leak pure nitroglycerine that is again sensitive to a slight shock.

Ethers, Aldehydes, and Ketones

An *ether* has a general formula of ROR′, and the best-known ether is diethylether. In a molecule of diethylether, both the R and the R′ are ethyl groups. Diethylether is a volatile, highly flammable liquid that was used as an anesthetic in the past. Today, it is used as an industrial and laboratory solvent.

Aldehydes and *ketones* both have a functional group of a carbon atom doubly bonded to an oxygen atom called a *carbonyl group*. The *aldehyde* has a hydrocarbon group, R (or a hydrogen in one case), and a hydrogen attached to the carbonyl group. A *ketone* has a carbonyl group with two hydrocarbon groups attached (Figure 12.17).

The simplest aldehyde is *formaldehyde*. Formaldehyde is soluble in water, and a 40 percent concentration called *formalin* has been used as an embalming agent and to preserve biological specimens. Formaldehyde is also a raw material used to make plastics such as Bakelite. All the aldehydes have odors, and the odors of some aromatic hydrocarbons include the odors of almonds, cinnamon, and vanilla. The simplest ketone is *acetone*. Acetone has a fragrant odor and is used as a solvent in paint removers and nail polish removers. By sketching the structural formulas you can see that ethers are isomers of alcohols while aldehydes and ketones are isomers of each other. Again, the physical and chemical properties are quite different.

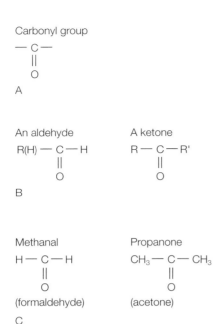

Carbonyl group

$$-\overset{\displaystyle \|}{\underset{\displaystyle O}{C}}-$$

A

An aldehyde

$$R(H)-\overset{\displaystyle \|}{\underset{\displaystyle O}{C}}-H$$

A ketone

$$R-\overset{\displaystyle \|}{\underset{\displaystyle O}{C}}-R'$$

B

Methanal

$$H-\overset{\displaystyle \|}{\underset{\displaystyle O}{C}}-H$$

(formaldehyde)

Propanone

$$CH_3-\overset{\displaystyle \|}{\underset{\displaystyle O}{C}}-CH_3$$

(acetone)

C

FIGURE 12.17

The carbonyl group (*A*) is present in both aldehydes and ketones, as shown in (*B*). (*C*) The simplest example of each, with the IUPAC name above and the common name below each formula.

Organic Acids and Esters

Mineral acids, such as hydrochloric and sulfuric acid, are made of inorganic materials. Acids that were derived from organisms are called *organic acids*. Because many of these organic acids can be formed from fats, they are sometimes called *fatty acids*. Chemically, they are known as the *carboxylic acids* because they contain the carboxyl functional group, −COOH, and have a general formula of RCOOH.

The simplest carboxylic acid has been known since the Middle Ages, when it was isolated by the distillation of ants. The Latin word *formica* means "ant," so this acid was given the name *formic acid* (Figure 12.18). Formic acid is

$$H-\overset{\displaystyle \|}{\underset{\displaystyle O}{C}}-O\,H$$

It is formic acid, along with other irritating materials, that causes the sting of bees, ants, and certain plants such as the stinging nettle.

Acetic acid, the acid of vinegar, has been known since antiquity. Acetic acid forms from the oxidation of ethanol. An oxidized bottle of wine contains acetic acid in place of the alcohol, which gives the wine a vinegar taste. Before wine is served in a restaurant, the person ordering is customarily handed the bottle cork and a glass with a small amount of wine. You first break the cork in half to make sure it is dry, which tells you that the wine has been sealed from oxygen. The

FIGURE 12.18

These red ants, like other ants, make the simplest of the organic acids, formic acid. The sting of bees, ants, and some plants contains formic acid, along with some other irritating materials. Formic acid is HCOOH.

small sip is to taste for vinegar before the wine is served. If the wine has been oxidized, the reaction is

$$H-\overset{\displaystyle H}{\underset{\displaystyle H}{C}}-\overset{\displaystyle H}{\underset{\displaystyle H}{C}}-OH \quad \xrightarrow{\text{oxidation}} \quad H-\overset{\displaystyle H}{\underset{\displaystyle H}{C}}-\overset{\displaystyle \|}{\underset{\displaystyle O}{C}}-OH$$

Ethanol   Acetic acid

Organic acids are common in many foods. The juice of citrus fruit, for example, contains citric acid, which relieves a thirsty feeling by stimulating the flow of saliva. Lactic acid is found in sour milk, buttermilk, sauerkraut, and pickles. Lactic acid also forms in your muscles as a product of carbohydrate metabolism, causing a feeling of fatigue. Citric and lactic acids are small molecules compared to some of the carboxylic acids that are formed from fats. Palmitic acid, for example, is $C_{16}H_{32}O_2$ and comes from palm oil. The structure of palmitic acid is a chain of fourteen CH_2 groups with CH_3- at one end and $-COOH$ at the other. Again, it is the functional carboxyl group, −COOH, that gives the molecule its acid properties. Organic acids are also raw materials used in the making of polymers of fabric, film, and paint.

Esters are common in both plants and animals, giving fruits and flowers their characteristic odor and taste. Esters are also used in perfumes and artificial flavorings. A few of the flavors that particular esters are responsible for are listed in Table 12.6. These liquid esters can be obtained from natural sources or they can be chemically synthesized. Whatever the source, amyl acetate, for example, is the chemical responsible for what you identify as the flavor of banana. Natural flavors, however, are complex mixtures of these esters along with other organic compounds. Lower molecular weight esters are fragrant-smelling liquids, but higher molecular weight esters are odorless oils and fats.

TABLE 12.6
Flavors and esters

Ester Name	Formula	Flavor
Amyl Acetate	$CH_3 - \overset{\displaystyle O}{\underset{\displaystyle \|}{C}} - O - C_5H_{11}$	Banana
Octyl Acetate	$CH_3 - \overset{\displaystyle O}{\underset{\displaystyle \|}{C}} - O - C_8H_{17}$	Orange
Ethyl Butyrate	$C_3H_7 - \overset{\displaystyle O}{\underset{\displaystyle \|}{C}} - O - C_2H_5$	Pineapple
Amyl Butyrate	$C_3H_7 - \overset{\displaystyle O}{\underset{\displaystyle \|}{C}} - O - C_5H_{11}$	Apricot
Ethyl Formate	$H - \overset{\displaystyle O}{\underset{\displaystyle \|}{C}} - O - C_2H_5$	Rum

Concepts Applied

Organic Products

Pick a household product that has ingredients sounding like they could be organic compounds. On a separate sheet of paper write the brand name of the product and the type of product (Example: Oil of Olay; skin moisturizer), then list the ingredients one under the other (column 1). In a second column beside each name put the type of compound if you can figure it out from its name or find it in any reference (Example: cetyl palmitate—an ester of cetyl alcohol and palmitic acid). In a third column, put the structural formula if you can figure it out or find it in any reference such as a CRC handbook or the Merck Index. Finally, in a fourth column put whatever you know or can find out about the function of that substance in the product.

Organic Compounds of Life

The chemical processes regulated by living organisms begin with relatively small organic molecules and water. The organism uses energy and matter from the surroundings to build large *macromolecules.* A **macromolecule** is a very large molecule that is a combination of many smaller, similar molecules joined together in a chainlike structure. Macromolecules have molecular weights of thousands or millions of atomic mass units. There are four main types of macromolecules: (1) proteins, (2) carbohydrates, (3) fats and oils, and (4) nucleic acids.

A living organism, even a single-cell organism such as a bacterium, contains six thousand or so different kinds of macromolecules. The basic unit of an organism is called a *cell.* Cells are made of macromolecules that are formed inside the cell. The cell decomposes organic molecules taken in as food and uses energy from the food molecules to build more macromolecules. The process of breaking down organic molecules and building up macromolecules is called *metabolism.* Through metabolism, the cell grows, then divides into two cells. Each cell is a generic duplicate of the other, containing the same number and kinds of macromolecules. Each new cell continues the process of growth, then reproduces again, making more cells. This is the basic process of life. The complete process is complicated and very involved, easily filling a textbook in itself, so the details will not be presented here. The following discussion will be limited to three groups of organic molecules involved in metabolic processes: proteins, carbohydrates, and fats and oils.

Proteins

Proteins are macromolecular polymers made up of smaller molecules called amino acids. These very large macromolecules have molecular weights that vary from about six thousand to fifty million. Some proteins are simple straight-chain polymers of amino acids, but others contain metal ions such as Fe^{2+} or parts of organic molecules derived from vitamins. Proteins serve as major structural and functional materials in living things. *Structurally,* proteins are major components of muscles, connective tissue, and the skin, hair, and nails. *Functionally,* some proteins are enzymes, which catalyze metabolic reactions; hormones, which regulate body activities; hemoglobin, which carries oxygen to cells; and antibodies, which protect the body.

Proteins are formed from twenty **amino acids,** which are organic molecules with acid and amino functional groups with the general formula of

Note the carbon atom labeled "alpha" in the general formula. The amino functional group (NH_2) is attached to this carbon atom, which is next to the carboxylic group (COOH). This arrangement is called an *alpha-amino acid,* and the building blocks of proteins are all alpha-amino acids. The twenty amino acids differ in the nature of the R group, also called the *side chain.* It is the linear arrangements of amino acids and their side chains that determine the properties of a protein.

Amino acids are linked to form a protein by a peptide bond between the amino group of one amino acid and the carboxyl group of a second amino acid. A polypeptide is a polymer formed from linking many amino acid molecules. If the polypeptide is involved in a biological structure or function, it is

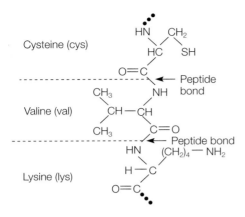

Cysteine (cys)

Valine (val)

Lysine (lys)

← Peptide bond

← Peptide bond

FIGURE 12.19

Part of a protein polypeptide made up of the amino acids cysteine (cys), valine (val), and lysine (lys). A protein can have from fifty to one thousand of these amino acid units; each protein has its own unique sequence.

called a *protein*. A protein chain can consist of different combinations of the twenty amino acids with hundreds or even thousands of amino acid molecules held together with peptide bonds (Figure 12.19). The arrangement or sequence of these amino acid molecules determines the structure that gives the protein its unique set of biochemical properties. Insulin, for example, is a protein hormone that biochemically regulates the blood sugar level. Insulin contains 86 amino acid molecules in the chain. Hemoglobin is the protein that carries oxygen in the bloodstream, and its biochemical characteristics are determined by its chain of 146 amino acid molecules.

Concepts Applied

Fats Foods

Pick a food product and write the number of grams of fats, proteins, and carbohydrates per serving according to the information on the label. Multiply the number of grams of proteins and carbohydrates each by 4 Cal/g and the number of grams of fat by 9 Cal/g. Add the total Calories per serving. Does your total agree with the number of Calories per serving given on the label? Also examine the given serving size. Is this a reasonable amount to be consumed at one time or would you probably eat two or three times this amount? Write the rest of the nutrition information (vitamins, minerals, sodium content, etc.) and then write the list of ingredients. Tell what ingredient you think is providing which nutrient. (Example: vegetable oil—source of fat; milk—provides calcium and vitamin A; MSG—source of sodium.)

Carbohydrates

Carbohydrates are an important group of organic compounds that includes sugars, starches, and cellulose, and they are important in plants and animals for structure, protection, and food. Cellulose is the skeletal substance of plants and plant materials, and chitin is a similar material that forms the hard, protective covering of insects and shellfish such as crabs and lobsters. *Glucose*, $C_6H_{12}O_6$, is the most abundant carbohydrate and serves as a food and a basic building block for other carbohydrates.

Carbohydrates were named when early studies found that water vapor was given off and carbon remained when sugar was heated. The name *carbohydrate* literally means "watered carbon," and the empirical formulas for most carbohydrates indeed indicate carbon (C) and water (H_2O). Glucose, for example, could be considered to be six carbons with six waters, or $C_6(H_2O)_6$. However, carbohydrate molecules are more complex than just water attached to a carbon atom. They are polyhydroxyl aldehydes and ketones, two of which are illustrated in Figure 12.20. The two carbohydrates in this illustration belong to a group of carbohydrates known as **monosaccharides,** or *simple sugars*. They are called simple sugars because they are the smallest units that have the characteristics of carbohydrates, and they can be combined to make larger complex carbohydrates. There are many kinds of simple sugars, but they are mostly 6-carbon molecules such as glucose and fructose. Glucose (also called dextrose) is found in the sap of plants, and in the human bloodstream it is called *blood sugar*. Corn syrup, which is often used as a sweetener, is mostly glucose. Fructose, as its name implies, is the sugar that occurs in fruits, and it is sometimes called *fruit sugar*. Both glucose and fructose have the same molecular formula, but glucose is an aldehyde sugar and fructose is a ketone sugar (Figure 12.20). A mixture of glucose and fructose is found in honey. This mixture also is formed when table sugar (sucrose) is reacted with water in the presence of an acid, a reaction that takes place in the preparation of canned fruit and candies. The

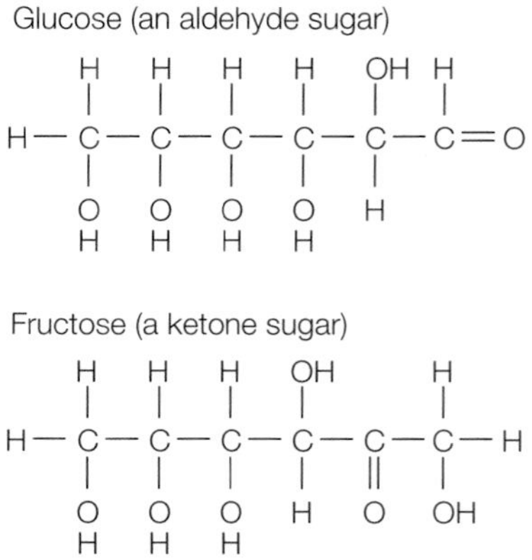

Glucose (an aldehyde sugar)

Fructose (a ketone sugar)

FIGURE 12.20

Glucose (blood sugar) is an aldehyde, and fructose (fruit sugar) is a ketone. Both have a molecular formula of $C_6H_{12}O_6$.

mixture of glucose and fructose is called *invert sugar.* Thanks to fructose, invert sugar is about twice as sweet to the taste as the same amount of sucrose.

Two monosaccharides are joined together to form **disaccharides** with the loss of a water molecule, for example,

$$C_6H_{12}O_6 + C_6H_{12}O_6 \longrightarrow C_{12}H_{22}O_{11} + H_2O$$

glucose   fructose   sucrose

The most common disaccharide is *sucrose,* or ordinary table sugar. Sucrose occurs in high concentrations in sugarcane and sugar beets. It is extracted by crushing the plant materials, then dissolving the sucrose from the materials with water. The water is evaporated and the crystallized sugar is decolorized with charcoal to produce white sugar. Other common disaccharides include *lactose* (milk sugar) and *maltose* (malt sugar). All three disaccharides have similar properties, but maltose tastes only about one-third as sweet as sucrose. Lactose tastes only about one-sixth as sweet as sucrose. No matter which disaccharide sugar is consumed (sucrose, lactose, or maltose), it is converted into glucose and transported by the bloodstream for use by the body.

Polysaccharides are polymers consisting of monosaccharide units joined together in straight or branched chains. Polysaccharides are the energy-storage molecules of plants and animals (starch and glycogen) and the structural molecules of plants (cellulose). **Starches** are a group of complex carbohydrates composed of many glucose units that plants use as a stored food source. Potatoes, rice, corn, and wheat store starch granules and serve as an important source of food for humans. The human body breaks down the starch molecules to glucose, which is transported by the bloodstream and utilized just like any other glucose. This digestive process begins with enzymes secreted with saliva in the mouth. You may have noticed a result of this enzyme-catalyzed reaction as you eat bread. If you chew the bread for a while it begins to taste sweet.

Plants store sugars in the form of starch polysaccharides, and animals store sugars in the form of the polysaccharide *glycogen.* Glycogen is a starchlike polysaccharide that is synthesized by the human body and stored in the muscles and liver. Glycogen, like starch, is a very high molecular weight polysaccharide, but it is more highly branched. These highly branched polysaccharides serve as a direct reserve source of energy in the muscles. In the liver, they serve as a reserve source to maintain the blood sugar level.

Cellulose is a polysaccharide that is abundant in plants, forming the fibers in cell walls that preserve the structure of plant materials (Figure 12.21). Cellulose molecules are straight chains, consisting of large numbers of glucose units. These glucose units are arranged in a way that is very similar to the arrangement of the glucose units of starch but with differences in the bonding arrangement that holds the glucose units together (Figure 12.22). This difference turns out to be an important one where humans are concerned because enzymes that break down starches do not affect cellulose. Humans do not have the necessary enzymes to break down the cellulose chain (digest it), so humans receive no food value from cellulose.

FIGURE 12.21

These plants and their flowers are made up of a mixture of carbohydrates that were manufactured from carbon dioxide and water, with the energy of sunlight. The simplest of the carbohydrates are the monosaccharides, simple sugars (fruit sugar) that the plant synthesizes. Food is stored as starches, which are polysaccharides made from the simpler monosaccharide glucose. The plant structure is held upright by fibers of cellulose, another form of a polysaccharide composed of glucose.

Cattle and termites that do utilize cellulose as a source of food have protozoa and bacteria (with the necessary enzymes) in their digestive systems. Cellulose is still needed in the human diet, however, for fiber and bulk.

Fats and Oils

Cereal grains and other plants also provide carbohydrates, the human body's preferred food for energy. When an excess amount of carbohydrates is consumed, the body begins to store some of its energy source in the form of glycogen in the muscles and liver. Beyond this storage for short-term needs, the body begins to store

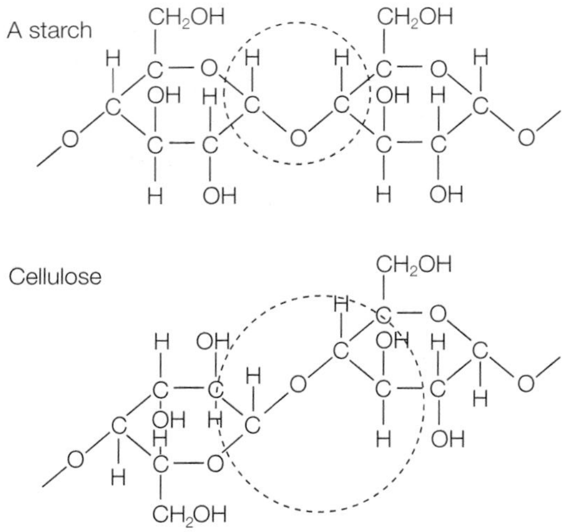

A starch

Cellulose

FIGURE 12.22

Starch and cellulose are both polymers of glucose, but humans cannot digest cellulose. The difference in the bonding arrangement might seem minor, but enzymes must fit a molecule very precisely. Thus, enzymes that break down starch do nothing to cellulose.

energy in a different chemical form for longer-term storage. This chemical form is called **fat** in animals and **oil** in plants. Fats and oils are esters formed from glycerol (1,2,3-trihydroxypropane) and three long-chain carboxylic acids (fatty acids). This ester is called a **triglyceride,** and its structural formula is shown in Figure 12.23. Fats are solids and oils are liquids at room temperature, but they both have this same general structure.

Fats and oils usually have two or three different fatty acids, and several are listed in Table 12.7. Animal fats can be either saturated or unsaturated, but most are saturated. Oils are liquids at room temperature because they contain a higher number of unsaturated units. These unsaturated oils (called "polyunsaturated" in news and advertisements), such as safflower and corn oils, are used as liquid cooking oils because unsaturated oils are believed to lead to lower cholesterol levels in the bloodstream. Saturated fats, along with

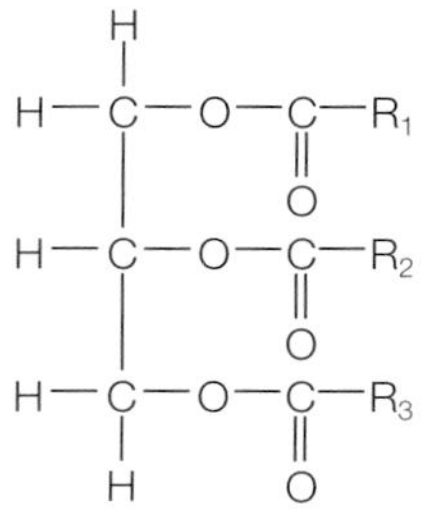

FIGURE 12.23

The triglyceride structure of fats and oils. Note the glycerol structure on the left and the ester structure on the right. Also notice that R_1, R_2, and R_3 are long-chained molecules of 12, 14, 16, 18, 20, 22, or 24 carbons that might be saturated or unsaturated.

TABLE 12.7
Some fatty acids occurring in fats

Common Name	Condensed Structure	Source
Lauric Acid	$CH_3(CH_2)_{10}COOH$	Coconuts
Palmitic Acid	$CH_3(CH_2)_{14}COOH$	Palm oil
Stearic Acid	$CH_3(CH_2)_{16}COOH$	Animal fats
Oleic Acid	$CH_3(CH_2)_7CH{=}CH(CH_2)_7COOH$	Corn oil
Linoleic Acid	$CH_3(CH_2)_4CH{=}CHCH_2{=}CH(CH_2)_7COOH$	Soybean oil
Linolenic Acid	$CH_3CH_2(CH{=}CHCH_2)_3(CH_2)_6COOH$	Fish oils

cholesterol, are believed to contribute to hardening of the arteries over time.

Cooking oils from plants, such as corn and soybean oil, are hydrogenated to convert the double bonds of the unsaturated oil to the single bonds of a saturated one. As a result, the liquid oils are converted to solids at room temperature. For example, one brand of margarine lists ingredients as "liquid soybean oil (nonhydrogenated) and partially hydrogenated cottonseed oil with water, salt, preservatives, and coloring." Complete hydrogenation would result in a hard solid, so the cottonseed oil is partially hydrogenated and then mixed with liquid soybean oil. Coloring is added because oleo is white, not the color of butter. Vegetable shortening is the very same product without added coloring. Reaction of a triglyceride with a strong base such as KOH or NaOH yields a fatty acid of salt and glycerol. A sodium or potassium fatty acid is commonly known as *soap.*

Excess food from carbohydrate, protein, or fat and oil sources is converted to fat for long-term energy storage in *adipose tissue,* which also serves to insulate and form a protective padding. In terms of energy storage, fats yield more than twice the energy per gram oxidized as carbohydrates or proteins.

Concepts Applied

Brand News

Pick two competing brands of a product you use (example: Tylenol and a store-brand acetaminophen) and write the following information: All ingredients, amount of each ingredient (if this is not given, remember that by law ingredients have to be listed in order of their percent by weight), and the cost per serving or dose. Comment on whether each of the listed ingredients is a single substance, and thus something with the same properties wherever it is found (example: salt, as a label ingredient, means sodium chloride no matter in what product it appears) or a mixture and thus possibly different in different products (example: tomatoes, as a ketchup ingredient, might be of better quality in one brand of ketchup than another). Then draw a reasonably informed conclusion as to whether there is any significant difference between the two brands or whether the more expensive one is worth the difference in price. Finally, do your own consumer test to check your prediction.

Synthetic Polymers

Polymers are huge, chainlike molecules made of hundreds or thousands of smaller, repeating molecular units called *monomers*. Polymers occur naturally in plants and animals. Cellulose, for example, is a natural plant polymer made of glucose monomers. Wool and silk are natural animal polymers made of amino acid monomers. *Synthetic polymers* are manufactured from a wide variety of substances, and you are familiar with these polymers as synthetic fibers such as nylon and the inexpensive light plastic used for wrappings and containers (Figure 12.24).

Name	Chemical unit	Uses	Name	Chemical unit	Uses
Polyethylene		Squeeze bottles, containers, laundry and trash bags, packaging	Polyvinyl acetate		Mixed with vinyl chloride to make vinylite; used as an adhesive and resin in paint
Polypropylene		Indoor-outdoor carpet, pipe valves, bottles	Styrene-butadiene rubber	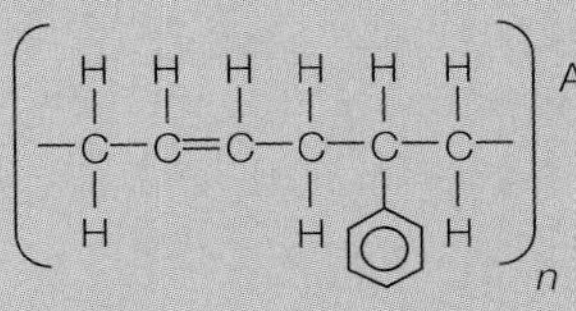	Automobile tires
Polyvinyl chloride (PVC)		Plumbing pipes, synthetic leather, plastic tablecloths, phonograph records, vinyl tile	Polychloroprene (Neoprene)	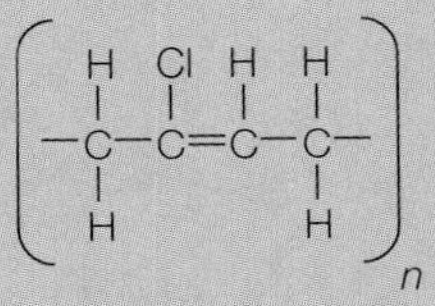	Shoe soles, heels
Polyvinylidene chloride (Saran)		Flexible food wrap	Polymethyl methacrylate (Plexiglas, Lucite)	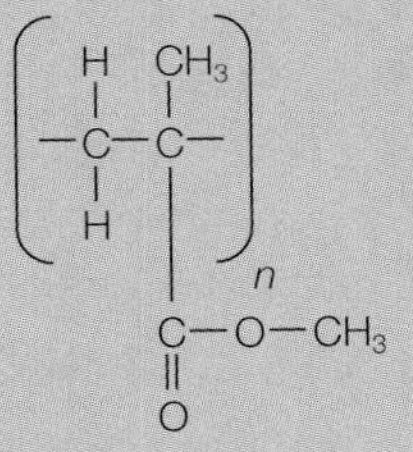	Moldings, transparent surfaces on furniture, lenses, jewelry, transparent plastic "glass"
Polystyrene (Styrofoam)		Coolers, cups, insulating foam, shock-resistant packing material, simulated wood furniture	Polycarbonate (Lexan)	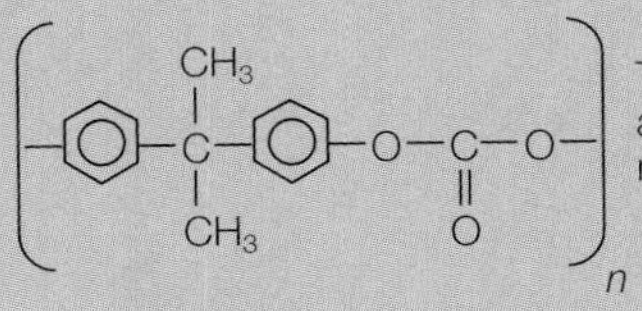	Tough, molded articles such as motorcycle helmets
Polytetrafluoroethylene (Teflon)		Gears, bearings, coating for nonstick surface of cooking utensils	Polyacrylonitrile (Orlon, Acrilan, Creslan)	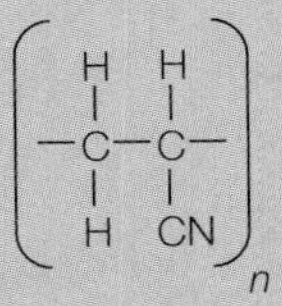	Textile fibers

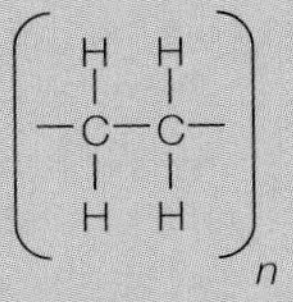

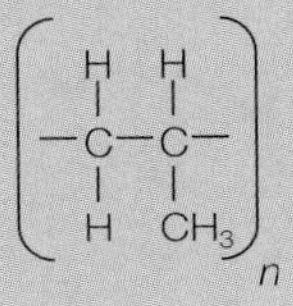

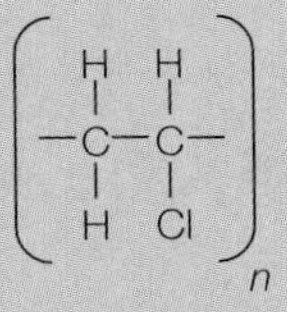

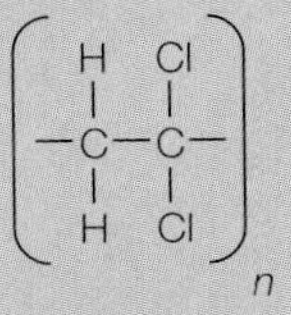

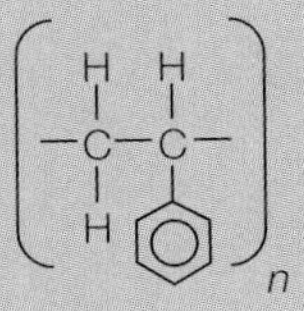

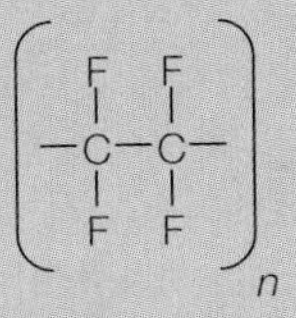

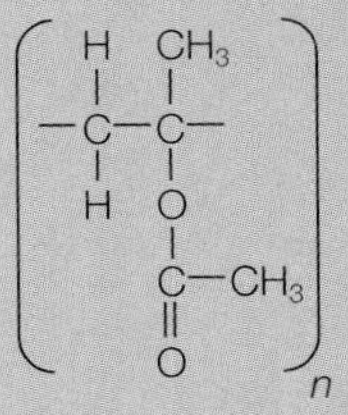

FIGURE 12.24

Synthetic polymers, the polymer unit, and some uses of each polymer.

Plastic containers are made of different types of plastic resins; some are suitable for recycling and some are not. How do you know which are suitable and how to sort them? Most plastic containers have a code stamped on the bottom. The code is a number in the recycling arrow logo, sometimes appearing with some letters. Here is what the numbers and letters mean in terms of (a) the plastic, (b) how it is used, and (c) if it is usually recycled or not.

a. Polyethylene terephthalate (PET)
b. Large soft-drink bottles, salad dressing bottles
c. Frequently recycled

a. High-density polyethylene (HDPE)
b. Milk jugs, detergent and bleach bottles, others
c. Frequently recycled

a. Polyvinyl chloride (PVC or PV)
b. Shampoos, hair conditioners, others
c. Rarely recycled

a. Low-density polyethylene (LDPE)
b. Plastic wrap, laundry and trash bags
c. Rarely recycled

a. Polypropylene (PP)
b. Food containers
c. Rarely recycled

a. Polystyrene (PS)
b. Styrofoam cups, burger boxes, plates
c. Occasionally recycled

a. Mixed resins
b. Catsup squeeze bottles, other squeeze bottles
c. Rarely recycled

The first synthetic polymer was a modification of the naturally existing cellulose polymer. Cellulose was chemically modified in 1862 to produce celluloid, the first *plastic*. The term "plastic" means that celluloid could be molded to any desired shape. Celluloid was produced by first reacting cotton with a mixture of nitric and sulfuric acids, which produced an ester of cellulose nitrate. This ester is an explosive compound known as "guncotton," or smokeless gunpowder. When made with ethanol and camphor, the product is less explosive and can be formed and molded into useful articles. This first plastic, celluloid, was used to make dentures, combs, eyeglass frames, and photographic film. Before the discovery of celluloid, many of these articles, including dentures, were made from wood. Today, only Ping-Pong balls are made from cellulose nitrate.

Cotton or other sources of cellulose reacted with acetic acid and sulfuric acid produce a cellulose acetate ester. This

Friedrich Wöhler (1800–1882)

Friedrich Wöhler was a German chemist who is generally credited with having carried out the first laboratory synthesis of an organic compound, although his main interest was inorganic chemistry.

Wöhler was born at Eschershein, near Frankfurt-am-Main, on July 31, 1800, the son of a veterinary surgeon in the service of the Crown Prince of Hesse-Kassel. He entered Marburg University in 1820 to study medicine, and after a year transferred to Heidelberg, where he studied in the laboratory of Leopold Gmelin (1788–1853). He gained his medical degree in 1823, but Gmelin had persuaded Wöhler to study chemistry, and so he spent the following year in Stockholm with Jöns Berzelius, beginning a lifelong association between the two chemists. From 1825 to 1831 he occupied a teaching position in a technical school in Berlin, and from 1831 to 1836 he held a similar post at Kassel. In 1836 he became Professor of Chemistry in the Medical Faculty of Göttingen University, as successor to Friedrich Strohmeyer (1776–1835), and he remained there for the rest of his career, making it one of the most prestigious teaching laboratories in Europe. He died in Göttingen on September 23, 1882.

In Wöhler's first research in 1827 he isolated metallic aluminum by heating its chloride with potassium; he then prepared many different aluminum salts. In 1828 he used the same procedure to isolate beryllium. Also in 1828 he carried out the reaction for which he is best known. He heated ammonium thiocyanate—a crystalline, inorganic substance—and converted it to urea (carbamide), an organic substance previously obtained only from natural sources. Until that time there had been a basic misconception in scientific thinking that the chemical changes undergone by substances in living organisms were not governed by the same laws as were inanimate substances; it was thought that these "vital" phenomena could not be described in ordinary chemical or physical terms. This theory gave rise to the original division between inorganic (nonvital) and organic (vital) chemistry, and its supporters were known as vitalists, who maintained that natural products formed by living organisms could never be synthesized by ordinary chemical means. Wöhler's synthesis of urea was a bitter blow to the vitalists and did much to overthrow their doctrine.

Wöhler worked with Justus von Liebig on a number of important investigations.

In 1830 they proved the polymerism of cyanates and fulminates, and two years later announced a series of studies of benzaldehyde (benzenecarbaldehyde) and the benzoyl (benzenecarboxyl) radical. In 1837 they investigated uric acid and its derivatives. Wöhler also discovered quinone (cyclohexadiene-1,4-dione), hydroquinone or quinol (benzene-1,4-diol) and quinhydrone (a molecular complex composed of equimolar amounts of quinone and hydroquinone).

In the inorganic field Wöhler isolated boron and silicon and prepared silicon nitride and hydride. He prepared phosphorus by the modern method, and discovered calcium carbide and showed that it can be reacted with water to produce acetylene (ethyne):

$$CaC_2 + 2H_2O \rightarrow Ca(OH)_2 + C_2H_2$$

He demonstrated the analogy between the compounds of carbon and silicon, and just missed being the first to discover vanadium and niobium. He also obtained pure titanium and showed the similarity between this element and carbon and silicon. He published little work after 1845 and concentrated on teaching.

Source: From the Hutchinson *Dictionary of Scientific Biography*. © Research Machines plc [2003] All Rights Reserved. Helicon Publishing is a division of Research Machines.

polymer, through a series of chemical reactions, produces viscose rayon filaments when forced through small holes. The filaments are twisted together to form viscose rayon thread. When the polymer is forced through a thin slit, a sheet is formed rather than filaments, and the transparent sheet is called *cellophane*. Both rayon and cellophane, like celluloid, are manufactured by modifying the natural polymer of cellulose.

The first truly synthetic polymer was produced in the early 1900s by reacting two chemicals with relatively small molecules rather than modifying a natural polymer. Phenol, an aromatic hydrocarbon, was reacted with formaldehyde, the simplest aldehyde, to produce the polymer named *Bakelite*. Bakelite is a *thermosetting* material that forms cross-links between the polymer chains. Once the links are formed during production, the plastic becomes permanently hardened and cannot be softened or made to flow. Some plastics are *thermoplastic* polymers and soften during heating and harden during cooling because they do not have cross-links.

Polyethylene is a familiar thermoplastic polymer used for vegetable bags, dry cleaning bags, grocery bags, and plastic squeeze bottles. Polyethylene is a polymer produced by a polymerization reaction of ethylene, which is derived from petroleum. Polyethylene was invented just before World War II and was used as an electrical insulating material during the war. Today, there are many variations of polyethylene that are produced by different reaction conditions or by the substitution of one or more hydrogen atoms in the ethylene molecule. When soft polyethylene near the melting point is rolled in alternating perpendicular directions or expanded and compressed as it is cooled, the polyethylene molecules become ordered in a way that improves the rigidity and tensile strength. This change in the microstructure produces *high-density polyethylene* with a superior rigidity and tensile strength compared to *low-density polyethylene*. High-density polyethylene is used in milk jugs, as liners in screw-on jar tops, in bottle caps, and as a material for toys.

The properties of polyethylene are changed by replacing one of the hydrogen atoms in a molecule of ethylene. If the hydrogen is replaced by a chlorine atom the compound is called vinyl chloride, and the polymer formed from vinyl chloride is

$$H_2C=CHCl$$

polyvinyl chloride (PVC). Polyvinyl chloride is used to make plastic water pipes, synthetic leather, and other vinyl products. It differs from the waxy plastic of polyethylene because of the chlorine atom that replaces hydrogen on each monomer.

The replacement of a hydrogen atom with a benzene ring makes a monomer called *styrene*. Styrene is

$$H_2C=CH-C_6H_5$$

and polymerization of styrene produces *polystyrene*. Polystyrene is puffed full of air bubbles to produce the familiar Styrofoam coolers, cups, and insulating materials.

If all hydrogens of an ethylene molecule are replaced with atoms of fluorine, the product is polytetrafluorethylene, a tough plastic that resists high temperatures and acts more like a metal than a plastic. Since it has a low friction it is used for bearings, gears, and as a nonsticking coating on frying pans. You probably know of this plastic by its trade name of *Teflon*.

There are many different polymers in addition to PVC, Styrofoam, and Teflon, and the monomers of some of these are shown in Figure 12.24. There are also polymers of isoprene, or synthetic rubber, in wide use. Fibers and fabrics may be polyamides (such as nylon), polyesters (such as Dacron), or polyacrylonitriles (Orlon, Acrilan, Creslan), which have a CN in place of a hydrogen atom on an ethylene molecule and are called acrylic materials. All of these synthetic polymers have added much to practically every part of your life. It would be impossible to list all of their uses here; however, they present problems because (1) they are manufactured from raw materials obtained from coal and a dwindling petroleum supply, and (2) they do not readily decompose when dumped into rivers, oceans, or other parts of the environment. However, research in the polymer sciences is beginning to reflect new understandings learned from research on biological tissues. This could lead to whole new molecular designs for synthetic polymers that will be more compatible with the ecosystems.

Summary

Organic chemistry is the study of compounds that have carbon as the principal element. Such compounds are called *organic compounds,* and all the rest are *inorganic compounds.* There are millions of organic compounds because a carbon atom can link with other carbon atoms as well as atoms of other elements.

A *hydrocarbon* is an organic compound consisting of hydrogen and carbon atoms. The simplest hydrocarbon is one carbon atom and four hydrogen atoms, or CH_4. All hydrocarbons larger than CH_4 have one or more carbon atoms bonded to another carbon atom. The bond can be single, double, or triple, and this forms a basis for classifying hydrocarbons. A second basis is whether the carbons are in a ring or not. The *alkanes* are hydrocarbons with single carbon-to-carbon bonds, the *alkenes* have a double carbon-to-carbon bond, and the *alkynes* have a triple carbon-to-carbon bond. The alkanes, alkenes, and alkynes can have straight- or branched-chain molecules. When the number of carbon atoms is greater than three, there are different arrangements that can occur for a particular number of carbon atoms. The different arrangements with the same molecular formula are called isomers. *Isomers* have different physical properties, so each isomer is given its own name. The name is determined by (1) identifying the longest continuous carbon chain as the base name, (2) locating the attachment of other atoms or hydrocarbon groups by counting from the direction that results in the smallest numbers, (3) identifying attached hydrocarbon groups by changing the "-ane" suffix of alkanes to "-yl," (4) identifying the number of these hydrocarbon groups with prefixes, and (5) identifying the location of the groups with the carbon atom number.

The alkanes have all the hydrogen atoms possible, so they are *saturated* hydrocarbons. The alkenes and the alkynes can add more hydrogens to the molecule, so they are *unsaturated* hydrocarbons. Unsaturated hydrocarbons are more chemically reactive than saturated molecules.

Hydrocarbons that occur in a ring or cycle structure are cyclohydrocarbons. A six-carbon cyclohydrocarbon with three double bonds has different properties than the other cyclohydrocarbons because the double bonds are not localized. This six-carbon molecule is *benzene,* the basic unit of the *aromatic hydrocarbons.*

Petroleum is a mixture of alkanes, cycloalkanes, and a few aromatic hydrocarbons that formed from the slow decomposition of buried marine plankton and algae. Petroleum from the ground, or *crude oil,* is distilled into petroleum products of *natural gas, LPG, petroleum ether, gasoline, kerosene, diesel fuel,* and *motor oils.* Each group contains a range of hydrocarbons and is processed according to use.

In addition to oxidation, hydrocarbons react by *substitution, addition,* and *polymerization* reactions. Reactions take place at sites of multiple bonds or lone pairs of electrons on the *functional groups.* The functional group determines the chemical properties of organic compounds. Functional group results in the *hydrocarbon derivatives* of *alcohols, ethers, aldehydes, ketones, organic acids, esters,* and *amines.*

Living organisms have an incredible number of highly organized chemical reactions that are catalyzed by *enzymes,* using food and energy to grow and reproduce. The process involves building large *macromolecules* from smaller molecules and units. The organic molecules involved in the process are proteins, carbohydrates, and fats and oils.

Proteins are macromolecular polymers of *amino acids* held together by *peptide bonds.* There are twenty amino acids that are used in various polymer combinations to build structural and functional proteins. *Structural proteins* are muscles, connective tissue, and the skin, hair, and nails of animals. *Functional proteins* are enzymes, hormones, and antibodies.

Carbohydrates are polyhydroxyl aldehydes and ketones that form three groups: the monosaccharides, disaccharides, and polysaccharides. The *monosaccharides* are simple sugars such as *glucose* and *fructose*. Glucose is *blood sugar,* a source of energy. The disaccharides are *sucrose* (table sugar), *lactose* (milk sugar), and *maltose* (malt sugar). The disaccharides are broken down (digested) to glucose for use by the body. The polysaccharides are polymers of glucose in straight or branched chains used as a near-term source of stored energy. Plants store the energy in the form of *starch,* and animals store it in the form of *glycogen. Cellulose* is a polymer similar to starch that humans cannot digest.

Fats and oils are esters formed from three fatty acids and glycerol into a *triglyceride. Fats* are usually solid triglycerides associated with animals, and *oils* are liquid triglycerides associated with plant life, but both represent a high-energy storage material.

Polymers are huge, chainlike molecules of hundreds or thousands of smaller, repeating molecular units called *monomers.* Polymers occur naturally in plants and animals, and many *synthetic polymers* are made today from variations of the ethylene-derived monomers. Among the more widely used synthetic polymers derived from ethylene are polyethylene, polyvinyl chloride, polystyrene, and Teflon. Problems with the synthetic polymers include that (1) they are manufactured from fossil fuels that are also used as the primary energy supply, and (2) they do not readily decompose and tend to accumulate in the environment.

KEY TERMS

alkanes (p. **301**)

alkenes (p. **305**)

alkyne (p. **306**)

amino acids (p. **313**)

aromatic hydrocarbons (p. **306**)

carbohydrates (p. **314**)

cellulose (p. **315**)

disaccharides (p. **315**)

fat (p. **316**)

functional group (p. **310**)

hydrocarbon (p. **300**)

hydrocarbon derivatives (p. **309**)

inorganic chemistry (p. **300**)

isomers (p. **302**)

macromolecule (p. **313**)

monosaccharides (p. **314**)

oil (p. **316**)

organic chemistry (p. **300**)

petroleum (p. **307**)

polymer (p. **309**)

polysaccharides (p. **315**)

proteins (p. **313**)

starches (p. **315**)

triglyceride (p. **316**)

APPLYING THE CONCEPTS

1. An organic compound is a compound that
   a. contains carbon and was formed only by a living organism.
   b. is a natural compound that has not been synthesized.
   c. contains carbon, whether it was formed by a living thing or not.
   d. was formed by a plant.

2. There are millions of organic compounds but only thousands of inorganic compounds because
   a. organic compounds were formed by living things.
   b. there is more carbon on the earth's surface than any other element.
   c. atoms of elements other than carbon never combine with themselves.
   d. carbon atoms can combine with up to four other atoms, including other carbon atoms.

3. You know for sure that the compound named decane has
   a. more than 10 isomers.
   b. 10 carbon atoms in each molecule.
   c. only single bonds.
   d. all of the above.

4. An alkane with 4 carbon atoms would have how many hydrogen atoms in each molecule?
   a. 4
   b. 8
   c. 10
   d. 16

5. Isomers are compounds with the same
   a. molecular formula with different structures.
   b. molecular formula with different atomic masses.
   c. atoms, but different molecular formulas.
   d. structures, but different formulas.

6. Isomers have
   a. the same chemical and physical properties.
   b. the same chemical, but different physical properties.
   c. the same physical, but different chemical properties.
   d. different physical and chemical properties.

7. The organic compound 2,2,4-trimethylpentane is an isomer of
   a. propane.
   b. pentane.
   c. heptane.
   d. octane.

8. The hydrocarbons with a double covalent carbon-carbon bond are called
   a. alkanes.
   b. alkenes.
   c. alkynes.
   d. none of the above.

9. According to their definitions, which of the following would not occur as unsaturated hydrocarbons?
   a. alkanes
   b. alkenes
   c. alkynes
   d. None of the above is correct.

10. Petroleum is believed to have formed mostly from the anaerobic decomposition of buried
    a. dinosaurs.
    b. fish.
    c. pine trees.
    d. plankton and algae.

11. The label on a container states that the product contains "petroleum distillates." Which of the following hydrocarbons is probably present?
    a. CH_4
    b. C_5H_{12}
    c. $C_{16}H_{34}$
    d. $C_{40}H_{82}$

12. Tetraethyl lead ("ethyl") was added to gasoline to increase the octane rating by
    a. increasing the power by increasing the heat of combustion.
    b. increasing the burning rate of the gasoline.
    c. absorbing the pings and knocks.
    d. decreasing the burning rate.

13. The reaction of $C_2H_2 + Br_2 \longrightarrow C_2H_2Br_2$ is a (an)
    a. substitution reaction.
    b. addition reaction.
    c. addition polymerization reaction.
    d. substitution polymerization reaction.

14. Ethylene molecules can add to each other in a reaction to form a long chain called a
    a. monomer.
    b. dimer.
    c. trimer.
    d. polymer.

15. Chemical reactions usually take place on an organic compound at the site of (a)
    a. double bond.
    b. lone pair of electrons.
    c. functional group.
    d. any of the above.

16. The R in ROH represents
    a. a functional group.
    b. a hydrocarbon group with a name ending in "-yl."
    c. an atom of an inorganic element.
    d. a polyatomic ion that does not contain carbon.

17. The OH in ROH represents
    a. a functional group.
    b. a hydrocarbon group with a name ending in "-yl."
    c. the hydroxide ion, which ionizes to form a base.
    d. the site of chemical activity in a strong base.

18. What is the proof of a "wine cooler" that is 5 percent alcohol by volume?
    a. 2.5 proof
    b. 5 proof
    c. 10 proof
    d. 50 proof

19. An alcohol with two hydroxyl groups per molecule is called
    a. ethanol.
    b. glycerol.
    c. glycerin.
    d. glycol.

20. A bottle of wine that has "gone bad" now contains
    a. CH_3OH.
    b. CH_3OCH_3.
    c. CH_3COOH.
    d. CH_3COOCH_3.

21. A protein is a polymer formed from the linking of many
    a. glucose units.
    b. DNA molecules.
    c. amino acid molecules.
    d. monosaccharides.

22. Which of the following is *not* converted to blood sugar by the human body?
    a. lactose
    b. dextrose
    c. cellulose
    d. glycogen

23. Fats from animals and oils from plants have the general structure of a (an)
    a. aldehyde.
    b. ester.
    c. amine.
    d. ketone.

24. Liquid oils from plants can be converted to solids by adding what to the molecule?
    a. metal ions
    b. carbon
    c. polyatomic ions
    d. hydrogen

25. The basic difference between a monomer of polyethylene and a monomer of polyvinyl chloride is
    a. the replacement of a hydrogen by a chlorine.
    b. the addition of four fluorines.
    c. the elimination of double bonds.
    d. the removal of all hydrogens.

26. Many synthetic polymers become a problem in the environment because they
    a. decompose to nutrients, which accelerates plant growth.
    b. do not readily decompose and tend to accumulate.
    c. do not contain vitamins as natural materials do.
    d. become a source of food for fish, but ruin the flavor of fish meat.

Answers

1. c 2. d 3. d 4. c 5. a 6. d 7. d 8. b 9. a 10. d 11. b 12. d 13. b 14. d
15. d 16. b 17. a 18. c 19. d 20. c 21. c 22. c 23. b 24. d 25. a 26. b

QUESTIONS FOR THOUGHT

1. What is an organic compound?
2. There are millions of organic compounds but only thousands of inorganic compounds. Explain why this is the case.

3. What is cracking and reforming? For what purposes are either or both used by the petroleum industry?

4. Is it possible to have an isomer of ethane? Explain.

5. Suggest a reason that ethylene is an important raw material used in the production of plastics but ethane is not.

6. What are (a) natural gas, (b) LPG, and (c) petroleum ether?

7. What does the octane number of gasoline describe? On what is the number based?

8. Why is unleaded gasoline more expensive than gasoline with lead additives?

9. What is a functional group? What is it about the nature of a functional group that makes it the site of chemical reactions?

10. Draw a structural formula for alcohol. Describe how alcohols are named.

11. A soft drink is advertised to "contain no sugar." The label lists ingredients of carbonated water, dextrose, corn syrup, fructose, and flavorings. Evaluate the advertising and the list of ingredients.

12. What are fats and oils? What are saturated and unsaturated fats and oils?

13. What is a polymer? Give an example of a naturally occurring plant polymer. Give an example of a synthetic polymer.

14. Explain why a small portion of wine is customarily poured before a bottle of wine is served. Sometimes the cork is handed to the person doing the ordering with the small portion of wine. What is the person supposed to do with the cork and why?

PARALLEL EXERCISES

The exercises in groups A and B cover the same concepts. Solutions to group A exercises are located in appendix D.

Group A

1. Draw the structural formulas for (a) *n*-pentane, and (b) an isomer of pentane with the maximum possible branching. (c) Give the IUPAC name of this isomer.

2. Write structural formulas for all the hexane isomers you can identify. Write the IUPAC name for each isomer.

3. Write structural formulas for
   a. 3,3,4-trimethyloctane.
   b. 2-methyl-1-pentene.
   c. 5,5-dimethyl-3-heptyne.

4. Write the IUPAC name for each of the following.

Group B

1. Write structural formulas for (a) *n*-octane, (b) an isomer of octane with the maximum possible branching. (c) Give the IUPAC name of this isomer.

2. Write the structural formulas for all the heptane isomers you can identify. Write the IUPAC name for each isomer.

3. Write structural formulas for
   a. 2,3-dimethylpentane.
   b. 1-butene.
   c. 3-ethyl-2-methyl-3-hexene.

4. Write the IUPAC name for each of the following.

—Continued top of next page.

```
            H
            |
      H—C—H
            |
  H   H     |   H
  |   |     |   |
H—C—C—C=C—C—H
  |   |     |   |
  H   H     H   H
          |
        H—C—H
          |
        H—C—H
          |
          H
```
C

```
  H   H         H
  |   |         |
H—C—C=C—C=C—C—H
  |   |     |   |   |
  H   Cl    H   H   H
```
C

```
       H  H  H  H
       |  |  |  |
Br—C—C=C—C—H
       |        |
       H        H
```
D

5. Which would have the higher octane rating, 2,2,3-trimethyl-butane or 2,2-dimethylpentane? Explain with an illustration.

6. Use the information in Table 12.5 to classify each of the following as an alcohol, ether, organic acid, ester, or amide.

5. Which would have the higher octane rating, 2-methyl-butane or dimethylpropane? Explain with an illustration.

6. Classify each of the following as an alcohol, ether, organic acid, ester, or amide.

A
```
      H   H   H
      |   |   |
  H—C—C—C—OH
      |   |   |
      H   H   H
```

A
```
      H   H   H
      |   |   |
  H—C—C—C—H
      |   |   |
      O   O   O
      H   H   H
```

B
```
      H   H   H
      |   |   |
  H—C—C—C—NH₂
      |   |   |
      H   H   H
```

B
```
      H   H
      |   |
  H—C—C—C—OH
      |   |   ‖
      H   H   O
```

C
```
      H   H   H       H   H   H
      |   |   |       |   |   |
  H—C—C—C—O—C—C—C—H
      |   |   |       |   |   |
      H   H   H       H   H   H
```

C
```
      H   H           H   H
      |   |           |   |
  H—C—C—O—C—C—H
      |   |           |   |
      H   H           H   H
```

D
```
      H   H           H   H   H
      |   |           |   |   |
  H—C—C—O—C—C—C—C—H
      |   |           ‖   |   |   |
      H   H           O   H   H   H
```

D
```
      H   H   H
      |   |   |
  H—C—C—C—NH₂
      |   |   |
      H   H   H
```

E
```
      H   H   H   H
      |   |   |   |
  H—C—C—C—C—C—OH
      |   |   |   |   ‖
      H   H   H   H   O
```

E
```
      H   H           H   H   H   H   H   H
      |   |           |   |   |   |   |   |
  H—C—C—O—C—C—C—C—C—C—H
      |   |           ‖   |   |   |   |   |   |
      H   H           O   H   H   H   H   H   H
```

With the top half of the steel vessel and control rods removed, fuel rod bundles can be replaced in the water-flooded nuclear reactor.

CHAPTER

13

Nuclear Reactions

The ancient alchemist dreamed of changing one element into another, such as lead into gold. The alchemist was never successful, however, because such changes were attempted with chemical reactions. Chemical reactions are reactions that involve only the electrons of atoms. Electrons are shared or transferred in chemical reactions, and the internal nucleus of the atom is unchanged. Elements thus retain their identity during the sharing or transferring of electrons. This chapter is concerned with a different kind of reaction, one that involves the *nucleus* of the atom. In nuclear reactions, the nucleus of the atom is often altered, changing the identity of the elements involved. The ancient alchemist's dream of changing one element into another was actually a dream of achieving a nuclear change, that is, a nuclear reaction.

Understanding nuclear reactions is important because although fossil fuels are the major source of energy today, there are growing concerns about (1) air pollution from fossil fuel combustion, (2) increasing levels of CO_2 from fossil fuel combustion, which may be warming the earth (the greenhouse effect), and (3) the dwindling fossil fuel supply itself, which cannot last forever. Energy experts see nuclear energy as a means of meeting rising energy demands in an environmentally acceptable way. However, the topic of nuclear energy is controversial, and discussions of it often result in strong emotional responses. Decisions about the use of nuclear energy require some understandings about nuclear reactions and some facts about radioactivity and radioactive materials (Figure 13.1). These understandings and facts are the topics of this chapter.

Natural Radioactivity

Natural **radioactivity** is the spontaneous emission of particles or energy from an atomic nucleus as it disintegrates. It was discovered in 1896 by Henri Becquerel, a French scientist who was very interested in the recent discovery of X rays. Becquerel was experimenting with fluorescent minerals, minerals that give off visible light after being exposed to sunlight. He wondered if fluorescent minerals emitted X rays in addition to visible light. From previous work with X rays, Becquerel knew that they would penetrate a wrapped, light-tight photographic plate, exposing it as visible light exposes an unprotected plate. Thus, Becquerel decided to place a fluorescent uranium mineral on a protected photographic plate while the mineral was exposed to sunlight. Sure enough, he found a silhouette of the mineral on the plate when it was developed. Believing the uranium mineral emitted X rays, he continued his studies until the weather turned cloudy. Storing a wrapped, protected photographic plate and the uranium mineral together during the cloudy weather, Becquerel returned to the materials later and developed the photographic plate to again find an image of the mineral (Figure 13.2). He concluded that the mineral was emitting an "invisible radiation" that was not induced by sunlight. Becquerel named the emission of invisible radiation *radioactivity*. Materials that have the property of radioactivity are called *radioactive* materials.

Becquerel's discovery led to the beginnings of the modern atomic theory and to the discovery of new elements. Ernest Rutherford studied the nature of radioactivity and found that there are three kinds, which are today known by the first three letters of the Greek alphabet—alpha (α), beta (β), and gamma (γ). These Greek letters were used at first before the nature of

FIGURE 13.1

Decisions about nuclear energy require some understanding of nuclear reactions and the nature of radioactivity. This is one of the three units of the Palo Verde Nuclear Generating Station in Arizona. With all three units running, enough power is generated to meet the electrical needs of nearly 4 million people.

A

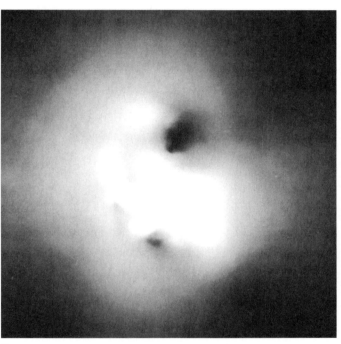

B

FIGURE 13.2

Radioactivity was discovered by Henri Becquerel when he exposed a light-tight photographic plate to a radioactive mineral, then developed the plate. (*A*) A photographic film is exposed to a uranite ore sample. (*B*) The film, developed normally after a four-day exposure to uranite. Becquerel found an image like this one and deduced that the mineral gave off invisible radiation that he called radioactivity.

the radiation was known. Today, an **alpha particle** (sometimes called an alpha ray) is known to be the nucleus of a helium atom, that is, two protons and two neutrons. A **beta particle** (or beta ray) is a high-energy electron. A **gamma ray** is electromagnetic radiation, as is light, but of very short wavelength (Figure 13.3).

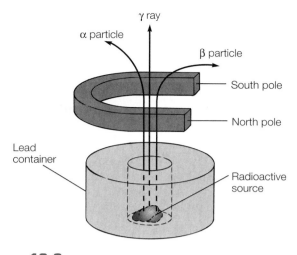

Figure 13.3

Radiation passing through a magnetic field shows that massive, positively charged alpha particles are deflected one way, and less massive beta particles with their negative charge are greatly deflected in the opposite direction. Gamma rays, like light, are not deflected.

At Becquerel's suggestion, Madame Marie Curie searched for other radioactive materials and in the process discovered two new elements: polonium and radium. More radioactive elements have been discovered since that time, and, in fact, all the isotopes of all the elements with an atomic number greater than 83 (bismuth) are radioactive. As a result of radioactive disintegration, the nucleus of an atom often undergoes a change of identity, becoming a simpler nucleus. The spontaneous disintegration of a given nucleus is a purely natural process and cannot be controlled or influenced. The natural spontaneous disintegration or decomposition of a nucleus is also called **radioactive decay.** Although it is impossible to know *when* a given nucleus will undergo radioactive decay, as you will see later, it is possible to deal with the *rate* of decay for a given radioactive material with precision.

Nuclear Equations

There are two main subatomic particles in the nucleus, the proton and the neutron. The proton and neutron are called **nucleons.** Recall that the number of protons, the *atomic number,* determines what element an atom is, and that all atoms of a given element have the same number of protons. The number of neutrons varies in *isotopes,* which are atoms with the same atomic number but different numbers of neutrons (Figure 13.4). The number of protons and neutrons together determines the *mass number,* so different isotopes of the same element are identified with their mass numbers. Thus, the two most common, naturally occurring isotopes of uranium are referred to as uranium-238 and uranium-235, and the 238 and 235 are the mass numbers of these isotopes. Isotopes are also represented by the following symbol:

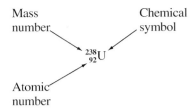

Subatomic particles involved in nuclear reactions are represented by symbols with the following form:

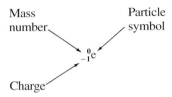

Symbols for these particles are illustrated in Table 13.1.

Symbols are used in an equation for a nuclear reaction that is written much like a chemical reaction with reactants and products. When a uranium-238 nucleus emits an alpha particle ($^4_2$He), for example, it loses two protons and two neutrons. The nuclear reaction is written in equation form as

$$^{238}_{92}\text{U} \rightarrow\; ^{234}_{90}\text{Th} +\; ^4_2\text{He}$$

The *products* of this nuclear reaction from the decay of a uranium-238 nucleus are (1) the alpha particle ($^4_2$He) given off and (2) the nucleus, which remains after the alpha particle leaves the original nucleus. What remains is easily determined since all nuclear equations must show conservation of charge and conservation of the total number of nucleons. In an alpha emission reaction, (1) the number of protons (positive charge) remains the same, and the sum of the subscripts (atomic

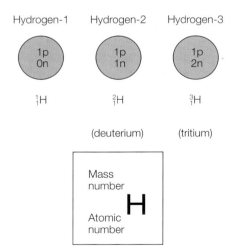

FIGURE 13.4

The three isotopes of hydrogen have the same number of protons but different numbers of neutrons. Hydrogen-1 is the most common isotope. Hydrogen-2, with an additional neutron, is named *deuterium,* and hydrogen-3 is called *tritium.* Neutrons and protons are called *nucleons* because they are in the nucleus.

TABLE 13.1

Names, symbols, and properties of particles in nuclear equations

Name	Symbol	Mass Number	Charge
Proton	^1_1H (or ^1_1p)	1	1+
Electron	$^0_{-1}\text{e}$ (or $^0_{-1}\beta$)	0	1−
Neutron	^1_0n	1	0
Gamma Photon	$^0_0\gamma$	0	0

number, or numbers of protons) in the reactants must equal the sum of the subscripts in the products; and (2) the total number of nucleons remains the same, and the sum of the superscripts (atomic mass, or number of protons plus neutrons) in the reactants must equal the sum of the superscripts in the products. The new nucleus remaining after the emission of an alpha particle, therefore, has an atomic number of 90 (92 − 2 = 90). According to the table of atomic numbers on the inside back cover of this text, this new nucleus is thorium (Th). The mass of the thorium isotope is 238 minus 4, or 234. The emission of an alpha particle thus decreases the number of protons by 2 and the mass number by 4. From the subscripts, you can see that the total charge is conserved (92 = 90 + 2). From the superscripts, you can see that the total number of nucleons is also conserved (238 = 234 + 4). The mass numbers (superscripts) and the atomic numbers (subscripts) are *balanced* in a correctly written nuclear equation. Such nuclear equations are considered to be independent of any chemical form or chemical reaction. Nuclear reactions are independent and separate from chemical reactions, whether or not the atom is in the pure element or in a compound. Each particle that is involved in nuclear reactions has its own symbol with a superscript indicating mass number and a subscript indicating the charge. These symbols, names, and numbers are given in Table 13.1.

EXAMPLE **13.1**

A plutonium-242 nucleus undergoes radioactive decay, emitting an alpha particle. Write the nuclear equation for this nuclear reaction.

Solution

Step 1: The table of atomic weights on the inside back cover gives the atomic number of plutonium as 94. Plutonium-242 therefore has a symbol of $^{242}_{94}\text{Pu}$. The symbol for an alpha particle is ($^4_2$He), so the nuclear equation so far is

$$^{242}_{94}\text{Pu} \rightarrow\; ^4_2\text{He} + \;?$$

Step 2: From the subscripts, you can see that 94 = 2 + 92, so the new nucleus has an atomic number of 92. The table of atomic weights identifies element 92 as uranium with a symbol of U.

Step 3: From the superscripts, you can see that the mass number of the uranium isotope formed is 242 − 4 = 238, so the product nucleus is $^{238}_{92}U$ and the complete nuclear equation is

$$^{242}_{94}Pu \rightarrow \, ^{4}_{2}He + \, ^{238}_{92}U$$

Step 4: Checking the subscripts (94 = 2 + 92) and the superscripts (242 = 4 + 238), you can see that the nuclear equation is balanced.

EXAMPLE **13.2**

What is the product nucleus formed when radium emits an alpha particle? (Answer: Radon-222, a chemically inert, radioactive gas)

The Nature of the Nucleus

The modern atomic theory does not picture the nucleus as a group of stationary protons and neutrons clumped together by some "nuclear glue." The protons and neutrons are understood to be held together by a **nuclear force,** a strong fundamental force of attraction that is functional only at very short distances, on the order of 10^{-15} m or less. At distances greater than about 10^{-15} m the nuclear force is negligible, and the weaker **electromagnetic force,** the force of repulsion between like charges, is the operational force. Thus, like-charged protons experience a repulsive force when they are farther apart than about 10^{-15} m. When closer together than 10^{-15} m, the short-range, stronger nuclear force predominates, and the protons experience a strong attractive force. This explains why the like-charged protons of the nucleus are not repelled by their like electric charges.

Observations of radioactive decay reactants and products and experiments with nuclear stability have led to a **shell model of the nucleus.** This model considers the protons and neutrons moving in energy levels, or shells, in the nucleus analogous to the orbital structure of electrons in the outermost part of the atom. As in the electron orbitals, there are certain configurations of nuclear shells that have a greater stability than others. Considering electrons, filled and half-filled orbitals are more stable than other arrangements, and maximum stability occurs with the noble gases and their 2, 10, 18, 36, 54, and 86 electrons. Considering the nucleus, atoms with 2, 8, 20, 28, 50, 82, or 126 protons or neutrons have a maximum nuclear stability. The stable numbers are not the same for electrons and nucleons because of differences in nuclear and electromagnetic forces.

Isotopes of uranium, radium, and plutonium, as well as other isotopes, emit an alpha particle during radioactive decay to a simpler nucleus. The alpha particle is a helium nucleus, $^{4}_{2}He$. The alpha particle contains two protons as well as

two neutrons, which is one of the nucleon numbers of stability, so you would expect the helium nucleus (or alpha particle) to have a stable nucleus, and it does. *Stable* means it does not undergo radioactive decay. Pairs of protons and pairs of neutrons have increased stability, just as pairs of electrons in a molecule do. As a result, nuclei with an *even number* of both protons and neutrons are, in general, more stable than nuclei with odd numbers of protons and neutrons. There are a little more than 150 stable isotopes with an even number of protons and an even number of neutrons, but there are only 5 stable isotopes with odd numbers of each. Just as in the case of electrons, there are other factors that come into play as the nucleus becomes larger and larger with increased numbers of nucleons.

The results of some of these factors are shown in Figure 13.5, which is a graph of the number of neutrons versus the number of protons in nuclei. As the number of protons increases, the neutron-to-proton ratio of the *stable nuclei* also increases in a **band of stability.** Within the band the neutron-to-proton ratio increases from about 1:1 at the bottom left to about 1½:1 at the top right. The increased ratio of neutrons is needed to produce a stable nucleus as the number of protons increases. Neutrons provide additional attractive *nuclear* (not electrical) forces, which counter the increased electrical repulsion

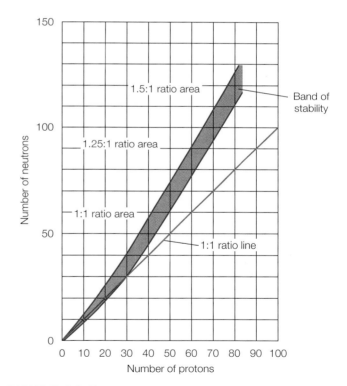

FIGURE 13.5

The shaded area indicates stable nuclei, which group in a band of stability according to their neutron-to-proton ratio. As the size of nuclei increases, so does the neutron-to-proton ratio that represents stability. Nuclei outside this band of stability are radioactive.

from a larger number of positively charged protons. Thus, more neutrons are required in larger nuclei to produce a stable nucleus. However, there is a limit to the additional attractive forces that can be provided by more and more neutrons, and all isotopes of all elements with more than 83 protons are unstable and thus undergo radioactive decay.

The generalizations about nuclear stability provide a means of predicting if a particular nucleus is radioactive. The generalizations are as follows:

1. All isotopes with an atomic number greater than 83 have an unstable nucleus.
2. Isotopes that contain 2, 8, 20, 28, 50, 82, or 126 protons or neutrons in their nuclei occur in more stable isotopes than those with other numbers of protons or neutrons.
3. Pairs of protons and pairs of neutrons have increased stability, so isotopes that have nuclei with even numbers of both protons and neutrons are generally more stable than nuclei with odd numbers of both protons and neutrons.
4. Isotopes with an atomic number less than 83 are stable when the ratio of neutrons to protons in the nucleus is about 1:1 in isotopes with up to 20 protons, but the ratio increases in larger nuclei in a band of stability (see Figure 13.5). Isotopes with a ratio to the left or right of this band are unstable and thus will undergo radioactive decay.

EXAMPLE 13.3

Would you predict the following isotopes to be radioactive or stable?

(a) $^{60}_{27}Co$

(b) $^{222}_{86}Rn$

(c) $^{3}_{1}H$

(d) $^{40}_{20}Ca$

Solution

(a) Cobalt-60 has 27 protons and 33 neutrons, both odd numbers, so you might expect $^{60}_{27}Co$ to be radioactive.

(b) Radon has an atomic number of 86, and all isotopes of all elements beyond atomic number 83 are radioactive. Radon-222 is therefore radioactive.

(c) Hydrogen-3 has an odd number of protons and an even number of neutrons, but its 2:1 neutron-to-proton ratio places it outside the band of stability. Hydrogen-3 is radioactive.

(d) Calcium-40 has an even number of protons and an even number of neutrons, containing 20 of each. The number 20 is a particularly stable number of protons or neutrons, and calcium-40 has 20 of each. In addition, the neutron-to-proton ratio is 1:1, placing it within the band of stability. All indications are that calcium-40 is stable, not radioactive.

EXAMPLE 13.4

Which of the following would you predict to be radioactive?

(a) $^{127}_{53}I$

(b) $^{131}_{53}I$

(c) $^{206}_{82}Pb$

(d) $^{214}_{82}Pb$

[Answer: (b) and (d)]

Types of Radioactive Decay

Through the process of radioactive decay, an unstable nucleus becomes a more stable one with less energy. The three more familiar types of radiation emitted—alpha, beta, and gamma— were introduced earlier. There are five common types of radioactive decay, and three of these involve alpha, beta, and gamma radiation.

1. *Alpha emission.* Alpha (α) emission is the expulsion of an alpha particle ($^{4}_{2}He$) from an unstable, disintegrating nucleus. The alpha particle, a helium nucleus, travels from 2 to 12 cm through the air, depending on the energy of emission from the source. An alpha particle is easily stopped by a sheet of paper close to the nucleus. As an example of alpha emission, consider the decay of a radon-222 nucleus,

$$^{222}_{86}Rn \rightarrow ^{218}_{84}Po + ^{4}_{2}He$$

The spent alpha particle eventually acquires two electrons and becomes an ordinary helium atom.

2. *Beta emission.* Beta (β^{-}) emission is the expulsion of a different particle, a beta particle, from an unstable disintegrating nucleus. A beta particle is simply an electron ($^{0}_{-1}e$) ejected from the nucleus at a high speed. The emission of a beta particle *increases the number of protons* in a nucleus. It is as if a neutron changed to a proton by emitting an electron, or

$$^{1}_{0}n \rightarrow ^{1}_{1}p + ^{0}_{-1}e$$

Carbon-14 is a carbon isotope that decays by beta emission:

$$^{14}_{6}C \rightarrow ^{14}_{7}N + ^{0}_{-1}e$$

Note that the number of protons increased from six to seven, but the mass number remained the same. The mass number is unchanged because the mass of the expelled electron (beta particle) is negligible.

Beta particles are more penetrating than alpha particles and may travel several hundred centimeters through the air. They can be stopped by a thin layer of metal close to the emitting nucleus, such as a 1 cm thick piece of aluminum. A spent beta particle may eventually join an ion to become part of an atom, or it may remain a free electron.

3. *Gamma emission.* Gamma (γ) emission is a high-energy burst of electromagnetic radiation from an excited nucleus. It is a burst of light (photon) of a wavelength much too short to be detected by the eye. Other types of radioactive decay, such as alpha or beta emission, sometimes leave the nucleus with an excess of energy,

a condition called an *excited state*. As in the case of excited electrons, the nucleus returns to a lower energy state by emitting electromagnetic radiation. From a nucleus, this radiation is in the high-energy portion of the electromagnetic spectrum. Gamma is the most penetrating of the three common types of nuclear radiation. Like X rays, gamma rays can pass completely through a person, but all gamma radiation can be stopped by a 5 cm thick piece of lead close to the source. As with other types of electromagnetic radiation, gamma radiation is absorbed by and gives its energy to materials. Since the product nucleus changed from an excited state to a lower energy state, there is no change in the number of nucleons. For example, radon-222 is an isotope that emits gamma radiation:

$$^{222}_{86}\text{Rn}^* \rightarrow ^{222}_{86}\text{Rn} + ^{0}_{0}\gamma$$

(*denotes excited state)

Radioactive decay by alpha, beta, and gamma emission is summarized in Table 13.2, which also lists the unstable nuclear conditions that lead to the particular type of emission. Just as electrons seek a state of greater stability, a nucleus undergoes radioactive decay to achieve a balance between nuclear attractions, electromagnetic repulsions, and a low quantum of nuclear shell energy. The key to understanding the types of reactions that occur is found in the band of stable nuclei illustrated in Figure 13.5. The isotopes within this band have achieved the state of stability, and other isotopes above, below, or beyond the band are unstable and thus radioactive.

Nuclei that have a neutron-to-proton ratio beyond the upper right part of the band are unstable because of an imbalance between the proton-proton electromagnetic repulsions and all the combined proton and neutron nuclear attractions. Recall that the neutron-to-proton ratio increases from about 1:1 to about 1½:1 in the larger nuclei. The additional neutron provided additional nuclear attractions to hold the nucleus together, but atomic number 83 appears to be the upper limit to this additional stabilizing contribution. Thus, all nuclei with an atomic number greater than 83 are outside the upper right limit of the band of stability. Emission of an alpha particle reduces the number of protons by two and the number of neutrons by two, moving the nucleus more toward the band of

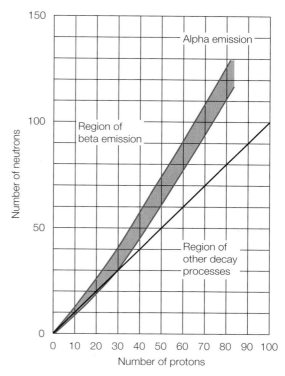

FIGURE 13.6

Unstable nuclei undergo different types of radioactive decay to obtain a more stable nucleus. The type of decay depends, in general, on the neutron-to-proton ratio, as shown.

stability. Thus, you can expect a nucleus that lies beyond the upper right part of the band of stability to be an alpha emitter (Figure 13.6).

A nucleus with a neutron-to-proton ratio that is too large will be on the left side of the band of stability. Emission of a beta particle decreases the number of neutrons and increases the number of protons, so a beta emission will lower the neutron-to-proton ratio. Thus, you can expect a nucleus with a large neutron-to-proton ratio, that is, one to the left of the band of stability, to be a beta emitter.

A nucleus that has a neutron-to-proton ratio that is too small will be on the right side of the band of stability. These nuclei can increase the number of neutrons and reduce the number of protons in the nucleus by other types of radioactive decay. As is usual when dealing with broad generalizations and trends, there are exceptions to the summarized relationships between neutron-to-proton ratios and radioactive decay.

EXAMPLE 13.5

Refer to Figure 13.6 and predict the type of radioactive decay for each of the following unstable nuclei:

(a) $^{131}_{53}\text{I}$

(b) $^{242}_{94}\text{Pu}$

TABLE 13.2			
Radioactive decay			
Unstable Condition	**Type of Decay**	**Emitted**	**Product Nucleus**
More than 83 protons	Alpha emission	$^{4}_{2}\text{He}$	Lost 2 protons and 2 neutrons
Neutron-to-proton ratio too large	Beta emission	$^{0}_{-1}\text{e}$	Gained 1 proton, no mass change
Excited nucleus	Gamma emission	$^{0}_{0}\gamma$	No change
Neutron-to-proton ratio too small	Other emission	$^{0}_{1}\text{e}$	Lost 1 proton, no mass change

Solution

(a) Iodine-131 has a nucleus with 53 protons and 131 minus 53, or 78 neutrons, so it has a neutron-to-proton ratio of 1.47:1. This places iodine-131 on the left side of the band of stability, with a high neutron-to-proton ratio that can be reduced by beta emission. The nuclear equation is

$$^{131}_{53}\text{I} \rightarrow {}^{131}_{54}\text{Xe} + {}^{0}_{-1}\text{e}$$

(b) Plutonium-242 has 94 protons and 242 minus 94, or 148 neutrons, in the nucleus. This nucleus is to the upper right, beyond the band of stability. It can move back toward stability by emitting an alpha particle, losing 2 protons and 2 neutrons from the nucleus. The nuclear equation is

$$^{242}_{94}\text{Pu} \rightarrow {}^{238}_{92}\text{U} + {}^{4}_{2}\text{He}$$

Radioactive Decay Series

A radioactive decay reaction produces a simpler, and eventually more stable nucleus than the reactant nucleus. As discussed in the previous section, large nuclei with an atomic number greater than 83 decay by alpha emission, giving up two protons and two neutrons with each alpha particle. A nucleus with an atomic number greater than 86, however, will emit an alpha particle and *still* have an atomic number greater than 83, which means the product nucleus will also be radioactive. This nucleus will also undergo radioactive decay, and the process will continue through a series of decay reactions until a stable nucleus is achieved. Such a series of decay reactions that (1) begins with one radioactive nucleus, which (2) decays to a second nucleus, which (3) then decays to a third nucleus, and so on until (4) a stable nucleus is reached is called a **radioactive decay series.** There are three naturally occurring radioactive decay series. One begins with thorium-232 and ends with lead-208, another begins with uranium-235 and ends with lead-207, and the third series begins with uranium-238 and ends with lead-206. Figure 13.7 shows the uranium-238 radioactive decay series.

As Figure 13.7 illustrates, the uranium-238 begins with uranium-238 decaying to thorium-234 by alpha emission. Thorium has a new position on the graph because it now has a new atomic number and a new mass number. Thorium-234 is unstable and decays to protactinium-234 by beta emission, which is also unstable and decays by beta emission to uranium-234. The process continues with five sequential alpha emissions,

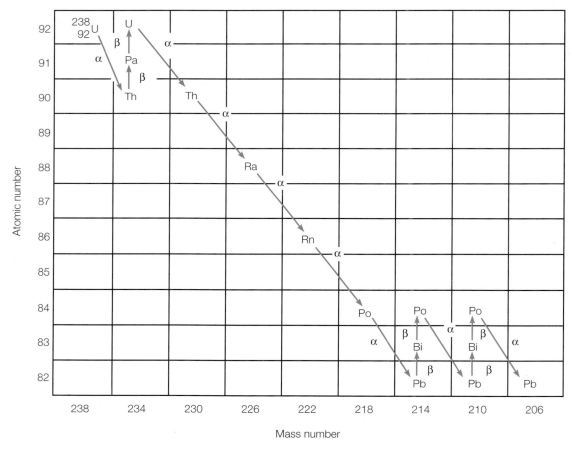

FIGURE 13.7

The radioactive decay series for uranium-238. This is one of three naturally occurring series.

TABLE 13.3
Half-lives of some radioactive isotopes

Isotope	Half-Life	Mode of Decay
$^{3}_{1}H$ (tritium)	12.26 years	Beta
$^{14}_{6}C$	5,730 years	Beta
$^{90}_{38}Sr$	28 years	Beta
$^{131}_{53}I$	8 days	Beta
$^{133}_{54}Xe$	5.27 days	Beta
$^{238}_{92}U$	4.51×10^{9} years	Alpha
$^{242}_{94}Pu$	3.79×10^{5} years	Alpha
$^{240}_{94}Pu$	6,760 years	Alpha
$^{239}_{94}Pu$	24,360 years	Alpha
$^{40}_{19}K$	1.3×10^{9} years	Alpha

then two beta-beta-alpha decay steps before the series terminates with the stable lead-206 nucleus.

The rate of radioactive decay is usually described in terms of its *half-life*. The **half-life** is the time required for one-half of the unstable nuclei to decay. Since each isotope has a characteristic decay constant, each isotope has its own characteristic half-life. Half-lives of some highly unstable isotopes are measured in fractions of seconds, and other isotopes have half-lives measured in seconds, minutes, hours, days, months, years, or billions of years. Table 13.3 lists half-lives of some of the isotopes, and the process is illustrated in Figure 13.8.

As an example of the half-life measure, consider a hypothetical isotope that has a half-life of one day. The half-life is independent of the amount of the isotope being considered, but suppose you start with a 1.0 kg sample of this element with a

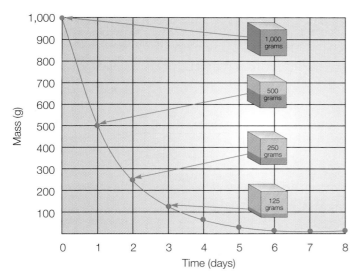

FIGURE 13.8
Radioactive decay of a hypothetical isotope with a half-life of one day. The sample decays each day by one-half to some other element. Actual half-lives may be in seconds, minutes, or any time unit up to billions of years.

half-life of one day. One day later, you will have half of the original sample, or 500 g. The other half did not disappear, but it is now the decay product, that is, some new element. During the next day half of the remaining nuclei will disintegrate, and only 250 g of the initial sample is still the original element. One-half of the remaining sample will disintegrate each day until the original sample no longer exists.

The half-life of uranium-238 is 4.5 billion years. Figure 13.9 gives the half-life for each step in the uranium-238 decay series.

Measurement of Radiation

The measurement of radiation is important in determining the half-life of radioactive isotopes. Radiation measurement is also important in considering biological effects, which will be discussed in the next section. As is the case with electricity, it is not possible to make direct measurements on things as small as electrons and other parts of atoms. Indirect measurement methods are possible, however, by considering the effects of the radiation.

Measurement Methods

As Becquerel discovered, radiation affects photographic film, exposing it as visible light does. Since the amount of film exposure is proportional to the amount of radiation, photographic film can be used as an indirect measure of radiation. Today, people who work around radioactive materials or X rays carry light-tight film badges. The film is replaced periodically and developed. The optical density of the developed film provides a record of the worker's exposure to radiation because the darkness of the developed film is proportional to the exposure.

There are also devices that indirectly measure radiation by measuring an effect of the radiation. An **ionization counter** is one type of device that measures ions produced by radiation. A second type of device is called a **scintillation counter.** *Scintillate* is a word meaning "sparks or flashes," and a scintillation counter measures the flashes of light produced when radiation strikes a phosphor.

The most common example of an ionization counter is known as a **Geiger counter** (Figure 13.10). The working components of a Geiger counter are illustrated in Figure 13.11. Radiation is received in a metal tube filled with an inert gas, such as argon. An insulated wire inside the tube is connected to the positive terminal of a direct current source. The metal cylinder around the insulated wire is connected to the negative terminal. There is not a current between the center wire and the metal cylinder because the gas acts as an insulator. When radiation passes through the window, however, it ionizes some of the gas atoms, releasing free electrons. These electrons are accelerated by the field between the wire and cylinder, and the accelerated electrons ionize more gas molecules, which results in an *avalanche* of free electrons. The avalanche creates a pulse of current that is amplified and then measured. More radiation means more avalanches, so the pulses are an indirect means of measuring radiation. When connected to a speaker or earphone, each avalanche produces a "pop" or "click."

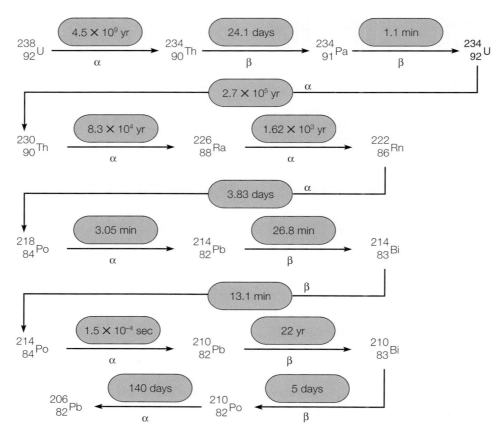

FIGURE 13.9
The half-life of each step in the uranium-238 radioactive decay series.

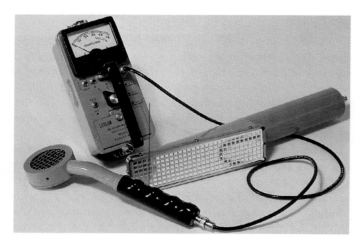

FIGURE 13.10
This is a beta-gamma probe, which can measure beta and gamma radiation in millirems per unit of time.

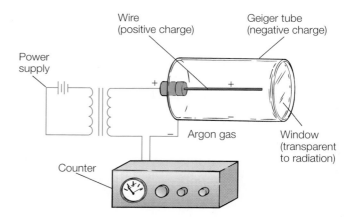

FIGURE 13.11
The working parts of a Geiger counter.

Some materials are *phosphors,* substances that emit a flash of light when excited by radiation. Zinc sulfide, for example, is used in television screens and luminous watches, and it was used by Rutherford to detect alpha particles. A luminous watch dial has a mixture of zinc sulfide and a small amount of radium sulfate. A zinc sulfide atom gives off a tiny flash of light when struck by radiation from a disintegrating radium nucleus. A scintillation counter measures the flashes of light through the photoelectric effect, producing free electrons that are accelerated to produce a pulse of current. Again, the pulses of current are used as an indirect means to measure radiation.

TABLE 13.4

Names, symbols, and conversion factors for radioactivity

Name	Symbol	To Obtain	Multiply by
Becquerel	Bq	Ci	2.7×10^{-11}
gray	Gy	rad	100
sievert	Sv	rem	100
curie	Ci	Bq	3.7×10^{10}
rem	rem	Sv	0.01
millirem	mrem	rem	0.001
rem	rem	millirem	1,000

TABLE 13.5

Approximate single dose, whole body effects of radiation exposure

Level	Comment
0.130 rem	Average annual exposure to natural background radiation
0.500 rem	Upper limit of annual exposure to general public
25.0 rem	Threshold for observable effects such as reduced blood cell count
100.0 rem	Fatigue and other symptoms of radiation sickness
200.0 rem	Definite radiation sickness, bone marrow damage, possibility of developing leukemia
500.0 rem	Lethal dose for 50 percent of individuals
1,000.0 rem	Lethal dose for all

Radiation Units

You have learned that *radioactivity* is a property of isotopes with unstable, disintegrating nuclei and *radiation* is emitted particles (alpha or beta) or energy traveling in the form of photons (gamma). Radiation can be measured (1) at the source of radioactivity or (2) at a place of reception, where the radiation is absorbed.

The *activity* of a radioactive source is a measure of the number of nuclear disintegrations per unit of time. The unit of activity at the source is called a **curie** (Ci), which is defined as 3.70×10^{10} nuclear disintegrations per second. Activities are usually expressed in terms of fractions of curies, for example, a *picocurie* (pCi), which is a millionth of a millionth of a curie. Activities are sometimes expressed in terms of so many picocuries per liter (pCi/L).

The International System of Units (SI) unit for radioactivity is the *Becquerel* (Bq), which is defined as one nuclear disintegration per second. The unit for reporting radiation in the United States is the curie, but the Becquerel is the internationally accepted unit. Table 13.4 gives the names, symbols, and conversion factors for units of radioactivity.

As radiation from a source moves out and strikes a material, it gives the material energy. The amount of energy released by radiation striking living tissue is usually very small, but it can cause biological damage nonetheless because chemical bonds are broken and free polyatomic ions are produced by radiation.

The amount of radiation received by a human is expressed in terms of radiological dose. Radiation dose is usually written in units of a **rem,** which takes into account the possible biological damage produced by different types of radiation. Doses are usually expressed in terms of fractions of the rem, for example, a *millirem* (mrem). A millirem is 1/1,000 of a rem and is the unit of choice when low levels of radiation are discussed. The SI unit for radiation dose is the *millisievert* (mSv). Both the millirem and the millisievert relate ionizing radiation and biological effect to humans. The natural radiation that people receive from nature in one day is about 1 millirem (0.01 millisievert). A single dose of 100,000 to 200,000 millirems (1,000 to 2,000 millisieverts) can cause radiation sickness in humans (Table 13.5). A single dose of 500,000 millirems (5,000 millisieverts) results in death about 50 percent of the time.

Another measure of radiation received by a material is the **rad.** The term *rad* is from <u>r</u>adiation <u>a</u>bsorbed <u>d</u>ose. The SI unit for radiation received by a material is the *gray* (Gy). One gray is equivalent to an exposure of 100 rad.

Overall, there are many factors and variables that affect the possible damage from radiation, including the distance from the source and what shielding materials are between a person and a source. A *millirem* is the unit of choice when low levels of radiation are discussed.

Radiation Exposure

Natural radioactivity is a part of your environment, and you receive between 100 and 500 millirems each year from natural sources. This radiation from natural sources is called **background radiation.** Background radiation comes from outer space in the form of cosmic rays and from unstable isotopes in the ground, building materials, and foods. Many activities and situations will increase your yearly exposure to radiation. For example, the atmosphere absorbs some of the cosmic rays from space, so the less atmosphere above you, the more radiation you will receive. You are exposed to one additional millirem per year for each 100 feet you live above sea level. You receive approximately 0.3 millirem for each hour spent on a jet flight. Airline crews receive an additional 300 to 400 millirems per year because they spend so much time high in the atmosphere. Additional radiation exposure comes from medical X rays, television sets, and luminous objects such as watch and clock dials. In general, the background radiation exposure for the average person is about 130 millirems per year.

What are the consequences of radiation exposure? Radiation can be a hazard to living organisms because it produces ionization along its path of travel. This ionization can (1) disrupt chemical bonds in essential macromolecules such as DNA and (2) produce molecular fragments, which are free polyatomic ions that can interfere with enzyme action and other essential cell functions. Tissues with rapidly dividing cells, such as blood-forming tissue, are more vulnerable to radiation damage than others. Thus, one of the symptoms of an excessive radiation exposure is a lowered red and white blood cell count. Table 13.5 compares the estimated results of various levels of acute radiation exposure.

Radiation is not a mysterious, unique health hazard. It is a hazard that should be understood and neither ignored nor exaggerated. Excessive radiation exposure should be avoided, just as you avoid excessive exposure to other hazards such as certain chemicals, electricity, or even sunlight. Everyone agrees that *excessive* radiation exposure should be avoided, but there is some controversy about long-term, low-level exposure and its possible role in cancer. Some claim that tolerable low-level exposure does not exist because that is not possible. Others point to many studies comparing high and low background radioactivity with cancer mortality data. For example, no cancer mortality differences could be found between people receiving 500 or more millirems a year and those receiving less than 100 millirems a year. The controversy continues, however, because of lack of knowledge about long-term exposure. Two models of long-term, low-level radiation exposure have been proposed: (1) a linear model and (2) a threshold model. The *linear model* proposes that any radiation exposure above zero is damaging and can produce cancer and genetic damage. The *threshold model* proposes that the human body can repair damage and get rid of damaging free polyatomic ions up to a certain exposure level called the threshold (Figure 13.12). The controversy over long-term, low-level radiation exposure will probably continue until there is clear evidence about which model is correct. Whichever is correct will not lessen the need for rational risks versus cost-benefit analyses of all energy alternatives.

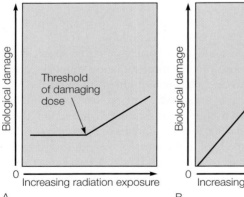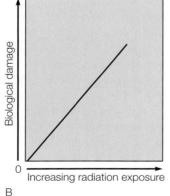

FIGURE 13.12

Graphic representation of the (A) threshold model and (B) linear model of low-level radiation exposure. The threshold model proposes that the human body can repair damage up to a threshold. The linear model proposes that any radiation exposure is damaging.

Nuclear Energy

As discussed, some nuclei are unstable because they are too large or because they have an unstable neutron-to-proton ratio. These unstable nuclei undergo radioactive decay, eventually forming products of greater stability. An example of this radioactive decay is the alpha emission reaction of uranium-238 to thorium-234,

$$^{238}_{92}U \rightarrow \, ^{234}_{90}Th + \, ^{4}_{2}He$$

$$238.0003 \text{ u} \rightarrow 233.9942 \text{ u} + 4.00150 \text{ u}$$

The numbers below the nuclear equation are the *nuclear* masses (u) of the reactant and products. As you can see, there seems to be a loss of mass in the reaction,

$$233.9942 + 4.00150 - 238.0003 = -0.0046 \text{ u}$$

This change in mass is related to the energy change according to the relationship that was formulated by Albert Einstein in 1905. The relationship is

$$E = mc^2$$

equation 13.1

where E is a quantity of energy, m is a quantity of mass, and c is a constant equal to the speed of light in a vacuum, 3.00×10^8 m/s. According to this relationship, matter and energy are the same thing, and energy can be changed to matter and vice versa.

The products of a mole of uranium-238 decaying to more stable products (1) have a lower energy of 4.14×10^{11} J and (2) lost a mass of 4.6×10^{-6} kg. As you can see, a very small amount of matter was converted into a large amount of energy in the process, forming products of lower energy.

The relationship between mass and energy explains why the mass of a nucleus is always *less* than the sum of the masses of the individual particles of which it is made. The difference between (1) the mass of the individual nucleons making up a nucleus and (2) the actual mass of the nucleus is called the **mass defect** of the nucleus. The explanation for the mass defect is again found in $E = mc^2$. When nucleons join to make a nucleus, energy is released as the more stable nucleus is formed.

The energy equivalent released when a nucleus is formed is the same as the **binding energy,** the energy required to break the nucleus into individual protons and neutrons. The binding energy of the nucleus of any isotope can be calculated from the mass defect of the nucleus.

The ratio of binding energy to nucleon number is a reflection of the stability of a nucleus (Figure 13.13). The greatest binding energy per nucleon occurs near mass number 56, with about 1.4×10^{-12} J per nucleon, then decreases for both more massive and less massive nuclei. This means that more massive nuclei can gain stability by splitting into smaller nuclei with the release of energy. It also means that less massive nuclei can gain stability by joining together with the release of energy. The slope also shows that more energy is released in the coming-together process than in the splitting process.

Radiation can be used to delay food spoilage and preserve foods by killing bacteria and other pathogens, just as heat is used to pasteurize milk. Foods such as wheat, flour, fruits, vegetables, pork, chicken, turkey, ground beef, and other uncooked meats are exposed to gamma radiation from cobalt-60 or cesium-137 isotopes, X rays, and electron beams. This kills insects, parasites such as *Trichinella spiralis* and tapeworms, and bacteria such as *E. coli, Listeria, Salmonellae,* and *Staphylococcus.* The overall effect is that many food-borne causes of human disease are eliminated and it is possible to store foods longer.

Food in the raw state, processed, or frozen is passed through a machine where it is irradiated while cold or frozen. This process does not make the food radioactive because it does not touch any radioactive substance. In addition, the radiation used in the process is not strong enough to disrupt the nuclei of atoms in food molecules, so it does not produce radioactivity, either.

In addition to killing the parasites and bacteria, the process might result in some nutritional loss, but no more than that which normally occurs in canning. Some new chemical products may be formed by the exposure to radiation, but studies in several countries have not been able to identify any health problems or ill effects from these compounds.

Treatment with radiation works better for some foods than others. Dairy products undergo some flavor changes that are undesirable, and some fruits such as peaches become soft. Irradiated strawberries, on the other hand, remain firm and last for weeks instead of a few days in the refrigerator. Foods that are sterilized with a stronger dose of radiation can be stored for years without refrigeration just like canned foods that have undergone heat pasteurization.

In the United States the Food and Drug Administration (FDA) regulates which products can be treated by radiation and the dosages used in the treatment. The U.S. Department of Agriculture (USDA) is responsible for the inspection of irradiated meat and poultry products. All foods that have undergone a radiation treatment must show the international logo for this and a statement. The logo, called a radura, is a stylized flower inside a circle with five openings on the top part (Box Figure 13.1).

BOX FIGURE 13.1

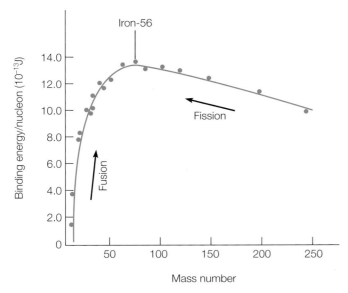

FIGURE 13.13

The maximum binding energy per nucleon occurs around mass number 56, then decreases in both directions. As a result, fission of massive nuclei and fusion of less massive nuclei both release energy.

The nuclear reaction of splitting a massive nucleus into more stable, less massive nuclei with the release of energy is **nuclear fission** (Figure 13.14). Nuclear fission occurs rapidly in an atomic bomb explosion and occurs relatively slowly in a nuclear reactor. The nuclear reaction of less massive nuclei, coming together to form more stable, and more massive, nuclei with the release of energy is **nuclear fusion.** Nuclear fusion occurs rapidly in a hydrogen bomb explosion and occurs continually in the sun, releasing the energy essential for the continuation of life on the earth.

Nuclear Fission

Nuclear fission was first accomplished in the late 1930s when researchers were attempting to produce isotopes by bombarding massive nuclei with neutrons. In 1938 two German scientists, Otto Hahn and Fritz Strassman, identified the element barium in a uranium sample that had been bombarded with neutrons. Where the barium came from was a puzzle at the time, but soon afterward Lise Meitner, an associate who had moved to Sweden, deduced that uranium nuclei had split, producing barium. The reaction might have been

$$\,_{0}^{1}\mathrm{n} + \,_{92}^{235}\mathrm{U} \rightarrow \,_{56}^{141}\mathrm{Ba} + \,_{36}^{92}\mathrm{Kr} + 3\,_{0}^{1}\mathrm{n}$$

Nuclear medicine had its beginnings in 1946 when radioactive iodine was first successfully used to treat thyroid cancer patients. Then physicians learned that radioactive iodine could also be used as a diagnostic tool, providing a way to measure the function of the thyroid and to diagnose thyroid disease. More and more physicians began to use nuclear medicine to diagnose thyroid disease as well as to treat hyperthyroidism and other thyroid problems. Nuclear medicine is a branch of medicine using radiation or radioactive materials to diagnose as well as treat diseases.

The development of new nuclear medicine technologies, such as new cameras, detection instruments, and computers, has led to a remarkable increase in the use of nuclear medicine as a diagnostic tool. Today, there are nearly 100 different nuclear medicine imaging procedures. These provide unique, detailed information about virtually every major organ system within the body, information that was unknown just years ago. Treatment of disease with radioactive materials continues to be a valuable part of nuclear medicine, too. The material that follows will consider some techniques of using nuclear medicine as a diagnostic tool, followed by a short discussion of the use of radioactive materials in the treatment of disease.

Nuclear medicine provides diagnostic information about organ function, compared to conventional radiology, which provides images about the structure. For example, a conventional X-ray image will show if a bone is broken or not, while a bone imaging nuclear scan will show changes caused by tumors, hairline fractures, or arthritis. There are procedures for making detailed structural X-ray pictures of internal organs such as the liver, kidney, or heart, but these images often cannot provide diagnostic information, showing only the structure. Nuclear medicine scans, on the other hand, can provide information about how much heart tissue is still alive after a heart attack or if a kidney is working, even when there are no detectable changes in organ appearance.

An X-ray image is produced when X rays pass through the body and expose photographic film on the other side. Some X-ray exams improve photographic contrast by introducing certain substances. A barium sulfate "milk shake," for example, can be swallowed to highlight the esophagus, stomach, and intestine. More information is provided if X rays are used in a CAT scan (CAT stands for "Computed Axial Tomography"). The CAT scan is a diagnostic test that combines the use of X rays with computer technology. The CAT scan shows organs of interest by making X-ray images from many different angles as the source of the X rays moves around the patient. Contrast-improving substances, such as barium sulfate, might also be used with a CAT scan. In any case, CAT scan images are assembled by a computer into a three-dimensional picture that can show organs, bones, and tissues in great detail.

The gamma camera is a key diagnostic imaging tool used in nuclear medicine. Its use requires a radioactive material, called a radiopharmaceutical, to be injected into or swallowed by the patient. A given radiopharmaceutical tends to go to a specific organ of the body; for example, radioactive iodine tends to go to the thyroid gland, and others go to other organs. Gamma-emitting radiopharmaceuticals are used with the gamma camera, and the gamma camera collects and processes these gamma rays to produce images. These images provide a way of studying the structure as well as measuring the function of the selected organ. Together, the structure and function provide a way of identifying tumors, areas of infection, or other problems. The patient experiences little or no discomfort and the radiation dose is small.

A SPECT scan (Single Photon Emission Computerized Tomography) is an imaging technique employing a gamma camera that rotates around the patient, measuring gamma rays and computing their point of origin. Cross-sectional images of a three-dimensional internal organ can be obtained from such data, resulting in images that have higher resolution and thus more diagnostic information than a simple gamma camera image. A Gallium radiopharmaceutical is often used in a scan to diagnose and follow the progression of tumors or infections. Gallium scans also can be used to evaluate the heart, lungs, or any other organ that may be involved with inflammatory disease.

Use of MRI (Magnetic Resonance Imaging) also produces images as an infinite number of projections through the body. Unlike CAT, gamma, or SPECT scans, MRI does not use any form of ionizing radiation. MRI uses magnetic fields, radio waves, and a computer to produce detailed images. As the patient enters an MRI scanner, his body is surrounded by a large magnet. The technique requires a very strong magnetic field, a field so strong that it aligns the nuclei of the person's atoms. The scanner sends a strong radio signal, temporarily knocking the nuclei out of alignment. When the radio signal stops, the nuclei return to the aligned position, releasing their own faint radio frequencies. These radio signals are read by the scanner, which uses them in a computer program to produce very detailed images of the human anatomy (Box Figure 13.2).

The PET scan (Positron Emission Tomography) produces 3D images superior to gamma camera images. This technique is built around a radiopharmaceutical that emits positrons (like an electron with a positive charge). Positrons collide with electrons, releasing a burst of energy in the form of photons. Detectors track the emissions and feed the information into a computer. The computer has a program to plot the source of radiation and translates the data into an image. Positron-emitting radiopharmaceuticals used in a PET scan can be low atomic weight elements like carbon, nitrogen, and oxygen. This is important for certain purposes since these are the same

—*Continued top of next page*

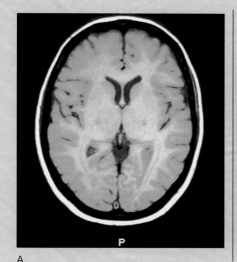

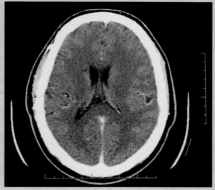

BOX FIGURE 13.2

(A) MRI scan of brain, (B) CAT scan of brain.

elements found in many biological substances like sugar, urea, or carbon dioxide. Thus a PET scan can be used to study processes in organs such as the brain and heart where glucose is being broken down or oxygen is being consumed. This diagnostic method can be used to detect epilepsy or brain tumors, among other problems.

Radiopharmaceuticals used for diagnostic examinations are selected for their affinity for certain organs, if they emit sufficient radiation to be easily detectable in the body, and if they have a rather short half-life, preferably no longer than a few hours. Useful radioisotopes that meet these criteria for diagnostic purposes are technetium-99, gallium-67, indium-111, iodine-123, iodine-131, thallium-201, and krypton-81.

The goal of therapy in nuclear medicine is to use radiation to destroy diseased or cancerous tissue while sparing adjacent healthy tissue. Few radioactive therapeutic agents are injected or swallowed, with the exception of radioactive iodine— mentioned earlier as a treatment for cancer of the thyroid. Useful radioisotopes for therapeutical purposes are iodine-131, phosphorus-32, iridium-192, and gold-198. The radioactive source placed in the body for local irradiation of a tumor is normally iridium-192. A nuclear pharmaceutical is a physiologically active carrier to which a radioisotope is attached. Today, it is possible to manufacture chemical or biological carriers that migrate to a particular part of the human body, and this is the subject of much ongoing medical research.

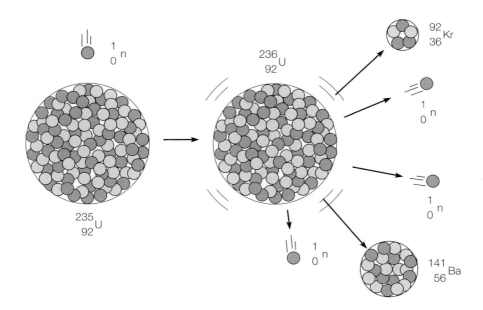

FIGURE 13.14

The fission reaction occurring when a neutron is absorbed by a uranium-235 nucleus. The deformed nucleus splits any number of ways into lighter nuclei, releasing neutrons in the process.

TABLE 13.6
Fragments and products from nuclear reactors using fission of uranium-235

Isotope	Major Mode of Decay	Half-Life	Isotope	Major Mode of Decay	Half-Life
Tritium	Beta	12.26 years	Cerium-144	Beta, gamma	285 days
Carbon-14	Beta	5,930 years	Promethium-147	Beta	2.6 years
Argon-41	Beta, gamma	1.83 hours	Samarium-151	Beta	90 years
Iron-55	Electron capture	2.7 years	Europium-154	Beta, gamma	16 years
Cobalt-58	Beta, gamma	71 days	Lead-210	Beta	22 years
Cobalt-60	Beta, gamma	5.26 years	Radon-222	Alpha	3.8 days
Nickel-63	Beta	92 years	Radium-226	Alpha, gamma	1,620 years
Krypton-85	Beta, gamma	10.76 years	Thorium-229	Alpha	7,300 years
Strontium-89	Beta	5.4 days	Thorium-230	Alpha	26,000 years
Strontium-90	Beta	28 years	Uranium-234	Alpha	2.48×10^5 years
Yttrium-91	Beta	59 days	Uranium-235	Alpha, gamma	7.13×10^8 years
Zirconium-93	Beta	9.5×10^5 years	Uranium-238	Alpha	4.51×10^9 years
Zirconium-95	Beta, gamma	65 days	Neptunium-237	Alpha	2.14×10^6 years
Niobium-95	Beta, gamma	35 days	Plutonium-238	Alpha	89 years
Technetium-99	Beta	2.1×10^5 years	Plutonium-239	Alpha	24,360 years
Ruthenium-106	Beta	1 year	Plutonium-240	Alpha	6,760 years
Iodine-129	Beta	1.6×10^7 years	Plutonium-241	Beta	13 years
Iodine-131	Beta, gamma	8 days	Plutonium-242	Alpha	3.79×10^5 years
Xenon-133	Beta, gamma	5.27 days	Americium-241	Alpha	458 years
Cesium-134	Beta, gamma	2.1 years	Americium-243	Alpha	7,650 years
Cesium-135	Beta	2×10^6 years	Curium-242	Alpha	163 days
Cesium-137	Beta	30 years	Curium-244	Alpha	18 years
Cerium-141	Beta	32.5 days			

The phrase "might have been" is used because a massive nucleus can split in many different ways, producing different products. About thirty-five different, less massive elements have been identified among the fission products of uranium-235. Some of these products are fission fragments, and some are produced by unstable fragments that undergo radioactive decay. Selected fission fragments are listed in Table 13.6, together with their major modes of radioactive decay and half-lives. Some of the isotopes are the focus of concern about nuclear wastes, the topic of the reading at the end of this chapter.

The fission of a uranium-235 nucleus produces two or three neutrons along with other products. These neutrons can each move to other uranium-235 nuclei where they are absorbed, causing fission with the release of more neutrons, which move to other uranium-235 nuclei to continue the process. A reaction where the products are able to produce more reactions in a self-sustaining series is called a **chain reaction.** A chain reaction is self-sustaining until all the uranium-235 nuclei have fissioned or until the neutrons fail to strike a uranium-235 nucleus (Figure 13.15).

You might wonder why all the uranium in the universe does not fission in a chain reaction. Natural uranium is mostly uranium-238, an isotope that does not fission easily. Only about 0.7 percent of natural uranium is the highly fissionable uranium-235. This low ratio of readily fissionable uranium-235 nuclei makes it unlikely that a stray neutron would be able to achieve a chain reaction.

In order to achieve a chain reaction, there must be (1) a sufficient mass with (2) a sufficient concentration of fissionable nuclei. When the mass and concentration are sufficient to sustain a chain reaction the amount is called a **critical mass.** Likewise, a mass too small to sustain a chain reaction is called a *subcritical mass.* A mass of sufficiently pure uranium-235 (or plutonium-239) that is large enough to produce a rapidly accelerating chain reaction is called a *supercritical mass.* An atomic bomb is simply a device that uses a small, conventional explosive to push subcritical masses of fissionable material into a supercritical mass. Fission occurs almost instantaneously in the supercritical mass, and tremendous energy is released in a violent explosion.

Nuclear Power Plants

The nuclear part of a nuclear power plant is the **nuclear reactor,** a steel vessel in which a controlled chain reaction of fissionable material releases energy (Figure 13.16). In the most popular design, called a pressurized light-water reactor, the fissionable material is enriched 3 percent uranium-235 and 97 percent

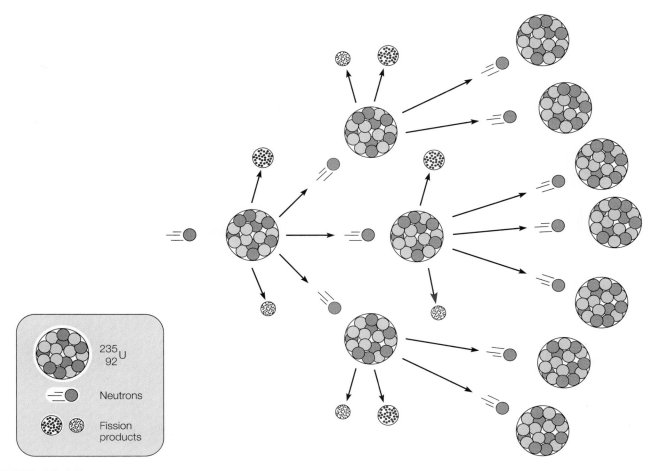

Legend (boxed key):
- $^{235}_{92}U$
- Neutrons
- Fission products

FIGURE 13.15

A schematic representation of a chain reaction. Each fissioned nucleus releases neutrons, which move out to fission other nuclei. The number of neutrons can increase quickly with each series.

uranium-238 that has been fabricated in the form of small ceramic pellets (Figure 13.17A). The pellets are encased in a long zirconium alloy tube called a **fuel rod.** The fuel rods are locked into a *fuel rod assembly* by locking collars, arranged to permit pressurized water to flow around each fuel rod (Figure 13.17B) and to allow the insertion of *control rods* between the fuel rods. **Control rods** are constructed of materials, such as cadmium, that absorb neutrons. The lowering or raising of control rods within the fuel rod assemblies slows or increases the chain reaction by varying the amount of neutrons absorbed. When they are lowered completely into the assembly, enough neutrons are absorbed to stop the chain reaction.

It is physically impossible for the low-concentration fuel pellets to form a supercritical mass. A nuclear reactor in a power plant can only release energy at a comparatively slow rate, and it is impossible for a nuclear power plant to produce a nuclear explosion. In a pressurized water reactor the energy released is carried away from the reactor by pressurized water in a closed pipe called the **primary loop** (Figure 13.18). The water is pressurized at about 150 atmospheres (about 2,200 lb/in²) to keep it from boiling, since its temperature may be 350°C (about 660°F).

In the pressurized light-water (ordinary water) reactor the circulating pressurized water acts as a coolant, carrying heat away from the reactor. The water also acts as a **moderator,** a substance that slows neutrons so they are more readily absorbed by uranium-235 nuclei. Other reactor designs use heavy water (dideuterium monoxide) or graphite as a moderator.

Water from the closed primary loop is circulated through a heat exchanger called a **steam generator** (Figure 13.18). The pressurized high-temperature water from the reactor moves through hundreds of small tubes inside the generator as *feed-water* from the **secondary loop** flows over the tubes. The water in the primary loop heats feedwater in the steam generator and then returns to the nuclear reactor to become heated again. The feedwater is heated to steam at about 235°C (455°F) with a pressure of about 68 atmospheres (1,000 lb/in²). This steam is piped to the turbines, which turn an electric generator (Figure 13.19).

After leaving the turbines, the spent steam is condensed back to liquid water in a second heat exchanger receiving water from the cooling towers. Again, the cooling water does not mix with the closed secondary loop water. The cooling-tower

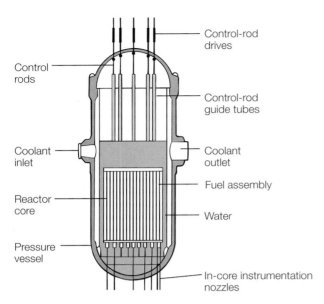

FIGURE 13.16

A schematic representation of the basic parts of a nuclear reactor. The largest commercial nuclear power plant reactors are nine- to eleven-inch-thick steel vessels with a stainless steel liner, standing about 40 feet high with a diameter of 16 feet. Such a reactor has four pumps, which move 440,000 gallons of water per minute through the primary loop.

water enters the condensing heat exchanger at about 32°C (90°F) and leaves at about 50°C (about 120°F) before returning to a cooling tower, where it is cooled by evaporation. The feedwater is preheated, then recirculated to the steam generator to start the cycle over again. The steam is condensed back to liquid water because of the difficulty of pumping and reheating steam.

After a period of time the production of fission products in the fuel rods begins to interfere with effective neutron transmission, so the reactor is shut down annually for refueling. During refueling about one-third of the fuel that had the longest exposure in the reactor is removed as "spent" fuel. New fuel rod assemblies are inserted to make up for the part removed (Figure 13.20). However, only about 4 percent of the "spent" fuel is unusable waste, about 94 percent is uranium-238, 0.8 percent is uranium-235, and about 0.9 percent is plutonium (Figure 13.21). Thus, "spent" fuel rods contain an appreciable amount of usable uranium and plutonium. For now, spent reactor fuel rods are mostly stored in cooling pools at the nuclear plant sites. In the future, a decision will be made either to reprocess the spent fuel, recovering the uranium and plutonium through chemical reprocessing, or put the fuel in terminal storage. Concerns about reprocessing are based on the fact that plutonium-239 and uranium-235 are fissionable and could possibly be used by terrorist groups to construct nuclear explosive devices. Six other countries do have reprocessing plants, however, and the spent fuel rods represent an energy source equivalent to more than 25 billion barrels of petroleum.

A

B

FIGURE 13.17

(*A*) These are uranium oxide fuel pellets that are stacked inside fuel rods, which are then locked together in a fuel rod assembly. (*B*) A fuel rod assembly. See also Figure 13.20, which shows a fuel rod assembly being loaded into a reactor.

Some energy experts say that it would be inappropriate to dispose of such an energy source.

The technology to dispose of fuel rods exists if the decision is made to do so. The longer half-life waste products are mostly alpha emitters. These metals could be converted to oxides, mixed with powdered glass (or a ceramic), melted, and then poured into stainless steel containers. The solidified canisters would then be buried in a stable geologic depository. The glass technology is used in France for disposal of high-level wastes. Buried at two-thousand- to three-thousand-foot depths in solid granite, the only significant means of the radioactive wastes reaching the surface would be through groundwater dissolving the stainless steel, glass, and waste products and then transporting them back to the surface. Many experts believe that if such groundwater dissolving were to take place it would require thousands of years. The radioactive isotopes would thus

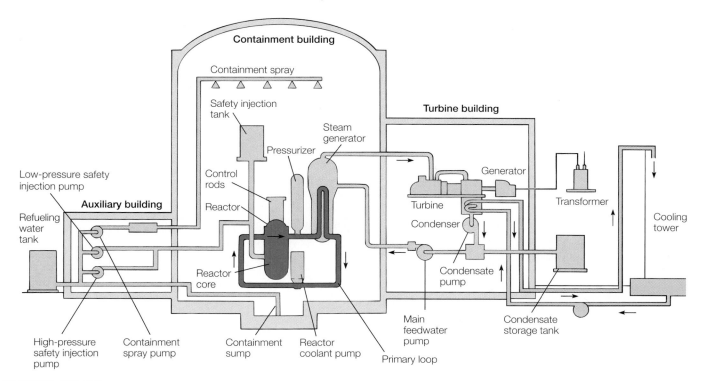

FIGURE 13.18

A schematic general system diagram of a pressurized water nuclear power plant, not to scale. The containment building is designed to withstand an internal temperature of 300°F at a pressure of 60 lb/in^2 and still maintain its leak-tight integrity.

FIGURE 13.19

The turbine deck of a nuclear generating station. There is one large generator in line with four steam turbines in this non-nuclear part of the plant. The large silver tanks are separators that remove water from the steam after it has left the high-pressure turbine and before it is recycled back into the low-pressure turbines.

FIGURE 13.20

Spent fuel rod assemblies are removed and new ones are added to a reactor head during refueling. This shows an initial fuel load to a reactor, which has the upper part removed and set aside for the loading.

undergo natural radioactive decay by the time they could reach the surface. Nonetheless, research is continuing on nuclear waste and its disposal. In the meantime, the question of whether it is best to reprocess fuel rods or place them in permanent storage remains unanswered.

What is the volume of nuclear waste under question? If all the spent fuel rods from all the commercial nuclear plants accumulated up to the year 2000 were reprocessed, then mixed with glass, the total amount of glassified waste would make a pile on one football field an estimated 4 m (about 13 ft) high.

Three Mile Island (downriver from Harrisburg, Pennsylvania, U.S.A.) and Chernobyl (former U.S.S.R., now Ukraine) are two nuclear power plants that became famous because of accidents. Here is a brief accounting of what happened.

Three Mile Island. It was March 28, 1979, and the 880-megawatt Three Mile Island Nuclear Plant, operated by Metropolitan Edison Co., was going full blast. At 4:00 A.M. that morning the main feedwater pump that pumps water to the steam generator failed for some unexplained reason (follow this description in Figure 13.18). Backup feedwater pumps kicked in, but the valves that should have been open were closed for maintenance, blocking the backup source of water for the steam generator. All of a sudden there was not a source of water for the steam generator that removes heat from the primary loop. Events began to happen quickly at this point.

The computer sensors registered that the steam generator was not receiving water, and it began to follow a shutdown procedure. First, the turbine was shut down as steam was vented from the steam line out through the turbine building. It sounded much like a large jet plane. Within six seconds the reactor was "scrammed," shut down with control rods dropped between the fuel rods in the reactor vessel. Fissioning began to slow, but the reactor was still hot.

Between three and six seconds a pressure relief valve opened on the primary loop, reducing the excess pressure that was generated because feedwater was not entering the steam generator to remove heat from the primary loop. The valve should have closed when the excess pressure was released, but it did not close. It was stuck in an open position. Pressurized water and steam was pouring from the primary loop into the containment building. As water was lost from the primary loop, temperatures inside the reactor began to climb. The loss of pressure resulted in high-temperature water flashing into steam. If an operator had pressed the right button in the control room

it would have closed the open valve, but this did not happen until thirty-two minutes later.

At this point the reactor could have recovered from the events. Two minutes after the initial shutdown the computer sensors noted the loss of pressure from the open valve and kicked in the emergency core cooling system, which pumps more water into the reactor vessel. However, for some unknown reason, the control room operators shut down one pump four and a half minutes after the initial event and the second pump six minutes later.

Water continued to move through the open pressure relief valve into the containment building. At seven and a half minutes after the accident began, the radioactive water on the floor was two feet deep and the sump pumps started pumping water into tanks in the auxiliary building. This water would become the source of the radioactivity that did escape the plant. It escaped because the pump seals leaked and spilled radioactive water. The filters had been removed from the auxiliary building air vents, allowing radioactive gases to escape.

Eleven minutes after the start of the accident, an operator restarted the emergency core cooling system that had been turned off. With the cooling water flowing again, the pressure in the reactor stopped falling. The fuel rods, some 36,000 in this reactor, had not yet suffered any appreciable damage. This would be taken care of by the next incredible event.

The next incredible event was that operators began turning off the emergency cooling pumps, perhaps because they were vibrating too much. In any case, with the pumps off, the water level in the reactor fell again, this time uncovering the fuel rods. Within minutes, the temperature was high enough to rupture fuel rods, dropping radioactive oxides into the bottom of the reactor vessel. The operators now had a general emergency.

It was eleven hours later that the operators decided to start the main reactor

coolant pump. This pump had shut down at the beginning of the series of events. Water again covered the fuel rods, and the pressure and temperature stabilized.

The consequences of this series of events were as follows:

1. Local residences did receive some radiation from the release of gases. They received 10 millirems (0.1 millisievert) in a low-exposure area and up to 25 millirems (0.25 millisievert) in a high-exposure area.
2. Cleaning up the damaged reactor vessel and core required more than ten years to cut it up, pack it into canisters, and ship everything to the Federal Nuclear Reservation at Idaho Falls.
3. The cost of the cleanup was more than $1 billion.
4. Changes at other nuclear power plants as a consequence: Pressure relief valves have been removed; operators can no longer turn off the emergency cooling system; and, operators must now spend about one-fourth of their time in training.

Chernobyl. The Soviet-designed Chernobyl reactor was a pressurized water reactor with individual fuel channels, which is very different from the pressurized water reactors used in the United States. The Chernobyl reactor was constructed with each fuel assembly in an individual pressure tube with circulating pressurized water. This heated water was circulated to a steam separator, and the steam was directed to turbines, which turned a generator to produce electricity. Graphite blocks surrounded the pressure tubes, serving as moderators to slow the neutrons involved in the chain reaction. The graphite was cooled by a mixture of helium and nitrogen. The reactor core was located in a concrete bunker that acted as a radiation shield. The top part was a steel cap and shield that supported the fuel assemblies. There were no containment buildings

—*Continued top of next page*

around the Soviet reactors as there are in the United States.

The Chernobyl accident was the result of a combination of a poorly engineered reactor design, poorly trained reactor operators, and serious mistakes made by the operators on the day of the accident.

The reactor design was flawed because at low power, steam tended to form pockets in the water-filled fuel channels, creating a condition of instability. Instability occurred because (1) steam is not as efficient at cooling as is liquid water, and (2) liquid water acts as a moderator and neutron absorber while steam does not. Excess steam therefore leads to overheating and increased power generation. Increased power can lead to increased steam generation, which leads to further increases in power. This coupled response is very difficult to control because it feeds itself.

On April 25, 1986, the operators of Chernobyl unit 4 started a test to find out how long the turbines would spin and supply power following the loss of electrical power. The operators disabled the automatic shutdown mechanisms and then started the test early on April 26, 1986. The plan was to stabilize the reactor at 1,000 MW, but an error was made; the power fell to about 30 MW and pockets of steam became a problem. Operators tried to increase the power by removing all the control rods. At 1:00 A.M., they were able to stabilize the reactor at 200 MW. Then instability returned, and the operators were making continuous adjustments to maintain a constant power. They reduced the feedwater to maintain steam pressure, and this created even more steam voids in the fuel channels. Power surged to a very high level and fuel elements ruptured. A steam explosion moved the reactor cap, exposing individual fuel channels and releasing fission products to the environment. A second steam explosion knocked a hole in the roof, exposing more of the reactor core, and the graphite, which served as a moderator, burst into flames. The graphite burned for nine days, releasing about 324 million Ci (12×10^{18} Bq) into the environment.

The fiery release of radioactivity was finally stopped by using a helicopter to drop sand, boron, lead, and other materials onto the burning graphite reactor. After the fire was out the remains of the reactor were covered with a large concrete shelter.

In addition to destroying the reactor, the accident killed 30 people, and 28 of these were from radiation exposure. Another 134 people were treated for acute radiation poisoning, and all recovered from the immediate effects.

Cleanup crews over the next year received about 10 rem (100 millisieverts) to 25 rem (250 millisieverts) and some received as much as 50 rem (500 millisieverts). In addition to this direct exposure, large expanses of Belarus, Ukraine, and Russia were contaminated by radioactive fallout from the reactor fire. Hundreds of thousands of people have been resettled into less contaminated areas. The World Health Organization and other international agencies have studied the data to understand the impact of radiation-related disease. These studies do confirm a rising incidence of thyroid cancer, but no increases in leukemia so far.

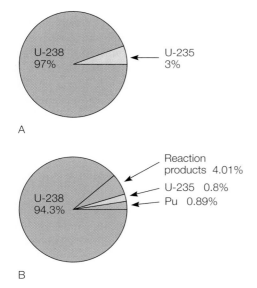

FIGURE 13.21

The composition of the nuclear fuel in a fuel rod (*A*) before and (*B*) after use over a three-year period in a nuclear reactor.

Nuclear Fusion

As the graph of nuclear binding energy versus mass numbers shows (see Figure 13.13), nuclear energy is released when (1) massive nuclei such as uranium-235 undergo fission and (2) when less massive nuclei come together to form more massive nuclei through nuclear fusion. Nuclear fusion is responsible for the energy released by the sun and other stars. At the present halfway point in the sun's life—with about 5 billion years to go—the core is now 35% hydrogen and 65% helium. Through fusion, the sun converts about 650 million tons of hydrogen to 645 million tons of helium every second. The other roughly 5 million tons of matter are converted into energy. Even at this rate, the sun has enough hydrogen to continue the process for an estimated 5 billion years. There are several fusion reactions that take place between hydrogen and helium isotopes, including the following:

$$\,_{1}^{1}\text{H} + \,_{1}^{1}\text{H} \rightarrow \,_{1}^{2}\text{H} + \,_{1}^{0}\text{e}$$

$$\,_{1}^{2}\text{H} + \,_{1}^{2}\text{H} \rightarrow \,_{2}^{3}\text{He} + \,_{0}^{1}\text{n}$$

$$\,_{2}^{3}\text{He} + \,_{2}^{3}\text{He} \rightarrow \,_{2}^{4}\text{He} + 2\,_{1}^{1}\text{H}$$

The fusion process would seem to be a desirable energy source on earth because (1) two isotopes of hydrogen, deuterium (^2_1H) and tritium (^3_1H), undergo fusion at a relatively low temperature; (2) the supply of deuterium is practically unlimited, with each gallon of seawater containing about a teaspoonful of heavy water; and (3) enormous amounts of energy are released with no radioactive by-products.

The oceans contain enough deuterium to generate electricity for the entire world for millions of years, and tritium can be constantly produced by a fusion device. Researchers know what needs to be done to tap this tremendous energy source. The problem is *how* to do it in an economical, continuous energy-producing fusion reactor. The problem, one of the most difficult engineering tasks ever attempted, is meeting three basic fusion reaction requirements of (1) temperature, (2) density, and (3) time (Figure 13.22):

1. *Temperature.* Nuclei contain protons and are positively charged, so they experience the electromagnetic repulsion of like charges. This force of repulsion can be overcome, moving the nuclei close enough to fuse together, by giving the nuclei sufficient kinetic energy. The fusion reaction of deuterium and tritium, which has the lowest temperature requirements of any fusion reaction known at the present time, requires temperatures on the order of 100 million°C.
2. *Density.* There must be a sufficiently dense concentration of heavy hydrogen nuclei, on the order of $10^{14}/cm^3$, so many reactions occur in a short time.
3. *Time.* The nuclei must be confined at the appropriate density up to a second or longer at pressures of at least 10 atmospheres to permit a sufficient number of reactions to take place.

The temperature, density, and time requirements of a fusion reaction are interrelated. A short time of confinement, for example, requires an increased density, and a longer confinement time requires less density. The primary problems of fusion research are the high-temperature requirements and confinement. No material in the world can stand up to a temperature of 100 million°C, and any material container would be instantly vaporized. Thus, research has centered on meeting the fusion reaction requirements without a material container. Two approaches are being tested, *magnetic confinement* and *inertial confinement.*

Magnetic confinement utilizes a **plasma,** a very hot gas consisting of atoms that have been stripped of their electrons because of the high kinetic energies. The resulting positively and negatively charged particles respond to electrical and magnetic forces, enabling researchers to develop a "magnetic bottle," that is, magnetic fields that confine the plasma and avoid the problems of material containers that would vaporize. A magnetically confined plasma is very unstable, however, and researchers have compared the problem to trying to carry a block of jello on a pair of rubber bands. Different magnetic field geometries and magnetic "mirrors" are the topics of research in attempts to stabilize the hot, wobbly plasma. Electric currents, injection of fast ions, and radio frequency (microwave) heating methods are also being studied.

Inertial confinement is an attempt to heat and compress small frozen pellets of deuterium and tritium with energetic laser beams or particle beams, producing fusion. The focus of this research is new and powerful lasers, light ion and heavy ion beams. If successful, magnetic or inertial confinement will provide a long-term solution for future energy requirements.

The Source of Nuclear Energy

When elements undergo the natural radioactive decay process, energy is released, and the decay products have less energy than the original reactant nucleus. When massive nuclei undergo fission, much energy is rapidly released along with fission products that continue to release energy through radioactive decay. What is the source of all this nuclear energy? The answer to this question is found in current theories about how the universe started and in theories about the life cycle of the stars. Theories about the life cycle of stars are discussed in chapters 14 and 15. For now, consider just a brief introduction to the life cycle of a star in order to understand the ultimate source of nuclear energy.

The current universe is believed to have started with a "big bang" of energy, which created a plasma of protons and neutrons. This primordial plasma cooled rapidly and, after several minutes, began to form hydrogen nuclei. Throughout the newly formed universe massive numbers of hydrogen atoms—on the order of 10^{57} nuclei—were gradually pulled together by gravity into masses that would become the stars. As the hydrogen atoms fell toward the center of each mass of gas, they accelerated, just like any other falling object. As they accelerated, the contracting mass began to heat up because the average kinetic energy of the atoms increased from acceleration. Eventually, after say ten million years or so of collapsing

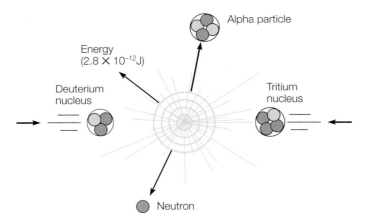

FIGURE 13.22

A fusion reaction between a tritium nucleus and a deuterium nucleus requires a certain temperature, density, and time of containment to take place.

There are two general categories of nuclear wastes: (1) low-level wastes and (2) high-level wastes. The *low-level wastes* are produced by hospitals, universities, and other facilities. They are also produced by the normal operation of a nuclear reactor. Radioactive isotopes sometimes escape from fuel rods in the reactor and in the spent fuel storage pools. These isotopes are removed from the water by ion-exchange resins and from the air by filters. The used resins and filters will contain the radioactive isotopes and will become low-level wastes. In addition, any contaminated protective clothing, tools, and discarded equipment also become low-level wastes.

Low-level liquid wastes are evaporated, mixed with cement, then poured into fifty-five-gallon steel drums. Solid wastes are compressed and placed in similar drums. The drums are currently disposed of by burial in government-licensed facilities. In general, low-level waste has an activity of less than 1.0 curie per cubic foot. Contact with the low-level waste could expose a person to up to 20 millirems per hour of contact.

High-level wastes from nuclear power plants are spent nuclear fuel rods. At the present time most of the commercial nuclear power plants have these rods in temporary storage at the plant sites. These rods are "hot" in the radioactive sense, producing about 100,000 curies per cubic foot. They are also hot in the thermal sense, continuing to generate heat for months after removal from the reactor. The rods are cooled by heat exchangers connected to storage pools; they could otherwise achieve an internal temperature as high as 800°C for several decades. In the future, these spent fuel rods will be reprocessed or disposed of through terminal storage.

Agencies of the United States federal government have also accumulated millions of gallons of high-level wastes from the manufacture of nuclear weapons and nuclear research programs. These liquid wastes are stored in million-gallon stainless steel containers that are surrounded by concrete. The future of this large amount of high-level wastes may be evaporation to a solid form or mixture with a glass or ceramic matrix, which is melted and poured into stainless steel containers. These containers would be buried in solid granite rock in a stable geologic depository. Such high-level wastes must be contained for thousands of years as they undergo natural radioactive decay (Box Figure 13.3). Burial at a depth of two thousand to three thousand feet in solid granite would provide protection from exposure by explosives, meteorite impact, or erosion. One major concern about this plan is that a hundred generations later, people might lose track of what is buried in the nuclear garbage dump.

BOX FIGURE 13.3

This is a standard warning sign for a possible radioactive hazard. Such warning signs would have to be maintained around a nuclear waste depository for thousands of years.

---

and heating, the mass of hydrogen condensed to a sphere with a diameter of 1.5 million miles or so, or about twice the size of the sun today. At the same time the interior temperature increased to millions of degrees, reaching the critical points of density, temperature, and containment for a fusion reaction to begin. Thus, a star was born as hydrogen nuclei fused into helium nuclei, releasing enough energy that the star began to shine.

Hydrogen nuclei in the newborn star had a higher energy per nucleon than helium nuclei, and helium nuclei had more energy per nucleon than other nuclei up to around iron. The fusion process continued for billions of years, releasing energy as heavier and heavier nuclei were formed. Eventually, the star materials were fused into nuclei around iron, the element with the lowest amount of energy per nucleon, and the star used up its energy source. Larger, more massive dying stars explode into supernovas (discussed in chapter 14). Such an explosion releases a flood of neutrons, which bombard medium-weight nuclei and build them up to more massive nuclei, all the way from iron up to uranium. Thus, the more massive elements were born from an exploding supernova, then spread into space as dust. In a process to be discussed later, this dust became the materials of which planets were made, including earth. The point for the present discussion, however, is that the energy of naturally radioactive elements, and the energy released during fission, can be traced back to the force of gravitational attraction, which provided the initial energy for the whole process.

Marie Curie (1867–1934)

Marie Curie was a Polish-born French scientist who, with her husband Pierre Curie (1859–1906), was an early investigator of radioactivity. The Curies discovered the radioactive elements polonium and radium, for which achievement they shared the 1903 Nobel Prize for Physics with Henri Becquerel. Madame Curie went on to study the chemistry and medical applications of radium, and was awarded the 1911 Nobel Prize for Chemistry in recognition of her work in isolating the pure metal.

Madame Curie's Polish maiden name was Manya Sklodowska. She was born in Warsaw on November 7, 1867, at a time when Poland was under Russian domination after the unsuccessful revolt of 1863. Her parents were teachers, and soon after Manya was born—their fifth child—they lost their teaching posts and had to take in boarders. Their young daughter worked long hours helping with the meals, but nevertheless won a medal for excellence at the local high school, where the examinations were held in Russian. No higher education was available, so Manya took a job as a governess, sending part of her savings to Paris to help to pay for her elder sister's medical studies. Her sister qualified and married a fellow doctor in 1891, and Manya went to join them in Paris. She entered the Sorbonne and studied physics and mathematics, graduating at the top of her class. In 1894 she met the French chemist Pierre Curie, and they were married the following year.

Pierre Curie was born in Paris on May 15, 1859, the son of a doctor. He was educated privately and at the Sorbonne, becoming an assistant there in 1878. He discovered the piezoelectric effect and, after being appointed head of the laboratory of the École de Physique et Chimie, went on to study magnetism and formulate Curie's law

(which states that magnetic susceptibility is inversely proportional to absolute temperature). In 1895 he discovered the Curie point, the critical temperature at which a paramagnetic substance becomes ferromagnetic. In the same year he married Manya Sklodowska.

From 1896 the Curies worked together on radioactivity, building on the results of Wilhelm Röntgen (who had discovered X rays) and Henri Becquerel (who had discovered that similar rays are emitted by uranium salts). Madame Curie discovered that thorium also emits radiation and found that the mineral pitchblende was even more radioactive than could be accounted for by any uranium and thorium content. The Curies then carried out an exhaustive search and in July 1898 announced the discovery of polonium, followed in December of that year with the discovery of radium. They eventually prepared 1 g/0.04 oz of pure radium

chloride from 8 metric tons of waste pitchblende from Austria. They also established that beta rays (now known to consist of electrons) are negatively charged particles.

In 1906 Pierre Curie was run down and killed by a horse-drawn wagon. Marie took over his post at the Sorbonne, becoming the first woman to teach there, and concentrated all her energies into research and caring for her daughters (one of whom, Irène, was to later marry Frédéric Joliot and become a famous scientist and Nobel prizewinner). In 1910 with André Debierne (1874–1949), who in 1899 had discovered actinium in pitchblende, she isolated pure radium metal.

At the outbreak of World War I in 1914 Madame Curie helped to equip ambulances with X-ray equipment, which she drove to the front lines. The International Red Cross made her head of its Radiological Service. Assisted by Iréne Curie and Martha Klein at the Radium Institute, she held courses for medical orderlies and doctors, teaching them how to use the new technique. By the late 1920s her health began to deteriorate: continued exposure to high-energy radiation had given her leukemia. She entered a sanatorium at Haute Savoie and died there on July 4, 1934, a few months after her daughter and son-in-law, the Joliot-Curies, had announced the discovery of artificial radioactivity.

Throughout much of her life Marie Curie was poor, and the painstaking radium extractions were carried out in primitive conditions. The Curies refused to patent any of their discoveries, wanting them freely to benefit everyone. The Nobel prize money and other financial rewards were used to finance further research. One of the outstanding applications of their work has been the use of radiation to treat cancer, one form of which cost Marie Curie her life.

Source: From the Hutchinson *Dictionary of Scientific Biography*. © Research Machines plc [2003] All Rights Reserved. Helicon Publishing is a division of Research Machines.

Shirley Ann Jackson (1946–)

Shirley Ann Jackson was born in Washington, D.C., in 1946. She received her B.S. from Massachusetts Institute of Technology in 1968 and her Ph.D. (Physics) in 1973. Shirley Jackson became the first African American female to receive a doctorate in Theoretical Solid State physics from MIT. Dr. Jackson became a Research Associate in Theoretical Physics at the Fermi National Accelerator Laboratory from 1973–1974 and served as a Visiting Science Associate at the European Organization for Nuclear Research (1974–1975). In 1975–1976 she returned to Fermi National Accelerator Laboratory as a Research Associate in Theoretical Physics. She spent 1976–1977 at the Stanford Linear Accelerator Center and Aspen Center for Physics. Dr. Jackson then served on the Technical Staff of Bell Telephone Laboratories in the-

oretical physics from 1976 until 1978. In 1978 she began working with the Technical Staff of the Scattering and Low Energy Physics Research Laboratory of Bell Telephone Laboratories. From 1976 to 1991 Dr. Jackson was appointed as Professor of Physics at Rutgers University in Piscataway, N.J. From 1991 to 1995 she served, concurrently with her professorship at Rutgers, as a consultant in semiconductor theory to AT&T Bell Laboratories in Murray Hill, N.J. Dr. Jackson was appointed as Commissioner of the Nuclear Regulatory Commission and assumed the Chairmanship on May 2, 1995.

Her research has focused on Landau theories of charge density waves in one- and two-dimensions. Dr. Jackson's research also touched on two-dimensional Yang-Mills gauge theories and neutrino reactions.

"I am interested in the electronic, optical, magnetic, and transport properties of novel semiconductor systems. Of special interest is the behavior of magnetic polarons in semimagnetic and dilute magnetic semiconductors, and the optical response properties of semiconductor quantum—wells and superlattices. My interests also include quantum dots, mesoscopic systems, and the role of antiferromagnetic fluctuations in correlated 2D electron systems."
—Professor Shirley Jackson

The Honorable Dr. Shirley Ann Jackson, chairman of the U.S. Nuclear Regulatory Commission, was also named the 18th president of Rensselaer Polytechnic Institute on July 1, 1999.

Source: Mitchell C. Brown, *The Faces of Science: African Americans in the Sciences.*

Summary

Radioactivity is the spontaneous emission of particles or energy from an unstable atomic nucleus. The modern atomic theory pictures the nucleus as protons and neutrons held together by a short-range *nuclear force* that has moving *nucleons* (protons and neutrons) in *energy shells* analogous to the shell structure of electrons. A graph of the number of neutrons to the number of protons in a nucleus reveals that stable nuclei have a certain neutron-to-proton ratio in a *band of stability.* Nuclei that are above or below the band of stability, and nuclei that are beyond atomic number 83, are radioactive and undergo *radioactive decay.*

Three common examples of radioactive decay involve the emission of an *alpha particle,* a *beta particle,* and a *gamma ray.* An alpha particle is a helium nucleus, consisting of two protons and two neutrons. A beta particle is a high-speed electron that is ejected from the nucleus. A gamma ray is a short-wavelength electromagnetic radiation from an excited nucleus. In general, nuclei with an atomic number of 83 or larger become more stable by alpha emission. Nuclei with a neutron-to-proton ratio that is too large become more stable by beta emission. Gamma ray emission occurs from a nucleus that was left in a high-energy state by the emission of an alpha or beta particle.

Each radioactive isotope has its own specific rate of nuclear disintegration. The rate is usually described in terms of *half-life,* the time required for one-half the unstable nuclei to decay.

Radiation is measured by (1) its effects on photographic film, (2) the number of ions it produces, or (3) the flashes of light produced

on a phosphor. It is measured at a source in units of a *curie,* defined as 3.70×10^{10} nuclear disintegrations per second. It is measured where received in units of a *rad,* defined as 1×10^{-5} J. A *rem* is a measure of radiation that takes into account the biological effectiveness of different types of radiation damage. In general, the natural environment exposes everyone to 100 to 500 millirems per year, an exposure called *background radiation.* Lifestyle and location influence the background radiation received, but the average is 130 millirems per year.

Energy and mass are related by Einstein's famous equation of $E = mc^2$, which means that *matter can be converted to energy and energy to matter.* The mass of a nucleus is always less than the sum of the masses of the individual particles of which it is made. This *mass defect* of a nucleus is equivalent to the energy released when the nucleus was formed according to $E = mc^2$. It is also the *binding energy,* the energy required to break the nucleus apart into nucleons.

When the binding energy is plotted against the mass number, the greatest binding energy per nucleon is seen to occur for an atomic number near that of iron. More massive nuclei therefore release energy by fission, or splitting to more stable nuclei. Less massive nuclei release energy by fusion, the joining of less massive nuclei to produce a more stable, more massive nucleus. Nuclear fission provides the energy for atomic explosions and nuclear power plants. Nuclear fusion is the energy source of the sun and other stars and also holds promise as a future energy source for humans. The source of the energy of a nucleus can be traced back to the gravitational attraction that formed a star.

Summary of Equations

energy = mass × the speed of light squared

$$E = mc^2$$

KEY TERMS

alpha particle (p. **327**)

background radiation (p. **335**)

band of stability (p. **329**)

beta particle (p. **327**)

binding energy (p. **336**)

chain reaction (p. **340**)

control rods (p. **341**)

critical mass (p. **340**)

curie (p. **335**)

electromagnetic force (p. **329**)

fuel rod (p. **341**)

gamma ray (p. **327**)

Geiger counter (p. **333**)

half-life (p. **333**)

ionization counter (p. **333**)

mass defect (p. **336**)

moderator (p. **341**)

nuclear fission (p. **337**)

nuclear force (p. **329**)

nuclear fusion (p. **337**)

nuclear reactor (p. **340**)

nucleons (p. **327**)

plasma (p. **346**)

primary loop (p. **341**)

rad (p. **335**)

radioactive decay (p. **327**)

radioactive decay series (p. **332**)

radioactivity (p. **326**)

rem (p. **335**)

scintillation counter (p. **333**)

secondary loop (p. **341**)

shell model of the nucleus (p. **329**)

steam generator (p. **341**)

APPLYING THE CONCEPTS

1. A high-speed electron ejected from a nucleus during radioactive decay is called a (an)
   a. alpha particle.
   b. beta particle.
   c. gamma ray.
   d. None of the above is correct.

2. The ejection of an alpha particle from a nucleus results in
   a. an increase in the atomic number by one.
   b. an increase in the atomic mass by four.
   c. a decrease in the atomic number by two.
   d. none of the above.

3. The emission of a gamma ray from a nucleus results in
   a. an increase in the atomic number by one.
   b. an increase in the atomic mass by four.
   c. a decrease in the atomic number by two.
   d. none of the above.

4. An atom of radon-222 loses an alpha particle to become a more stable atom of
   a. radium.
   b. bismuth.
   c. polonium.
   d. radon.

5. The nuclear force is
   a. attractive when nucleons are closer than 10^{-15} m.
   b. repulsive when nucleons are closer than 10^{-15} m.
   c. attractive when nucleons are farther than 10^{-15} m.
   d. repulsive when nucleons are farther than 10^{-15} m.

6. Which of the following is most likely to be radioactive?
   a. nuclei with an even number of protons and neutrons
   b. nuclei with an odd number of protons and neutrons
   c. nuclei with the same number of protons and neutrons
   d. Number of protons and neutrons have nothing to do with radioactivity.

7. Which of the following isotopes is most likely to be radioactive?
   a. magnesium-24
   b. calcium-40
   c. astatine-210
   d. ruthenium-101

8. Hydrogen-3 is a radioactive isotope of hydrogen. Which type of radiation would you expect an atom of this isotope to emit?
   a. an alpha particle
   b. a beta particle
   c. either of the above
   d. neither of the above

9. A sheet of paper will stop a (an)
   a. alpha particle.
   b. beta particle.
   c. gamma ray.
   d. none of the above.

10. The most penetrating of the three common types of nuclear radiation is the
    a. alpha particle.
    b. beta particle.
    c. gamma ray.
    d. All have equal penetrating ability.

11. An atom of an isotope with an atomic number greater than 83 will probably emit a (an)
    a. alpha particle.
    b. beta particle.
    c. gamma ray.
    d. none of the above.

12. An atom of an isotope with a large neutron-to-proton ratio will probably emit a (an)
    a. alpha particle.
    b. beta particle.
    c. gamma ray.
    d. none of the above.

13. All of the naturally occurring radioactive decay series end when the radioactive elements have decayed to
    a. lead.
    b. bismuth.
    c. uranium.
    d. hydrogen.

14. The rate of radioactive decay can be increased by increasing the
    a. temperature.
    b. pressure.
    c. size of the sample.
    d. None of the above is correct.

15. Isotope A has a half-life of seconds, and isotope B has a half-life of millions of years. Which isotope is more radioactive?
    a. It depends on the sample size.
    b. isotope A
    c. isotope B
    d. Unknown, from the information given.

16. A Geiger counter indirectly measures radiation by measuring
    a. ions produced.
    b. flashes of light.
    c. speaker static.
    d. curies.

17. A measure of radioactivity at the *source* is (the)
    a. curie.
    b. rad.
    c. rem.
    d. any of the above.

18. A measure of radiation received that considers the biological effect resulting from the radiation is (the)
    a. curie.
    b. rad.
    c. rem.
    d. any of the above.

19. The mass of a nucleus is always _____ the sum of the masses of the individual particles of which it is made.
    a. equal to
    b. less than
    c. more than
    d. Unable to say without more information.

20. When protons and neutrons join together to make a nucleus, energy is
    a. released.
    b. absorbed.
    c. neither released nor absorbed.
    d. unpredictably absorbed or released.

21. Used fuel rods from a nuclear reactor contain about
    a. 96% usable uranium and plutonium.
    b. 33% usable uranium and plutonium.
    c. 4% usable uranium and plutonium.
    d. 0% usable uranium and plutonium.

22. The source of energy from the sun is
    a. chemical (burning).
    b. fission.
    c. fusion.
    d. radioactive decay.

23. The energy released by radioactive decay and the energy released by nuclear reactions can be traced back to the energy that isotopes acquired from
    a. fusion.
    b. the sun.
    c. gravitational attraction.
    d. the big bang.

Answers

1. b 2. c 3. d 4. c 5. a 6. b 7. c 8. b 9. a 10. c 11. a 12. b 13. a 14. c 15. b 16. a 17. a 18. c 19. b 20. a 21. a 22. c 23. c

QUESTIONS FOR THOUGHT

1. How is a radioactive material different from a material that is not radioactive?

2. What is radioactive decay? Describe how the radioactive decay rate can be changed if this is possible.

3. Describe three kinds of radiation emitted by radioactive materials. Describe what eventually happens to each kind of radiation after it is emitted.

4. How are positively charged protons able to stay together in a nucleus since like charges repel?

5. What is half-life? Give an example of the half-life of an isotope, describing the amount remaining and the time elapsed after five half-life periods.

6. Would you expect an isotope with a long half-life to be more, the same, or less radioactive than an isotope with a short half-life? Explain.

7. What is (a) a curie? (b) a rad? (c) a rem?

8. What is meant by background radiation? What is the normal radiation dose for the average person from background radiation?

9. Why is there controversy about the effects of long-term, low levels of radiation exposure?

10. What is a mass defect? How is it related to the binding energy of a nucleus? How can both be calculated?

11. Compare and contrast nuclear fission and nuclear fusion.

The exercises in groups A and B cover the same concepts. Solutions to group A exercises are located in appendix D.

Group A

Note: *You will need the table of atomic weights inside the back cover of this text.*

1. Give the number of protons and the number of neutrons in the nucleus of each of the following isotopes:
   (a) cobalt-60
   (b) potassium-40
   (c) neon-24
   (d) lead-208

2. Write the nuclear symbols for each of the nuclei in exercise 1.

3. Predict if the nuclei in exercise 1 are radioactive or stable, giving your reasoning behind each prediction.

4. Write a nuclear equation for the decay of the following nuclei as they give off a beta particle:
   (a) $^{56}_{26}Fe$
   (b) $^{7}_{4}Be$
   (c) $^{64}_{29}Cu$
   (d) $^{24}_{11}Na$
   (e) $^{214}_{82}Pb$
   (f) $^{32}_{15}P$

5. Write a nuclear equation for the decay of the following nuclei as they undergo alpha emission:
   (a) $^{235}_{92}U$
   (b) $^{226}_{88}Ra$
   (c) $^{239}_{94}Pu$
   (d) $^{214}_{83}Bi$
   (e) $^{230}_{90}Th$
   (f) $^{210}_{84}Po$

6. The half-life of iodine-131 is 8 days. How much of a 1.0 oz sample of iodine-131 will remain after 32 days?

Group B

Note: *You will need the table of atomic weights inside the back cover of this text.*

1. Give the number of protons and the number of neutrons in the nucleus of each of the following isotopes:
   (a) aluminum-25
   (b) technetium-95
   (c) tin-120
   (d) mercury-200

2. Write the nuclear symbols for each of the nuclei in exercise 1.

3. Predict if the nuclei in exercise 1 are radioactive or stable, giving your reasoning behind each prediction.

4. Write a nuclear equation for the beta emission decay of each of the following:
   (a) $^{14}_{6}C$
   (b) $^{60}_{27}Co$
   (c) $^{24}_{11}Na$
   (d) $^{241}_{94}Pu$
   (e) $^{131}_{53}I$
   (f) $^{210}_{82}Pb$

5. Write a nuclear equation for each of the following alpha emission decay reactions:
   (a) $^{241}_{95}Am$
   (b) $^{232}_{90}Th$
   (c) $^{223}_{88}Ra$
   (d) $^{234}_{92}U$
   (e) $^{242}_{96}Cm$
   (f) $^{237}_{93}Np$

6. If the half-life of cesium-137 is 30 years, approximately how much time will be required to reduce a 1 kg sample to about 1 g?

Appendix A
Mathematical Review

Working with Equations

Many of the problems of science involve an equation, a short-hand way of describing patterns and relationships that are observed in nature. Equations are also used to identify properties and to define certain concepts, but all uses have well-established meanings, symbols that are used by convention, and allowed mathematical operations. This appendix will assist you in better understanding equations and the reasoning that goes with the manipulation of equations in problem-solving activities.

Background

In addition to a knowledge of rules for carrying out mathematical operations, an understanding of certain quantitative ideas and concepts can be very helpful when working with equations. Among these helpful concepts are (1) the meaning of inverse and reciprocal, (2) the concept of a ratio, and (3) fractions.

The term *inverse* means the opposite, or reverse, of something. For example, addition is the opposite, or inverse, of subtraction, and division is the inverse of multiplication. A *reciprocal* is defined as an inverse multiplication relationship between two numbers. For example, if the symbol n represents any number (except zero), then the reciprocal of n is $1/n$. The reciprocal of a number ($1/n$) multiplied by that number (n) always gives a product of 1. Thus, the number multiplied by 5 to give 1 is $1/5$ ($5 \times 1/5 = 5/5 = 1$). So $1/5$ is the reciprocal of 5, and 5 is the reciprocal of $1/5$. Each number is the *inverse* of the other.

The fraction $1/5$ means 1 divided by 5, and if you carry out the division it gives the decimal 0.2. Calculators that have a $1/x$ key will do the operation automatically. If you enter 5, then press the $1/x$ key, the answer of 0.2 is given. If you press the $1/x$ key again, the answer of 5 is given. Each of these numbers is a reciprocal of the other.

A *ratio* is a comparison between two numbers. If the symbols m and n are used to represent any two numbers, then the ratio of the number m to the number n is the fraction m/n. This expression means to divide m by n. For example, if m is 10 and n is 5, the ratio of 10 to 5 is $10/5$, or $2:1$.

Working with *fractions* is sometimes necessary in problem-solving exercises, and an understanding of these operations is needed to carry out unit calculations. It is helpful in many of these operations to remember that a number (or a unit) divided by itself is equal to 1, for example,

$$\frac{5}{5} = 1 \qquad \frac{\text{inch}}{\text{inch}} = 1 \qquad \frac{5 \text{ inches}}{5 \text{ inches}} = 1$$

When one fraction is divided by another fraction, the operation commonly applied is to "invert the denominator and multiply." For example, $2/5$ divided by $1/2$ is

$$\frac{\frac{2}{5}}{\frac{1}{2}} = \frac{2}{5} \times \frac{2}{1} = \frac{4}{5}$$

What you are really doing when you invert the denominator of the larger fraction and multiply is making the denominator ($1/2$) equal to 1. Both the numerator ($2/5$) and the denominator ($1/2$) are multiplied by $2/1$, which does not change the value of the overall expression. The complete operation is

$$\frac{\frac{2}{5}}{\frac{1}{2}} \times \frac{\frac{2}{1}}{\frac{2}{1}} = \frac{\frac{2}{5} \times \frac{2}{1}}{\frac{1}{2} \times \frac{2}{1}} = \frac{\frac{4}{5}}{\frac{2}{2}} = \frac{\frac{4}{5}}{1} = \frac{4}{5}$$

Symbols and Operations

The use of symbols seems to cause confusion for some students because it seems different from their ordinary experiences with arithmetic. The rules are the same for symbols as they are for numbers, but you cannot do the operations with the symbols until you know what values they represent. The operation signs, such as $+$, $\div$, $\times$, and $-$ are used with symbols to indicate the operation that you *would* do if you knew the values. Some of the mathematical operations are indicated several ways. For example, $a \times b$, $a \cdot b$, and ab all indicate the same thing, that a is to be multiplied by b. Likewise, $a \div b$, a/b, and $a \times 1/b$ all indicate that a is to be divided by b. Since it is not possible to carry out the operations on symbols alone, they are called *indicated operations*.

Operations in Equations

An equation is a shorthand way of expressing a simple sentence with symbols. The equation has three parts: (1) a left side, (2) an equal sign ($=$), which indicates the equivalence of the two sides, and (3) a right side. The left side has the same value and units as the right side, but the two sides may have a very different appearance. The two sides may also have the symbols that indicate mathematical operations ($+$, $-$, $\times$, and so forth) and may be in certain forms that indicate operations (a/b, ab, and so forth). In any case, the equation is a complete expression that states the left side has the same value and units as the right side.

Equations may contain different symbols, each representing some unknown quantity. In science, the term "solve the equation" means to perform certain operations with one symbol (which represents some variable) by itself on one side of the equation. This single symbol is usually, but not necessarily, on the left side and is not present on the other side. For example, the equation $F = ma$ has the symbol F on the left side. In science, you would say that this equation is solved for F. It could also be solved for m or for a, which will be considered shortly. The equation $F = ma$ is solved for F, and the *indicated operation* is to multiply m by a because they are in the form ma, which means the same thing as $m \times a$. This is the only indicated operation in this equation.

A solved equation is a set of instructions that has an order of indicated operations. For example, the equation for the relationship between a Fahrenheit and Celsius temperature, solved for °C, is $C = 5/9(F - 32)$. A list of indicated operations in this equation is as follows:

1. Subtract 32° from the given Fahrenheit temperature.
2. Multiply the result of (1) by 5.
3. Divide the result of (2) by 9.

Why are the operations indicated in this order? Because the bracket means 5/9 of the *quantity* $(F - 32°)$. In its expanded form, you can see that $5/9(F - 32°)$ actually means $5/9(F) - 5/9(32°)$. Thus, you cannot multiply by 5 or divide by 9 until you have found the quantity $(F - 32°)$. Once you have figured out the order of operations, finding the answer to a problem becomes almost routine as you complete the needed operations on both the numbers and the units.

Solving Equations

Sometimes it is necessary to rearrange an equation to move a different symbol to one side by itself. This is known as solving an equation for an unknown quantity. But you cannot simply move a symbol to one side of an equation. Since an equation is a statement of equivalence, the right side has the same value as the left side. If you move a symbol, you must perform the operation in a way that the two sides remain equivalent. This is accomplished by "canceling out" symbols until you have the unknown on one side by itself. One key to understanding the canceling operation is to remember that a fraction with the same number (or unit) over itself is equal to 1. For example, consider the equation $F = ma$, which is solved for F. Suppose you are considering a problem in which F and m are given, and the unknown is a. You need to solve the equation

for a so it is on one side by itself. To eliminate the m, you do the *inverse* of the indicated operation on m, dividing both sides by m. Thus,

$$F = ma$$

$$\frac{F}{m} = \frac{ma}{m}$$

$$\frac{F}{m} = a$$

Since m/m is equal to 1, the a remains by itself on the right side. For convenience, the whole equation may be flipped to move the unknown to the left side,

$$a = \frac{F}{m}$$

Thus, a quantity that indicated a multiplication (ma) was removed from one side by an inverse operation of dividing by m.

Consider the following inverse operations to "cancel" a quantity from one side of an equation, moving it to the other side:

If the Indicated Operation of the Symbol You Wish to Remove Is:	Perform This Inverse Operation on Both Sides of the Equation
multiplication	division
division	multiplication
addition	subtraction
subtraction	addition
squared	square root
square root	square

EXAMPLE A.1

The equation for finding the kinetic energy of a moving body is $KE = 1/2mv^2$. You need to solve this equation for the velocity, v.

Solution

The order of indicated operations in the equation is as follows:

1. Square v.
2. Multiply v^2 by m.
3. Divide the result of (2) by 2.

To solve for v, this order is *reversed* as the "canceling operations" are used:

Step 1: Multiply both sides by 2

$$KE = \frac{1}{2}mv^2$$

$$2KE = \frac{2}{2}mv^2$$

$$2KE = mv^2$$

Step 2: Divide both sides by m

$$\frac{2KE}{m} = \frac{mv^2}{m}$$

$$\frac{2KE}{m} = v^2$$

Step 3: Take the square root of both sides

$$\sqrt{\frac{2KE}{m}} = \sqrt{v^2}$$

$$\sqrt{\frac{2KE}{m}} = v$$

or

$$v = \sqrt{\frac{2KE}{m}}$$

The equation has been solved for v, and you are now ready to substitute quantities and perform the needed operations (see example 1.3 in chapter 1 for information on this topic).

Significant Figures

The numerical value of any measurement will always contain some uncertainty. Suppose, for example, that you are measuring one side of a square piece of paper as shown in Figure A.1. You

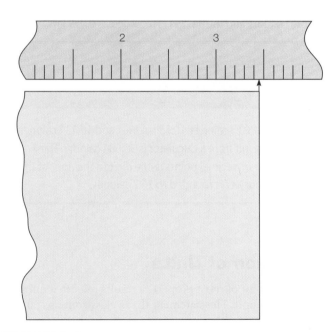

FIGURE A.1

How wide is this sheet of paper? Write your answer before reading the text _____.

could say that the paper is *about* 3.5 cm wide and you would be correct. This measurement, however, would be unsatisfactory for many purposes. It does not approach the true value of the length and contains too much uncertainty. It seems clear that the paper width is larger than 3.4 cm but shorter than 3.5 cm. But how much larger than 3.4 cm? You cannot be certain if the paper is 3.44, 3.45, or 3.46 cm wide. As your best estimate, you might say that the paper is 3.45 cm wide. Everyone would agree that you can be certain about the first two numbers (3.4) and they should be recorded. The last number (0.05) has been estimated and is not certain. The two certain numbers, together with one uncertain number, represent the greatest accuracy possible with the ruler being used. The paper is said to be 3.45 cm wide.

A *significant figure* is a number that is believed to be correct with some uncertainty only in the last digit. The value of the width of the paper, 3.45 cm, represents three significant figures. As you can see, the number of significant figures can be determined by the degree of accuracy of the measuring instrument being used. But suppose you need to calculate the area of the paper. You would multiply 3.45 cm $\times$ 3.45 cm and the product for the area would be 11.9025 cm^2. This is a greater precision than you were able to obtain with your measuring instrument. The result of a calculation can be no more accurate than the values being treated. Because the measurement had only three significant figures (two certain, one uncertain), then the answer can have only three significant figures. The area is correctly expressed as 11.9 cm^2.

There are a few simple rules that will help you determine how many significant figures are contained in a reported measurement:

1. All digits reported as a direct result of a measurement are significant.
2. Zero is significant when it occurs between nonzero digits. For example, 607 has three significant figures, and the zero is one of the significant figures.
3. In figures reported as *larger than the digit 1*, the digit zero is not significant when it follows a nonzero digit to indicate place. For example, in a report that "23,000 people attended the rock concert," the digits 2 and 3 are significant but the zeros are not significant. In this situation the 23 is the measured part of the figure, and the three zeros tell you an estimate of how many attended the concert, that is, 23 thousand. If the figure is a measurement rather than an estimate, then it is written *with a decimal point after the last zero* to indicate that the zeros *are* significant. Thus 23,000 has *two* significant figures (2 and 3), but 23,000. has *five* significant figures. The figure 23,000 means "about 23 thousand," but 23,000. means 23,000. and not 22,999 or 23,001.
4. In figures reported as *smaller than the digit 1*, zeros after a decimal point that come before nonzero digits *are not* significant and serve only as place holders. For example, 0.0023 has two significant figures: 2 and 3. Zeros alone after a decimal point or zeros after a nonzero digit indicate a measurement, however, so these zeros *are* significant. The figure 0.00230, for example, has three significant figures since the 230 means 230 and not 229 or 231. Likewise, the figure 3.000 cm has four significant figures because the presence of the three zeros means that the measurement was actually 3.000 and not 2.999 or 3.001.

Multiplication and Division

When you multiply or divide measurement figures, the answer may have no more significant figures than the *least* number of significant figures in the figures being multiplied or divided. This simply means that an answer can be no more accurate than the least accurate measurement entering into the calculation, and that you cannot improve the accuracy of a measurement by doing a calculation. For example, in multiplying 54.2 mi/h × 4.0 h to find out the total distance traveled, the first figure (54.2) has three significant figures but the second (4.0) has only two significant figures. The answer can contain only two significant figures since this is the weakest number of those involved in the calculation. The correct answer is therefore 220 mi, not 216.8 mi. This may seem strange since multiplying the two numbers together gives the answer of 216.8 mi. This answer, however, means a greater accuracy than is possible, and the accuracy cannot be improved over the weakest number involved in the calculation. Since the weakest number (4.0) has only two significant figures the answer must also have only two significant figures, which is 220 mi.

The result of a calculation is *rounded* to have the same least number of significant figures as the least number of a measurement involved in the calculation. When rounding numbers the last significant figure is increased by 1 if the number after it is 5 or larger. If the number after the last significant figure is 4 or less, the nonsignificant figures are simply dropped. Thus, if two significant figures are called for in the answer of the previous example, 216.8 is rounded up to 220 because the last number after the two significant figures is 6 (a number larger than 5). If the calculation result had been 214.8, the rounded number would be 210 miles.

Note that *measurement figures* are the only figures involved in the number of significant figures in the answer. Numbers that are counted or **defined** are not included in the determination of significant figures in an answer. For example, when dividing by 2 to find an average, the 2 is ignored when considering the number of significant figures. Defined numbers are defined exactly and are not used in significant figures. Since 1 kilogram is *defined* to be exactly 1,000 grams, such a conversion is not a measurement.

Addition and Subtraction

Addition and subtraction operations involving measurements, as with multiplication and division, cannot result in an answer that implies greater accuracy than the measurements had before the calculation. Recall that the last digit to the right in a measurement is uncertain, that is, it is the result of an estimate. The answer to an addition or subtraction calculation can have this uncertain number *no farther from the decimal place than it was in the weakest number involved in the calculation.* Thus, when 8.4 is added to 4.926, the weakest number is 8.4, and the uncertain number is .4, one place to the right of the decimal. The sum of 13.326 is therefore rounded to 13.3, reflecting the placement of this weakest doubtful figure.

The rules for counting zeros tell us that the numbers 203 and 0.200 both have three significant figures. Likewise, the numbers 230 and 0.23 only have two significant figures. Once you remember the rules, the counting of significant figures is straightforward. On the other hand, sometimes you find a number that seems to make it impossible to follow the rules. For example, how would you write 3,000 with two significant figures? There are several special systems in use for taking care of problems such as this, including the placing of a little bar over the last significant digit. One of the convenient ways of showing significant figures for difficult numbers is to use scientific notation, which is also discussed elsewhere in this appendix. The convention for writing significant figures is to display one digit to the left of the decimal. The exponents are not considered when showing the number of significant figures in scientific notation. Thus if you want to write three thousand showing one significant figure, you would write 3×10^3. To show two significant figures it is 3.0×10^3 and for three significant figures it becomes 3.00×10^3. As you can see, the correct use of scientific notation leaves little room for doubt about how many significant figures are intended.

EXAMPLE **A.2**

In a problem it is necessary to multiply 0.0039 km by 15.0 km. The result from a calculator is 0.0585 km². The least number of significant figures involved in this calculation is *two* (0.0039 is two significant figures; 15.0 is three—read the rules again to see why). The calculator result is therefore rounded off to have only two significant figures, and the answer is recorded as 0.059 km².

EXAMPLE **A.3**

The quantities of 10.3 calories, 10.15 calories, and 16.234 calories are added. The result from a calculator is 36.684 calories. The smallest number of decimal points is *one* digit to the right of the decimal, so the answer is rounded to 16.7 calories.

Conversion of Units

The measurement of most properties results in both a numerical value and a unit. The statement that a glass contains 50 cm³ of a liquid conveys two important concepts—the numerical value of 50 and the referent unit of cubic centimeters. Both the numerical value and the unit are necessary to communicate correctly the volume of the liquid.

When working with calculations involving measurement units, *both* the numerical value and the units are treated mathematically. As in other mathematical operations, there are general rules to follow.

1. Only properties with like units may be added or subtracted. It should be obvious that adding quantities such as 5 dollars and 10 dimes is meaningless. You must first convert to like units before adding or subtracting.

2. Like or unlike units may be multiplied or divided and treated in the same manner as numbers. You have used this rule when dealing with area (length $\times$ length = length^2, for example, cm $\times$ cm = cm^2) and when dealing with volume (length $\times$ length $\times$ length = length^3, for example, cm $\times$ cm $\times$ cm = cm^3).

You can use these two rules to create a *conversion ratio* that will help you change one unit to another. Suppose you need to convert 2.3 kg to grams. First, write the relationship between kilograms and grams:

$$1{,}000 \text{ g} = 1 \text{ kg}$$

Next, divide both sides by what you wish to convert *from* (kilograms in this example):

$$\frac{1{,}000 \text{ g}}{1} = \frac{1 \text{ kg}}{1 \text{ kg}}$$

One kilogram divided by 1 kg equals 1, just as 10 divided by 10 equals 1. Therefore, the right side of the relationship becomes 1 and the equation is:

$$\frac{1{,}000 \text{ g}}{1 \text{ kg}} = 1$$

The 1 is usually understood, that is, not stated, and the operation is called *canceling*. Canceling leaves you with the fraction 1,000 g/1 kg, which is a conversion ratio that can be used to convert from kilograms to grams. You simply multiply the conversion ratio by the numerical value and unit you wish to convert:

$$= 2.3 \text{ kg} \times \frac{1{,}000 \text{ g}}{1 \text{ kg}}$$

$$= \frac{2.3 \times 1{,}000}{1} \frac{\text{kg} \times \text{g}}{\text{kg}}$$

$$= \boxed{2{,}300 \text{ g}}$$

The kilogram units cancel. Showing the whole operation with units only, you can see how you end up with the correct unit of grams:

$$\text{kg} \times \frac{\text{g}}{\text{kg}} = \frac{\text{kg} \cdot \text{g}}{\text{kg}} = \text{g}$$

Since you did obtain the correct unit, you know that you used the correct conversion ratio. If you had blundered and used an inverted conversion ratio, you would obtain

$$2.3 \text{ kg} \times \frac{1 \text{ kg}}{1{,}000 \text{ g}} = .0023 \frac{\text{kg}^2}{\text{g}}$$

which yields the meaningless, incorrect units of kg^2/g. Carrying out the mathematical operations on the numbers and the units will always tell you whether or not you used the correct conversion ratio.

EXAMPLE A.4

A distance is reported as 100.0 km, and you want to know how far this is in miles.

Solution

First, you need to obtain a *conversion factor* from a textbook or reference book, which usually lists the conversion factors by properties in a table. Such a table will show two conversion factors for kilometers and miles: (1) 1 km = 0.621 mi and (2) 1 mi = 1.609 km. You select the factor that is in the same form as your problem; for example, your problem is 100.0 km = ? mi. The conversion factor in this form is 1 km = 0.621 mi.

Second, you convert this conversion factor into a *conversion ratio* by dividing the factor by what you wish to convert *from:*

conversion factor:	1 km = 0.621
divide factor by what you want to convert from:	$\dfrac{1 \text{ km}}{1 \text{ km}} = \dfrac{0.621 \text{ mi}}{1 \text{ km}}$
resulting conversion rate:	$\dfrac{0.621 \text{ mi}}{\text{km}}$

Note that if you had used the 1 mi = 1.609 km factor, the resulting units would be meaningless. The conversion ratio is now multiplied by the numerical value *and unit* you wish to convert:

$$100.0 \text{ km} \times \frac{0.621 \text{ mi}}{\text{km}}$$

$$(100.0)(0.621) \frac{\text{km} \cdot \text{mi}}{\text{km}}$$

$$62.1 \text{ mi}$$

EXAMPLE A.5

A service station sells gasoline by the liter, and you fill your tank with 72 liters. How many gallons is this? (Answer: 19 gal)

Scientific Notation

Most of the properties of things that you might measure in your everyday world can be expressed with a small range of numerical values together with some standard unit of measure. The range of numerical values for most everyday things can be dealt with by using units (1s), tens (10s), hundreds (100s), or perhaps thousands (1,000s). But the actual universe contains some objects of incredibly large size that require some very big numbers to describe. The sun, for example, has a mass of about

1,970,000,000,000,000,000,000,000,000,000 kg. On the other hand, very small numbers are needed to measure the size and parts of an atom. The radius of a hydrogen atom, for example, is about 0.00000000005 m. Such extremely large and small numbers are cumbersome and awkward since there are so many zeros to keep track of, even if you are successful in carefully counting all the zeros. A method does exist to deal with extremely large or small numbers in a more condensed form. The method is called *scientific notation,* but it is also sometimes called *powers of ten* or *exponential notation,* since it is based on exponents of 10. Whatever it is called, the method is a compact way of dealing with numbers that not only helps you keep track of zeros but provides a simplified way to make calculations as well.

In algebra you save a lot of time (as well as paper) by writing $(a \times a \times a \times a \times a)$ as a^5. The small number written to the right and above a letter or number is a superscript called an *exponent.* The exponent means that the letter or number is to be multiplied by itself that many times, for example, a^5 means a multiplied by itself five times, or $a \times a \times a \times a \times a$. As you can see, it is much easier to write the exponential form of this operation than it is to write it out in the long form. Scientific notation uses an exponent to indicate the power of the base 10. The exponent tells how many times the base, 10, is multiplied by itself. For example,

$$10,000 = 10^4$$
$$1,000 = 10^3$$
$$100 = 10^2$$
$$10 = 10^1$$
$$1 = 10^0$$
$$0.1 = 10^{-1}$$
$$0.01 = 10^{-2}$$
$$0.001 = 10^{-3}$$
$$0.0001 = 10^{-4}$$

This table could be extended indefinitely, but this somewhat shorter version will give you an idea of how the method works. The symbol 10^4 is read as "ten to the fourth power" and means $10 \times 10 \times 10 \times 10$. Ten times itself four times is 10,000, so 10^4 is the scientific notation for 10,000. It is also equal to the number of zeros between the 1 and the decimal point, that is, to write the longer form of 10^4 you simply write 1, then move the decimal point four places to the *right;* 10 to the fourth power is 10,000.

The power of ten table also shows that numbers smaller than 1 have negative exponents. A negative exponent means a reciprocal:

$$10^{-1} = \frac{1}{10} = 0.1$$

$$10^{-2} = \frac{1}{100} = 0.01$$

$$10^3 = \frac{1}{1,000} = 0.001$$

To write the longer form of 10^{-4}, you simply write 1 then move the decimal point four places to the *left;* 10 to the negative fourth power is 0.0001.

Scientific notation usually, but not always, is expressed as the product of two numbers: (1) a number between 1 and 10 that is called the *coefficient* and (2) a power of ten that is called the *exponent.* For example, the mass of the sun that was given in long form earlier is expressed in scientific notation as

$$1.97 \times 10^{30} \text{ kg}$$

and the radius of a hydrogen atom is

$$5.0 \times 10^{-11} \text{ m}$$

In these expressions, the coefficients are 1.97 and 5.0, and the power of ten notations are the exponents. Note that in both of these examples, the exponent tells you where to place the decimal point if you wish to write the number all the way out in the long form. Sometimes scientific notation is written without a coefficient, showing only the exponent. In these cases the coefficient of 1.0 is understood, that is, not stated. If you try to enter a scientific notation in your calculator, however, you will need to enter the understood 1.0, or the calculator will not be able to function correctly. Note also that 1.97×10^{30} kg and the expressions 0.197×10^{31} kg and 19.7×10^{29} kg are all correct expressions of the mass of the sun. By convention, however, you will use the form that has one digit to the left of the decimal.

EXAMPLE **A.6**

What is 26,000,000 in scientific notation?

Solution

Count how many times you must shift the decimal point until one digit remains to the left of the decimal point. For numbers larger than the digit 1, the number of shifts tells you how much the exponent is increased, so the answer is

$$26 \times 10^7$$

which means the coefficient 2.6 is multiplied by 10 seven times.

EXAMPLE **A.7**

What is 0.000732 in scientific notation? (Answer: 7.32×10^{-4})

It was stated earlier that scientific notation provides a compact way of dealing with very large or very small numbers, but it provides a simplified way to make calculations as well. There are a few mathematical rules that will describe how the use of scientific notation simplifies these calculations.

To *multiply* two scientific notation numbers, the coefficients are multiplied as usual, and the exponents are *added* algebraically. For example, to multiply (2×10^2) by (3×10^3), first separate the coefficients from the exponents,

$$(2 \times 3) \times (10^2 \times 10^3),$$

then multiply the coefficients and add the exponents,

$$6 \times 10^{(2 + 3)} = 6 \times 10^5$$

Adding the exponents is possible because $10^2 \times 10^3$ means the same thing as $(10 \times 10) \times (10 \times 10 \times 10)$, which equals $(100) \times (1,000)$, or 100,000, which is expressed as 10^{-5} in scientific notation. Note that two negative exponents add algebraically, for example $10^{-2} \times 10^{-3} = 10^{[(-2) + (-3)]} = 10^{-5}$. A negative and a positive exponent also add algebraically, as in $10^5 \times 10^{-3} = 10^{[(+5) + (-3)]} = 10^2$.

If the result of a calculation involving two scientific notation numbers does not have the conventional one digit to the left of the decimal, move the decimal point so it does, changing the exponent according to which way and how much the decimal point is moved. Note that the exponent increases by one number for each decimal point moved to the left. Likewise, the exponent decreases by one number for each decimal point moved to the right. For example, $938. \times 10^3$ becomes 9.38×10^5 when the decimal point is moved two places to the left.

To *divide* two scientific notation numbers, the coefficients are divided as usual and the exponents are *subtracted*. For example, to divide (6×10^6) by (3×10^2), first separate the coefficients from the exponents,

$$(6 \div 3) \times (10^6 \div 10^2)$$

then divide the coefficients and subtract the exponents,

$$2 \times 10^{(6 - 2)} = 2 \times 10^4$$

Note that when you subtract a negative exponent, for example, $10^{[(3) - (-2)]}$, you change the sign and add, $10^{(3 + 2)} = 10^5$.

EXAMPLE **A.8**

Solve the following problem concerning scientific notation:

$$\frac{(2 \times 10^4) \times (8 \times 10^{-6})}{8 \times 10^4}$$

Solution

First, separate the coefficients from the exponents,

$$\frac{2 \times 8}{8} \times \frac{10^4 \times 10^{-6}}{10^4}$$

then multiply and divide the coefficients and add and subtract the exponents as the problem requires,

$$2 \times 10^{\{[(4) + (-6)] - [4]\}}$$

Solving the remaining additions and subtractions of the coefficients gives

$$2 \times 10^{-6}$$

The Simple Line Graph

An equation describes a relationship between variables, and a graph helps you picture this relationship. A line graph pictures how changes in one variable correspond with changes in a second variable, that is, how the two variables change together. Usually one variable can be easily manipulated. The other variable is caused to change in value by the manipulation of the first variable. The **manipulated variable** is known by various names (*independent, input,* or *cause variable*), and the **responding variable** is known by various related names (*dependent, output,* or *effect variable*). The manipulated variable is usually placed on the horizontal axis, or *x-*axis, of the graph, so you could also identify it as the *x-variable.* The responding variable is placed on the vertical axis, or *y-*axis. This variable is identified as the *y-variable.*

Figure A.2 shows the mass of different volumes of water at room temperature. Volume is placed on the *x-*axis because the volume of water is easily manipulated, and the mass values change as a consequence of changing the values of volume. Note that both variables are named and that the measuring unit for each is identified on the graph.

Figure A.2 also shows a number *scale* on each axis that represents changes in the values of each variable. The scales are usually, but not always, linear. A **linear scale** has equal intervals that represent equal increases in the value of the variable. Thus a certain distance on the *x-*axis to the right represents a certain

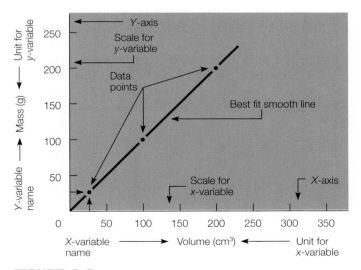

FIGURE A.2

The parts of a graph. On this graph, volume is placed on the *x-*axis and mass on the *y-*axis.

increase in the value of the x-variable. Likewise, certain distances up the y-axis represent certain increases in the value of the y-variable. The **origin** is the only point where both the x- and y-variables have a value of zero at the same time.

Figure A.2 shows three **data points.** A data point represents measurements of two related variables that were made at the same time. For example, a volume of 25 cm³ of water was found to have a mass of 25 g. Locate 25 cm³ on the x-axis and imagine a line moving straight up from this point on the scale. Locate 25 g on the y-axis and imagine a line moving straight out from this point on the scale. Where the lines meet is the data point for the 25 cm³ and 25 g measurements. A data point is usually indicated with a small dot or x (dots are used in the graph in Figure A.2).

A "best fit" smooth line is drawn through all the data points as close to them as possible. If it is not possible to draw the straight line *through* all the data points, then a straight line is drawn that has the same number of data points on both sides of the line. Such a line will represent a "best approximation" of the relationship between the two variables. The origin is also used as a data point in this example because a volume of zero will have a mass of zero.

The smooth line tells you how the two variables get larger together. With the same x- and y-axis scale, a 45° line means that they are increasing in an exact direct proportion. A more flat or more upright line means that one variable is increasing faster than the other. The more you work with graphs, the easier it will become for you to analyze what the "picture" means. There are more exact ways to extract information from a graph, and one of these techniques is discussed next.

One way to determine the relationship between two variables that are graphed with a straight line is to calculate the **slope.** The slope is a *ratio* between the changes in one variable and the changes in the other. The ratio is between the change in the value of the x-variable and the change in the value of the y-variable. Recall that the symbol Δ (Greek letter delta) means "change in," so the symbol Δx means the "change in x." The first step in calculating the slope is to find out how much the x-variable is changing (Δx) in relation to how much the y-variable is changing (Δy). You can find this relationship by first drawing a dashed line to the right of the straight *line* (not the data points), so that the x-variable has increased by some convenient unit (Figure A.3). Where you start or end this dashed line will not matter since the ratio between the variables will be the same everywhere on the graph line. The Δx is determined by subtracting the initial value of the x-variable on the dashed line (x_i) from the final value of the x-variable on the dashed line x_f, or $\Delta x = x_f - x_i$. In Figure A.3, the dashed line has an x_f of 200 cm³ and an x_i of 100 cm³, so Δx is 200 cm³ − 100 cm³, or 100 cm³. Note that Δx has both a number value and a unit.

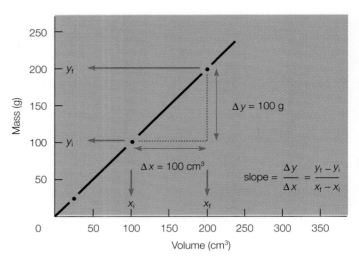

FIGURE A.3

The slope is a ratio between the changes in the y-variable and the changes in the x-variable, or $\Delta y/\Delta x$.

Now you need to find Δy. The example in Figure A.3 shows a dashed line drawn back up to the graph line from the x-variable dashed line. The value of Δy is $y_f - y_i$. In the example, $\Delta y = 200$ g − 100 g, or 100 g.

The slope of a straight graph line is the ratio of Δy to Δx, or

$$\text{slope} = \frac{\Delta y}{\Delta x}$$

In the example,

$$\text{slope} = \frac{100 \text{ g}}{100 \text{ cm}^3}$$

$$= 1 \frac{\text{g}}{\text{cm}^3} \text{ or } 1 \text{ g/cm}^3$$

Thus the slope is 1 g/cm³, and this tells you how the variables change together. Since g/cm³ is also the unit of density, you know that you have just calculated the density of water from a graph.

Note that the slope can be calculated only for two variables that are increasing together, that is, for variables that are in direct proportion and have a line that moves upward and to the right. If variables change in any other way, mathematical operations must be performed to change the variables *into* this relationship. Examples of such necessary changes include taking the inverse of one variable, squaring one variable, taking the inverse square, and so forth.

Appendix B
Solubilities Chart

	Acetate	Bromide	Carbonate	Chloride	Fluoride	Hydroxide	Iodide	Nitrate	Oxide	Phosphate	Sulfate	Sulfide
Aluminum	S	S	—	S	s	i	S	S	i	i	S	d
Ammonium	S	S	S	S	S	S	S	S	—	S	S	S
Barium	S	S	i	S	s	S	S	S	S	i	i	d
Calcium	S	S	i	S	i	s	S	S	s	i	s	d
Copper (I)	—	s	i	s	i	—	i	—	i	—	d	i
Copper (II)	S	S	i	S	S	i	S	S	i	i	S	i
Iron (II)	S	S	i	S	s	i	S	S	i	i	S	i
Iron (III)	S	S	i	S	s	i	S	S	i	i	S	d
Lead	S	s	i	s	i	i	s	S	i	i	i	i
Magnesium	S	S	i	S	i	i	S	S	i	i	S	d
Mercury (I)	s	i	i	i	d	d	i	S	i	i	i	i
Mercury (II)	S	s	i	S	d	i	i	S	i	i	i	i
Potassium	S	S	S	S	S	S	S	S	S	S	S	i
Silver	s	i	i	i	S	—	i	S	i	i	i	i
Sodium	S	S	S	S	S	S	S	S	d	S	S	S
Strontium	S	S	s	S	i	s	S	S	—	i	i	i
Zinc	S	S	i	S	S	i	S	S	i	i	S	i

S–soluble
i–insoluble
s–slightly soluble
d–decomposes

Appendix C
Relative Humidity Table

Dry-Bulb Temperature (°C)	Difference between Wet-Bulb and Dry-Bulb Temperatures (°C)																			
	1	2	3	4	5	6	7	8	9	10	11	12	13	14	15	16	17	18	19	20
0	81	64	46	29	13															
1	83	66	49	33	17															
2	84	68	52	37	22	7														
3	84	70	55	40	26	12														
4	86	71	57	43	29	16														
5	86	72	58	45	33	20	7													
6	86	73	60	48	35	24	11													
7	87	74	62	50	38	26	15													
8	87	75	63	51	40	29	19	8												
9	88	76	64	53	42	32	22	12												
10	88	77	66	55	44	34	24	15	6											
11	89	78	67	56	46	36	27	18	9											
12	89	78	68	58	48	39	29	21	12											
13	89	79	69	59	50	41	32	23	15	7										
14	90	79	70	60	51	42	34	26	18	10										
15	90	80	71	61	53	44	36	27	20	13	6									
16	90	81	71	63	54	46	38	30	23	15	8									
17	90	81	72	64	55	47	40	32	25	18	11									
18	91	82	73	65	57	49	41	34	27	20	14	7								
19	91	82	74	65	58	50	43	36	29	22	16	10								
20	91	83	74	66	59	51	44	37	31	24	18	12	6							
21	91	83	75	67	60	53	46	39	32	26	20	14	9							
22	92	83	76	68	61	54	47	40	34	28	22	17	11	6						
23	92	84	76	69	62	55	48	42	36	30	24	19	13	8						
24	92	84	77	69	62	56	49	43	37	31	26	20	15	10	5					
25	92	84	77	70	63	57	50	44	39	33	28	22	17	12	8					
26	92	85	78	71	64	58	51	46	40	34	29	24	19	14	10	5				
27	92	85	78	71	65	58	52	47	41	36	31	26	21	16	12	7				
28	93	85	78	72	65	59	53	48	42	37	32	27	22	18	13	9	5			
29	93	86	79	72	66	60	54	49	43	38	33	28	24	19	15	11	7			
30	93	86	79	73	67	61	55	50	44	39	35	30	25	21	17	13	9	5		
31	93	86	80	73	67	61	56	51	45	40	36	31	27	22	18	14	11	7		
32	93	86	80	74	68	62	57	51	46	41	37	32	28	24	20	16	12	9	5	
33	93	87	80	74	68	63	57	52	47	42	38	33	29	25	21	17	14	10	7	
34	93	87	81	75	69	63	58	53	48	43	39	35	30	28	23	19	15	12	8	5
35	94	87	81	75	69	64	59	54	49	44	40	36	32	28	24	20	17	13	10	7

Appendix D
Solutions for Group A Parallel Exercises

NOTE: Solutions that involve calculations of measurements are rounded up or down to conform to the rules for significant figures described in Appendix A.

1.1. Answers will vary but should have the relationship of 100 cm in 1 m, for example, 178 cm = 1.78 m.

1.2. Since mass density is given by the relationship $\rho = m/V$, then

$$\rho = \frac{m}{V} = \frac{272 \text{ g}}{20.0 \text{ cm}^3}$$

$$= \frac{272}{20.0} \frac{\text{g}}{\text{cm}^3}$$

$$= \boxed{13.6 \frac{\text{g}}{\text{cm}^3}}$$

1.3. The volume of a sample of lead is given and the problem asks for the mass. From the relationship of $\rho = m/V$, solving for the mass (m) tells you that the mass density (ρ) times the volume (V), or $m = \rho V$. The mass density of lead, 11.4 g/cm³, can be obtained from Table 1.4, so

$$\rho = \frac{m}{V}$$

$$V\rho = \frac{m\cancel{V}}{\cancel{V}}$$

$$m = \rho V$$

$$m = \left(11.4 \frac{\text{g}}{\text{cm}^3}\right)(10.0 \text{ cm}^3)$$

$$11.4 \times 10.0 \frac{\text{g}}{\text{cm}^3} \times \text{cm}^3$$

$$114 \frac{\text{g} \cdot \cancel{\text{cm}^3}}{\cancel{\text{cm}^3}}$$

$$= \boxed{114 \text{ g}}$$

1.4. Solving the relationship $\rho = m/V$ for volume gives $V = m/\rho$, and

$$\rho = \frac{m}{V}$$

$$V\rho = \frac{m\cancel{V}}{\cancel{V}}$$

$$\frac{V\cancel{\rho}}{\cancel{\rho}} = \frac{m}{\rho}$$

$$V = \frac{m}{\rho}$$

$$V = \frac{600 \text{ g}}{3.00 \dfrac{\text{g}}{\text{cm}^3}}$$

$$= \frac{600}{3.00} \frac{\text{g}}{1} \times \frac{\text{cm}^3}{\text{g}}$$

$$= 200 \frac{\cancel{\text{g}} \cdot \text{cm}^3}{\cancel{\text{g}}}$$

$$= \boxed{200 \text{ cm}^3}$$

1.5. A 50.0 cm³ sample with a mass of 34.0 grams has a density of

$$\rho = \frac{m}{V} = \frac{34.0 \text{ g}}{50.0 \text{ cm}^3}$$

$$= \frac{34.0}{50.0} \frac{\text{g}}{\text{cm}^3}$$

$$= \boxed{0.680 \frac{\text{g}}{\text{cm}^3}}$$

According to Table 1.4, 0.680 g/cm³ is the mass density of gasoline, so the substance must be gasoline.

1.6. The problem asks for a mass and gives a volume, so you need a relationship between mass and volume. Table 1.4 gives the mass density of water as 1.00 g/cm^3, which is a density that is easily remembered. The volume is given in liters (L), which should first be converted to cm^3 because this is the unit in which density is expressed. The relationship of $\rho = m/V$ solved for mass is ρV, so the solution is

$$\rho = \frac{m}{V} \quad \therefore \quad m = \rho V$$

$$m = \left(1.00\,\frac{g}{cm^3}\right)(40,000\,cm^3)$$

$$= 1.00 \times 40,000\,\frac{g}{cm^3} \times cm^3$$

$$= 40,000\,\frac{g \cdot cm^3}{cm^3}$$

$$= 40,000\,g$$

$$= \boxed{40\,kg}$$

1.7. From Table 1.4, the mass density of aluminum is given as 2.70 g/cm^3. Converting 2.1 kg to the same units as the density gives 2,100 g. Solving $\rho = m/V$ for the volume gives

$$V = \frac{m}{\rho} = \frac{2,100\,g}{2.70\,\dfrac{g}{cm^3}}$$

$$= \frac{2,100}{2.70}\,\frac{g}{1} \times \frac{cm^3}{g}$$

$$= 777.78\,\frac{g \cdot cm^3}{g}$$

$$= \boxed{780\,cm^3}$$

1.8. The length of one side of the box is 0.1 m. Reasoning: Since the density of water is 1.00 g/cm^3, then the volume of 1,000 g of water is 1,000 cm^3. A cubic box with a volume of 1,000 cm^3 is 10 cm (since $10 \times 10 \times 10 = 1,000$). Converting 10 cm to m units, the cube is 0.1 m on each edge.

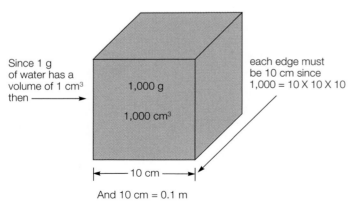

Since 1 g of water has a volume of 1 cm³ then ⟶

1,000 g

1,000 cm³

each edge must be 10 cm since 1,000 = 10 × 10 × 10

├── 10 cm ──┤

And 10 cm = 0.1 m

FIGURE A.4

Visualize the reasoning in 1.8.

1.9. The relationship between mass, volume, and density is $\rho = m/V$. The problem gives a volume, but not a mass. The mass, however, can be assumed to remain constant during the compression of the bread so the mass can be obtained from the original volume and density, or

$$\rho = \frac{m}{V} \quad \therefore \quad m = \rho V$$

$$m = \left(0.2\,\frac{g}{cm^3}\right)(3,000\,cm^3)$$

$$= 0.2 \times 3,000\,\frac{g}{cm^3} \times cm^3$$

$$= 600\,\frac{g \cdot cm^3}{cm^3}$$

$$= 600\,g$$

A mass of 600 g and the new volume of 1,500 cm^3 means that the new density of the crushed bread is

$$\rho = \frac{m}{V}$$

$$= \frac{600\,g}{1,500\,cm^3}$$

$$= \frac{600}{1,500}\,\frac{g}{cm^3}$$

$$= \boxed{0.4\,\frac{g}{cm^3}}$$

1.10. According to Table 1.4, lead has a density of 11.4 g/cm^3. Therefore, a 1.00 cm^3 sample of lead would have a mass of

$$\rho = \frac{m}{V} \quad \therefore \quad m = \rho V$$

$$m = \left(11.4\,\frac{g}{cm^3}\right)(1.00\,cm^3)$$

$$= 11.4 \times 1.00\,\frac{g}{cm^3} \times cm^3$$

$$= 11.4\,\frac{g \cdot cm^3}{cm^3}$$

$$= 11.4\,g$$

Also according to Table 1.4, copper has a density of 8.96 g/cm^3. To balance a mass of 11.4 g of lead, a volume of this much copper would be required:

$$\rho = \frac{m}{V} \quad \therefore \quad V = \frac{m}{\rho}$$

$$V = \frac{11.4\,g}{8.96\,\dfrac{g}{cm^3}}$$

$$= \frac{11.4}{8.96}\,\frac{g}{1} \times \frac{cm^3}{g}$$

$$= 1.27\,\frac{g \cdot cm^3}{g}$$

$$= \boxed{1.27\,cm^3}$$

2.1. Listing these quantities given in this problem, with their symbols, we have

$$d = 22 \text{ km}$$
$$t = 15 \text{ min}$$
$$\bar{v} = ?$$

The usual units for a speed problem are km/h or m/s, and the problem specifies that the answer should be in km/h. We see that 15 minutes is 15/60, or 1/4, or 0.25 of an hour. We will now make a new list of the quantities with the appropriate units:

$$d = 22 \text{ km}$$
$$t = 0.25 \text{ h}$$
$$\bar{v} = ?$$

These quantities are related in the average speed equation, which is already solved for the unknown average velocity:

$$\bar{v} = \frac{d}{t}$$

Substituting the known quantities, we have

$$\bar{v} = \frac{22 \text{ km}}{0.25 \text{ h}}$$

$$= \boxed{88 \frac{\text{km}}{\text{h}}}$$

2.2. Listing the quantities with their symbols:

$$\bar{v} = 3.0 \times 10^8 \text{ m/s}$$
$$t = 20.0 \text{ min}$$
$$d = ?$$

We see that the velocity units are meters per second, but the time units are minutes. We need to convert minutes to seconds, and:

$$\bar{v} = 3.0 \times 10^8 \text{ m/s}$$
$$t = 1.20 \times 10^3 \text{ s}$$
$$d = ?$$

These relationships can be found in the average speed equation, which can be solved for the unknown:

$$\bar{v} = \frac{d}{t} \qquad \therefore \qquad d = \bar{v}t$$

$$d = \left(3.0 \times 10^8 \frac{\text{m}}{\text{s}}\right)(1.20 \times 10^3 \text{ s})$$

$$= (3.0)(1.20) \times 10^{8+3} \frac{\text{m}}{\text{s}} \times \text{s}$$

$$= 3.6 \times 10^{11} \text{ m}$$

$$= \boxed{3.6 \times 10^8 \text{ km}}$$

2.3. Listing the quantities with their symbols, we can see the problem involves the quantities found in the definition of average speed:

$$\bar{v} = 350.0 \text{ m/s}$$
$$t = 5.00 \text{ s}$$
$$d = ?$$

$$v = \frac{d}{t} \qquad \therefore \qquad d = vt$$

$$d = \left(350.0 \frac{\text{m}}{\text{s}}\right)(5.00 \text{ s})$$

$$= (350.0)(5.00) \frac{\text{m}}{\text{s}} \times \text{s}$$

$$= \boxed{1,750 \text{ m}}$$

2.4. Note that the two speeds given (100.0 km/h and 50.0 km/h) are *average* speeds for two different legs of a trip. They are not initial and final speeds of an accelerating object, so you cannot add them together and divide by 2. The average speed for the total (entire) trip can be found from the definition of average speed, that is, average speed is the *total* distance covered divided by the *total* time elapsed. So, we start by finding the distance covered for each of the two legs of the trip:

$$\bar{v} = \frac{d}{t} \qquad \therefore \qquad d = \bar{v}t$$

$$\text{Leg 1 distance} = \left(100.0 \frac{\text{km}}{\text{h}}\right)(2.00 \text{ h})$$

$$= 200.0 \text{ km}$$

$$\text{Leg 2 distance} = \left(50.0 \frac{\text{km}}{\text{h}}\right)(1.00 \text{ h})$$

$$= 50.0 \text{ km}$$

Total distance (leg 1 plus leg 2) = 250.0 km
Total time = 3.00 h

$$\bar{v} = \frac{d}{t} = \frac{250.0 \text{ km}}{3.00 \text{ h}} = \boxed{83.3 \text{ km/h}}$$

2.5. The initial velocity, final velocity, and time are known and the problem asked for the acceleration. Listing these quantities with their symbols, we have

$$v_i = 0$$
$$v_f = 15.0 \text{ m/s}$$
$$t = 10.0 \text{ s}$$
$$a = ?$$

These are the quantities involved in the acceleration equation, which is already solved for the unknown:

$$a = \frac{v_f - v_i}{t}$$

$$a = \frac{15.0 \text{ m/s} - 0 \text{ m/s}}{10.0 \text{ s}}$$

$$= \frac{15.0}{10.0} \frac{m}{s} \times \frac{1}{s}$$

$$= \boxed{1.50 \frac{m}{s^2}}$$

2.6. The initial velocity, final velocity, and acceleration are known and the problem asked for the time. Listing these quantities with their symbols, we have

$$v_i = 20.0 \text{ m/s}$$

$$v_f = 25.0 \text{ m/s}$$

$$a = 3.0 \text{ m/s}^2$$

$$t = ?$$

These are the quantities involved in the acceleration equation, which must first be solved for the unknown time:

$$a = \frac{v_f - v_i}{t} \qquad \therefore \qquad t = \frac{v_f - v_i}{a}$$

$$t = \frac{25.0 \frac{m}{s} - 20.0 \frac{m}{s}}{3.0 \frac{m}{s^2}}$$

$$= \frac{5.00 \frac{m}{s}}{3.0 \frac{m}{s^2}}$$

$$= \frac{5.00}{3.0} \frac{m}{s} \times \frac{s \cdot s}{m}$$

$$= \boxed{1.7 \text{ s}}$$

2.7. The relationship between average velocity ($\bar{v}$), distance (d), and time (t) can be solved for time:

$$\bar{v} = \frac{d}{t}$$

$$\bar{v}t = d$$

$$t = \frac{d}{\bar{v}}$$

$$t = \frac{5,280 \text{ ft}}{2,360 \frac{ft}{s}}$$

$$= \frac{5,280}{2,360} \frac{ft}{1} \times \frac{s}{ft}$$

$$= 2.24 \frac{ft \cdot s}{ft}$$

$$= \boxed{2.24 \text{ s}}$$

2.8. The relationship between average velocity ($\bar{v}$), distance (d), and time (t) can be solved for distance:

$$\bar{v} = \frac{d}{t} \qquad \therefore \qquad d = \bar{v}t$$

$$d = \left(40.0 \frac{m}{s}\right)(0.4625 \text{ s})$$

$$= 40.0 \times 0.4625 \frac{m \cdot s}{s}$$

$$= \boxed{18.5 \text{ m}}$$

2.9. "How many minutes . . .," is a question about time and the distance is given. Since the distance is given in km and the speed in m/s, a unit conversion is needed. The easiest thing to do is to convert km to m. There are 1,000 m in a km, and

$$(1.50 \times 10^8 \text{ km}) \times (1 \times 10^3 \text{ m/km}) = 1.50 \times 10^{11} \text{ m}$$

The relationship between average velocity ($\bar{v}$), distance (d), and time (t) can be solved for time:

$$\bar{v} = \frac{d}{t} \qquad \therefore \qquad t = \frac{d}{\bar{v}}$$

$$t = \frac{1.50 \times 10^{11} \text{ m}}{3.00 \times 10^8 \frac{m}{s}}$$

$$= \frac{1.50}{3.00} \times 10^{11-8} \frac{m}{1} \times \frac{s}{m}$$

$$= 0.500 \times 10^3 \frac{m \cdot s}{m}$$

$$= 5.00 \times 10^2 \text{ s}$$

$$\frac{500 \text{ s}}{60 \frac{s}{min}} = \frac{500}{60} \frac{s}{1} \times \frac{min}{s}$$

$$= 8.33 \frac{s \cdot min}{s}$$

$$= \boxed{8.33 \text{ min}}$$

[Information on how to use scientific notation (also called powers of ten or exponential notation) is located in the Mathematical Review of appendix A.]
All significant figures are retained here because the units are defined exactly, without uncertainty.

2.10. The initial velocity (v_i) is given as 100.0 m/s, the final velocity (v_f) is given as 51.0 m/s, and the time is given as 5.00 s. Acceleration, including a deceleration or negative acceleration, is found from a change of velocity during a given time. Thus,

$$a = \frac{v_f - v_i}{t}$$

$$= \frac{\left(51.0\,\frac{m}{s}\right) - \left(100.0\,\frac{m}{s}\right)}{5.00\ s}$$

$$= \frac{-49.0\,\frac{m}{s}}{5.00\ s}$$

$$= -9.80\,\frac{m}{s} \times \frac{1}{s}$$

$$= \boxed{-9.80\,\frac{m}{s^2}}$$

(The negative sign means a negative acceleration, or deceleration.)

2.11. A ball thrown straight up decelerates to a velocity of zero, then accelerates back to the surface, just as a dropped ball would do from the height reached. Thus, the time required to decelerate upward is the same as the time required to accelerate downward. The ball returns to the surface with the same velocity with which it was thrown (neglecting friction). Therefore:

$$a = \frac{v_f - v_i}{t}$$

$$at = v_f - v_i$$

$$v_f = at + v_i$$

$$= \left(9.8\,\frac{m}{s^2}\right)(3.0\ s)$$

$$= (9.8)(3.0)\,\frac{m}{s^2} \times s$$

$$= 29\,\frac{m \cdot s}{s \cdot s}$$

$$= \boxed{29\ m/s}$$

2.12. These three questions are easily answered by using the three sets of relationships, or equations, that were presented in this chapter:

(a) $v_f = at + v_i$, and when v_i is zero,

$$v_f = at$$

$$v_f = \left(9.8\,\frac{m}{s^2}\right)(4.00\ s)$$

$$= 9.8 \times 4.00\,\frac{m}{s^2} \times s$$

$$= 39\,\frac{m \cdot s}{s \cdot s}$$

$$= \boxed{39\ m/s}$$

(b) $\bar{v} = \dfrac{v_f + v_i}{2} = \dfrac{39\ m/s + 0}{2} = 19.5\ m/s$

(c) $\bar{v} = \dfrac{d}{t}$ $\quad \therefore \quad$ $d = \bar{v}t = \left(19.5\,\dfrac{m}{s}\right)(4.00\ s)$

$$= 19.5 \times 4.00\,\frac{m}{s} \times s$$

$$= 78\,\frac{m \cdot s}{s}$$

$$= \boxed{78\ m}$$

2.13. Note that this problem can be solved with a series of three steps as in the previous problem. It can also be solved by the equation that combines all the relationships into one step. Either method is acceptable, but the following example of a one-step solution reduces the possibilities of error since fewer calculations are involved:

$$d = \frac{1}{2}gt^2 = \frac{1}{2}\left(9.8\,\frac{m}{s^2}\right)(5.00\ s)^2$$

$$= \frac{1}{2}\left(9.8\,\frac{m}{s^2}\right)(25.0\ s^2)$$

$$= \left(\frac{1}{2}\right)(9.8)(25.0)\,\frac{m}{s^2} \times s^2$$

$$= 4.90 \times 25.0\,\frac{m \cdot s^2}{s^2}$$

$$= 122.5\ m$$

$$= \boxed{120\ m}$$

2.14. Listing the known and unknown quantities:

$$F = 100\ N$$
$$m = 5\ kg$$
$$a = ?$$

These are the quantities of Newton's second law of motion, $F = ma$, and

$$F = ma \quad \therefore \quad a = \frac{F}{m}$$

$$= \frac{100\,\frac{kg \cdot m}{s^2}}{5\ kg}$$

$$= \frac{100}{5}\,\frac{kg \cdot m}{s^2} \times \frac{1}{kg}$$

$$= \boxed{20\,\frac{m}{s^2}}$$

2.15. Listing the known and unknown quantities:

$$m = 100\ kg$$
$$v = 6\ m/s$$
$$p = ?$$

These are the quantities found in the equation for momentum, $p = mv$, which is already solved for momentum (p). Thus,

$$p = mv$$

$$= (100 \text{ kg})\left(6\frac{m}{s}\right)$$

$$= \boxed{600 \frac{\text{kg·m}}{\text{s}}}$$

2.16 Listing the known and unknown quantities:

$$w = 13{,}720 \text{ N}$$

$$v = 91 \text{ km/h}$$

$$p = ?$$

The equation for momentum is $p = mv$, which is already solved for momentum (p). The weight unit must be first converted to a mass unit:

$$w = mg \qquad \therefore \qquad m = \frac{w}{g}$$

$$= \frac{13{,}720 \frac{\text{kg·m}}{s^2}}{9.8 \frac{m}{s^2}}$$

$$= \frac{13{,}720}{9.8} \frac{\text{kg·m}}{s^2} \times \frac{s^2}{m}$$

$$= 1{,}400 \text{ kg}$$

The km/h unit should next be converted to m/s. Using the conversion factor from inside the front cover:

$$\frac{0.2778 \frac{m}{s}}{1 \frac{\text{km}}{h}} \times 91 \frac{\text{km}}{h}$$

$$0.2778 \times 91 \frac{m}{s} \times \frac{h}{\text{km}} \times \frac{\text{km}}{h}$$

$$25 \frac{m}{s}$$

Now, listing the converted known and unknown quantities:

$$m = 1{,}400 \text{ kg}$$

$$v = 25 \text{ m/s}$$

$$p = ?$$

and solving for momentum (p),

$$p = mv$$

$$= (1{,}400 \text{ kg})\left(25\frac{m}{s}\right)$$

$$= \boxed{35{,}000 \frac{\text{kg·m}}{\text{s}}}$$

2.17. Listing the known and unknown quantities:

Bullet → $m = 0.015$ kg   Rifle → $m = 6$ kg

Bullet → $v = 200$ m/s   Rifle → $v = ?$ m/s

Note the mass of the bullet was converted to kg. This is a conservation of momentum question, where the bullet and rifle can be considered as a system of interacting objects:

$$\text{Bullet momentum} = -\text{rifle momentum}$$

$$(mv)_b = -(mv)_r$$

$$(mv)_b - (mv)_r = 0$$

$$(0.015 \text{ kg})\left(200\frac{m}{s}\right) - (6 \text{ kg})v_r = 0$$

$$\left(3 \text{ kg} \cdot \frac{m}{s}\right) - (6 \text{ kg·}v_r) = 0$$

$$\left(3 \text{ kg} \cdot \frac{m}{s}\right) = (6 \text{ kg·}v_r)$$

$$v_r = \frac{3 \text{ kg} \cdot \frac{m}{s}}{6 \text{ kg}}$$

$$= \frac{3}{6} \frac{\text{kg}}{1} \times \frac{1}{\text{kg}} \times \frac{m}{s}$$

$$= \boxed{0.5 \frac{m}{s}}$$

The rifle recoils with a velocity of 0.5 m/s.

2.18. Listing the known and unknown quantities:

Astronaut → $w = 2{,}156$ N   Wrench → $m = 5.0$ kg

Astronaut → $v = ?$ m/s   Wrench → $v = 5.0$ m/s

Note that the astronaut's weight is given, but we need mass for the conservation of momentum equation. Mass can be found because the weight on earth was given, where we know $g = 9.8$ m/s². Thus the mass is

$$w = mg \qquad \therefore \qquad m = \frac{w}{g}$$

$$= \frac{2{,}156 \frac{\text{kg·m}}{s^2}}{9.8 \frac{m}{s^2}}$$

$$= \frac{2{,}156}{9.8} \frac{\text{kg·m}}{s^2} \times \frac{s^2}{m}$$

$$= 220 \text{ kg}$$

So the converted known and unknown quantities are:

Astronaut → $m = 220$ kg   Wrench → $m = 5.0$ kg

Astronaut → $v = ?$ m/s   Wrench → $v = 5.0$ m/s

This is a conservation of momentum question, where the astronaut and wrench can be considered as a system of interacting objects:

$$\text{Wrench momentum} = -\text{astronaut momentum}$$

$$(mv)_w = -(mv)_a$$

$$(mv)_w - (mv)_a = 0$$

$$(5.0 \text{ kg})\left(5.0 \frac{m}{s}\right) - (220 \text{ kg})v_a = 0$$

$$\left(25 \text{ kg} \cdot \frac{m}{s}\right) - (220 \text{ kg} \cdot v_a) = 0$$

$$v_a = \frac{25 \text{ kg} \cdot \dfrac{m}{s}}{220 \text{ kg}}$$

$$= \frac{25}{220} \frac{\text{kg}}{1} \times \frac{1}{\text{kg}} \times \frac{m}{s}$$

$$= \boxed{0.11 \frac{m}{s}}$$

The astronaut moves away with a velocity of 0.11 m/s.

2.19.   (a) Weight (w) is a downward force from the acceleration of gravity (g) on the mass (m) of an object. This relationship is the same as Newton's second law of motion, $F = ma$, and

$$w = mg = (1.25 \text{ kg})\left(9.8 \frac{m}{s^2}\right)$$

$$= (1.25)(9.8) \text{ kg} \times \frac{m}{s^2}$$

$$= 12.25 \frac{\text{kg} \cdot m}{s^2}$$

$$= \boxed{12 \text{ N}}$$

   (b) First, recall that a force (F) is measured in newtons (N) and a newton has units of $N = \dfrac{\text{kg} \cdot m}{s^2}$. Second, the relationship between force (F), mass (m), and acceleration (a) is given by Newton's second law of motion, force = mass times acceleration, or $F = ma$. Thus,

$$F = ma \qquad \therefore \qquad a = \frac{F}{m} = \frac{10.0 \dfrac{\text{kg} \cdot m}{s^2}}{1.25 \text{ kg}}$$

$$= \frac{10.0}{1.25} \frac{\text{kg} \cdot m}{s^2} \times \frac{1}{\text{kg}}$$

$$= 8.00 \frac{\text{kg} \cdot m}{\text{kg} \cdot s^2}$$

$$= \boxed{8.00 \frac{m}{s^2}}$$

(Note how the units were treated mathematically in this solution and why it is necessary to show the units for a newton of force. The resulting unit in the answer

is a unit of acceleration, which provides a check that the problem was solved correctly.

2.20.
$$F = ma = (1.25 \text{ kg})\left(5.00 \frac{m}{s^2}\right)$$

$$= (1.25)(5.00) \text{ kg} \times \frac{m}{s^2}$$

$$= 6.25 \frac{\text{kg} \cdot m}{s^2}$$

$$= \boxed{6.25 \text{ N}}$$

(Note that the solution is correctly reported in *newton* units of force rather than kg·m/s².)

2.21.   The bicycle tire exerts a backward force on the road, and the equal and opposite reaction force of the road on the bicycle produces the forward motion. (The motion is always in the direction of the applied force.) Therefore,

$$F = ma = (70.0 \text{ kg})\left(2.0 \frac{m}{s^2}\right)$$

$$= (70.0)(2.0) \text{ kg} \times \frac{m}{s^2}$$

$$= 140 \frac{\text{kg} \cdot m}{s^2}$$

$$= \boxed{140 \text{ N}}$$

2.22.   The question requires finding a force in the metric system, which is measured in newtons of force. Since newtons of force are defined in kg, m, and s, unit conversions are necessary, and these should be done first.

$$1 \frac{\text{km}}{h} = \frac{1,000 \text{ m}}{3,600 \text{ s}} = 0.2778 \frac{m}{s}$$

Dividing both sides of this conversion factor by what you are converting *from* gives the conversion ratio of

$$\frac{0.2778 \dfrac{m}{s}}{\dfrac{\text{km}}{h}}$$

Multiplying this conversion ratio times the two velocities in km/h will convert them to m/s as follows:

$$\left(\frac{0.2778 \dfrac{m}{s}}{\dfrac{\text{km}}{h}}\right)\left(80.0 \frac{\text{km}}{h}\right)$$

$$= (0.2778)(80.0) \frac{m}{s} \times \frac{h}{\text{km}} \times \frac{\text{km}}{h}$$

$$= 22.2 \frac{m}{s}$$

$$\left(\frac{0.2778 \dfrac{m}{s}}{\dfrac{\text{km}}{h}}\right)\left(44.0 \frac{\text{km}}{h}\right)$$

$$= (0.2778)(44.0) \frac{m}{s} \times \frac{h}{km} \times \frac{km}{h}$$

$$= 12.2 \frac{m}{s}$$

Now you are ready to find the appropriate relationship between the quantities involved. This involves two separate equations: Newton's second law of motion and the relationship of quantities involved in acceleration. These may be combined as follows:

$$F = ma \text{ and } a = \frac{v_f - v_i}{t} \qquad \therefore \qquad F = m\left(\frac{v_f - v_i}{t}\right)$$

Now you are ready to substitute quantities for the symbols and perform the necessary mathematical operations:

$$= (1,500 \text{ kg})\left(\frac{22.2 \text{ m/s} - 12.2 \text{ m/s}}{10.0 \text{ s}}\right)$$

$$= (1,500 \text{ kg})\left(\frac{10.0 \text{ m/s}}{10.0 \text{ s}}\right)$$

$$= 1,500 \times 1.00 \frac{kg \cdot \frac{m}{s}}{s}$$

$$= 1,500 \frac{kg \cdot m}{s} \times \frac{1}{s}$$

$$= 1,500 \frac{kg \cdot m}{s \cdot s}$$

$$= 1,500 \frac{kg \cdot m}{s^2}$$

$$= 1,500 \text{ N} = \boxed{1.5 \times 10^3 \text{ N}}$$

2.23. A unit conversion is needed as in the previous problem:

$$\left(90.0 \frac{km}{h}\right)\left(0.2778 \frac{\frac{m}{s}}{\frac{km}{h}}\right) = 25.0 \text{ m/s}$$

(a) $F = ma \qquad \therefore \qquad m = \frac{F}{a} \text{ and } a = \frac{v_f - v_i}{t}$, so

$$m = \frac{F}{\frac{v_f - v_i}{t}} = \frac{5,000.0 \frac{kg \cdot m}{s^2}}{\frac{25.0 \text{ m/s} - 0}{5.0 \text{ s}}}$$

$$= \frac{5,000.0 \frac{kg \cdot m}{s^2}}{5.0 \frac{m}{s^2}}$$

$$= \frac{5,000.0}{5.0} \frac{kg \cdot m}{s^2} \times \frac{s^2}{m}$$

$$= 1,000 \frac{kg \cdot m \cdot s^2}{m \cdot s^2}$$

$$= \boxed{1.0 \times 10^3 \text{ kg}}$$

(b) $w = mg$

$$= (1.0 \times 10^3 \text{ kg})\left(9.8 \frac{m}{s^2}\right)$$

$$= (1.0 \times 10^3)(9.8) \text{ kg} \times \frac{m}{s^2}$$

$$= 9.8 \times 10^3 \frac{kg \cdot m}{s^2}$$

$$= \boxed{9.8 \times 10^3 \text{ N}}$$

2.24. $w = mg$

$$= (70.0 \text{ kg})\left(9.8 \frac{m}{s^2}\right)$$

$$= 70.0 \times 9.8 \text{ kg} \times \frac{m}{s^2}$$

$$= 686 \frac{kg \cdot m}{s^2}$$

$$= \boxed{690 \text{ N}}$$

2.25. $F = \frac{mv^2}{r}$

$$= \frac{(0.20 \text{ kg})\left(3.0 \frac{m}{s}\right)^2}{1.5 \text{ m}}$$

$$= \frac{(0.20 \text{ kg})\left(9.0 \frac{m^2}{s^2}\right)}{1.5 \text{ m}}$$

$$= \frac{0.20 \times 9.0}{1.5} \frac{kg \cdot m^2}{s^2} \times \frac{1}{m}$$

$$= 1.2 \frac{kg \cdot m \cdot m}{s^2 \cdot m}$$

$$= \boxed{1.2 \text{ N}}$$

2.26. **(a)** Newton's laws of motion consider the resistance to a change of motion, or mass, and not weight. The astronaut's mass is

$$w = mg \qquad \therefore \qquad m = \frac{w}{g} = \frac{1,960.0 \frac{kg \cdot m}{s^2}}{9.8 \frac{m}{s^2}}$$

$$= \frac{1,960.0}{9.8} \frac{kg \cdot m}{s^2} \times \frac{s^2}{m} = 200 \text{ kg}$$

(b) From Newton's second law of motion, you can see that the 100 N rocket gives the 200 kg astronaut an acceleration of:

$$F = ma \qquad \therefore \qquad a = \frac{F}{m} = \frac{100 \frac{kg \cdot m}{s^2}}{200 \text{ kg}}$$

$$= \frac{100 \text{ kg} \cdot m}{200 \text{ s}^2} \times \frac{1}{kg} = 0.5 \text{ m/s}^2$$

(c) An acceleration of 0.5 m/s² for 2.0 s will result in a final velocity of

$$a = \frac{v_f - v_i}{t} \qquad \therefore \qquad v_f = at + v_i$$

$$= (0.5 \text{ m/s}^2)(2.0 \text{ s}) + 0 \text{ m/s}$$

$$= \boxed{1 \text{ m/s}}$$

EXERCISES Chapter 3

3.1. Listing the known and unknown quantities:

$$F = 200 \text{ N}$$

$$d = 3 \text{ m}$$

$$W = ?$$

These are the quantities found in the equation for work, $W = Fd$, which is already solved for work (W). Thus,

$$W = Fd$$

$$= \left(200 \frac{\text{kg·m}}{\text{s}^2}\right)(3 \text{ m})$$

$$= (200)(3) \text{ N·m}$$

$$= \boxed{600 \text{ J}}$$

3.2. Listing the known and unknown quantities:

$$F = 440 \text{ N}$$

$$d = 5.0 \text{ m}$$

$$w = 880 \text{ N}$$

$$W = ?$$

These are the quantities found in the equation for work, $W = Fd$, which is already solved for work (W). As you can see in the equation, the force exerted and the distance the box was moved are the quantities used in determining the work accomplished. The weight of the box is a different variable, and one that is not used in this equation. Thus,

$$W = Fd$$

$$= \left(440 \frac{\text{kg·m}}{\text{s}^2}\right)(5.0 \text{ m})$$

$$= 2,200 \text{ N·m}$$

$$= \boxed{2,200 \text{ J}}$$

3.3. Note that 10.0 kg is a mass quantity, and not a weight quantity. Weight is found from $w = mg$, a form of Newton's second law of motion. Thus the force that must

be exerted to lift the backpack is its weight, or (10.0 kg) × (9.8 m/s²) which is 98 N. Therefore, a force of 98 N was exerted on the backpack through a distance of 1.5 m, and

$$W = Fd$$

$$= \left(98 \frac{\text{kg·m}}{\text{s}^2}\right)(1.5 \text{ m})$$

$$= 147 \text{ N·m}$$

$$= \boxed{150 \text{ J}}$$

3.4. Weight is defined as the force of gravity acting on an object, and the greater the force of gravity the harder it is to lift the object. The force is proportional to the mass of the object, as the equation $w = mg$ tells you. Thus the force you exert when lifting is $F = w = mg$, so the work you do on an object you lift must be $W = mgh$.

You know the mass of the box and you know the work accomplished. You also know the value of the acceleration due to gravity, g, so the list of known and unknown quantities is:

$$m = 102 \text{ kg}$$

$$g = 9.8 \text{ m/s}^2$$

$$W = 5,000 \text{ J}$$

$$h = ?$$

The equation $W = mgh$ is solved for work, so the first thing to do is to solve it for h, the unknown height in this problem (note that height is also a distance):

$$W = mgh \qquad \therefore \qquad h = \frac{W}{mg}$$

$$= \frac{5,000 \frac{\text{kg·m}}{\text{s}^2} \times \text{m}}{(102 \text{ kg})\left(9.8 \frac{\text{m}}{\text{s}^2}\right)}$$

$$= \frac{5,000.0}{102 \times 9.8} \frac{\text{kg·m}}{\text{s}^2} \times \frac{\text{m}}{1} \times \frac{1}{\text{kg}} \times \frac{\text{s}^2}{\text{m}}$$

$$= \frac{5,000}{999.6} \text{ m}$$

$$= \boxed{5 \text{ m}}$$

3.5. A student running up the stairs has to lift herself, so her weight is the required force needed. Thus the force exerted is $F = w = mg$, and the work done is $W = mgh$. You know the mass of the student, the height, and the time. You also know the value of the acceleration due to gravity, g, so the list of known and unknown quantities is:

$$m = 60.0 \text{ kg}$$

$$g = 9.8 \text{ m/s}^2$$

$$h = 5.00 \text{ m}$$

$$t = 3.94 \text{ s}$$

$$P = ?$$

The equation $P = \dfrac{mgh}{t}$ is already solved for power, so:

$$P = \dfrac{mgh}{t}$$

$$= \dfrac{(60.0 \text{ kg})\left(9.8 \dfrac{\text{m}}{\text{s}^2}\right)(5.00 \text{ m})}{3.92 \text{ s}}$$

$$= \dfrac{(60.0)(9.8)(5.00)}{(3.94)} \dfrac{\left(\dfrac{\text{kg·m}}{\text{s}^2}\right) \times \text{m}}{\text{s}}$$

$$= \dfrac{2{,}940}{3.92} \dfrac{\text{N·m}}{\text{s}}$$

$$= 750 \dfrac{\text{J}}{\text{s}}$$

$$= \boxed{750 \text{ W}}$$

3.6.

(a)

$$\dfrac{1.00 \text{ hp}}{746 \text{ W}} \times 1{,}400 \text{ W}$$

$$\dfrac{1{,}400}{746} \dfrac{\text{hp·W}}{\text{W}}$$

$$\boxed{1.9 \text{ hp}}$$

(b)

$$\dfrac{746 \text{ W}}{1 \text{ hp}} \times 3.5 \text{ hp}$$

$$746 \times 3.5 \dfrac{\text{W·hp}}{\text{hp}}$$

$$2{,}611 \text{ W}$$

$$\boxed{2{,}600 \text{ hp}}$$

3.7. Listing the known and unknown quantities:

$$m = 2{,}000 \text{ kg}$$

$$v = 72 \text{ km/h}$$

$$KE = ?$$

These are the quantities found in the equation for kinetic energy, $KE = 1/2mv^2$, which is already solved. However, note that the velocity is in units of km/h, which must be changed to m/s before doing anything else (it must be m/s because all energy and work units are in units of the joule [J]. A joule is a newton-meter, and a newton is a kg·m/s²). Using the conversion factor from inside the front cover of your text,

$$\dfrac{0.2778 \dfrac{\text{m}}{\text{s}}}{1.0 \dfrac{\text{km}}{\text{h}}} \times 72 \dfrac{\text{km}}{\text{h}}$$

$$(0.2778)(72) \dfrac{\text{m}}{\text{s}} \times \dfrac{\text{h}}{\text{km}} \times \dfrac{\text{km}}{\text{h}}$$

$$20 \dfrac{\text{m}}{\text{s}}$$

and

$$KE = \dfrac{1}{2}mv^2$$

$$= \dfrac{1}{2}(2{,}000 \text{ kg})\left(20 \dfrac{\text{m}}{\text{s}}\right)^2$$

$$= \dfrac{1}{2}(2{,}000 \text{ kg})\left(400 \dfrac{\text{m}^2}{\text{s}^2}\right)$$

$$= \dfrac{1}{2} \times 2{,}000 \times 400 \dfrac{\text{kg·m}^2}{\text{s}^2}$$

$$= 400{,}000 \dfrac{\text{kg·m}}{\text{s}^2} \times \text{m}$$

$$= 40{,}000 \text{ N·m}$$

$$= \boxed{4 \times 10^5 \text{ J}}$$

Scientific notation is used here to simplify a large number and to show one significant figure.

3.8. Recall the relationship between work and energy—that you do work on an object when you throw it, giving it kinetic energy, and the kinetic energy it has will do work on something else when stopping. Because of the relationship between work and energy you can calculate (1) the work you do, (2) the kinetic energy a moving object has as a result of your work, and (3) the work it will do when coming to a stop, and all three answers should be the same. Thus you do not have a force or a distance to calculate the work needed to stop a moving car, but you can simply calculate the kinetic energy of the car. Both answers should be the same.

Before you start, note that the velocity is in units of km/hr, which must be changed to m/s before doing anything else (it must be m/s because all energy and work units are in units of the joule [J]. A joule is a newton-meter, and a newton is a kg·m/s²). Using the conversion factor from inside the front cover,

$$\dfrac{0.2778 \dfrac{\text{m}}{\text{s}}}{1.0 \dfrac{\text{km}}{\text{h}}} \times 54.0 \dfrac{\text{km}}{\text{h}}$$

$$0.2778 \times 54.0 \dfrac{\text{m}}{\text{s}} \times \dfrac{\text{h}}{\text{km}} \times \dfrac{\text{km}}{\text{h}}$$

$$15.0 \dfrac{\text{m}}{\text{s}}$$

and

$$KE = \dfrac{1}{2}mv^2$$

$$= \dfrac{1}{2}(1{,}000.0 \text{ kg})\left(15.0 \dfrac{\text{m}}{\text{s}}\right)^2$$

$$= \dfrac{1}{2}(1{,}000.0 \text{ kg})\left(225 \dfrac{\text{m}^2}{\text{s}^2}\right)$$

$$= \dfrac{1}{2} \times 1{,}000.0 \times 225 \dfrac{\text{kg·m}^2}{\text{s}^2}$$

$$= 112{,}500 \ \frac{\text{kg} \cdot \text{m}}{\text{s}^2} \times \text{m}$$

$$= 112{,}500 \ \text{N} \cdot \text{m}$$

$$= \boxed{1.13 \times 10^5 \ \text{J}}$$

Scientific notation is used here to simplify a large number and to easily show three significant figures. The answer could likewise be expressed as 113 kJ.

3.9. **(a)** $\qquad W = Fd$

$$= (10 \ \text{lb})(5 \ \text{ft})$$

$$= (10)(5) \ \text{ft} \times \text{lb}$$

$$= 50 \ \text{ft} \cdot \text{lb}$$

(b) The distance of the bookcase from some horizontal reference level did not change, so the gravitational potential energy does not change.

3.10. The force (F) needed to lift the book is equal to the weight (w) of the book, or $F = w$. Since $w = mg$, then $F = mg$. Work is defined as the product of a force moved through a distance, or $W = Fd$. The work done in lifting the book is therefore $W = mgd$, and:

(a) $\qquad W = mgd$

$$= (2.0 \ \text{kg})(9.8 \ \text{m/s}^2)(2.00 \ \text{m})$$

$$= (2.0)(9.8)(2.00) \ \frac{\text{kg} \cdot \text{m}}{\text{s}^2} \times \text{m}$$

$$= 39.2 \ \frac{\text{kg} \cdot \text{m}^2}{\text{s}^2}$$

$$= 39.2 \ \text{J} = \boxed{39 \ \text{J}}$$

(b) $\qquad PE = mgh = \boxed{39 \ \text{J}}$

(c) $\qquad PE_{\text{lost}} = KE_{\text{gained}} = mgh = \boxed{39 \ \text{J}}$

$$\text{(or)}$$

$$v = \sqrt{2gh} = \sqrt{(2)(9.8 \ \text{m/s}^2)(2.00 \ \text{m})}$$

$$= \sqrt{39.2 \ \text{m}^2/\text{s}^2}$$

$$= 6.26 \ \text{m/s}$$

$$KE = \frac{1}{2}mv^2 = \left(\frac{1}{2}\right)(2.0 \ \text{kg})(6.26 \ \text{m/s})^2$$

$$= \left(\frac{1}{2}\right)(2.0 \ \text{kg})(39.2 \ \text{m}^2/\text{s}^2)$$

$$= (1.0)(39.2) \ \frac{\text{kg} \cdot \text{m}^2}{\text{s}^2}$$

$$= \boxed{39 \ \text{J}}$$

3.11. Note that the gram unit must be converted to kg to be consistent with the definition of a newton-meter, or joule unit of energy:

$$KE = \frac{1}{2}mv^2 = \left(\frac{1}{2}\right)(0.15 \ \text{kg})(30.0 \ \text{m/s})^2$$

$$= \left(\frac{1}{2}\right)(0.15 \ \text{kg})(900 \ \text{m}^2/\text{s}^2)$$

$$= \left(\frac{1}{2}\right)(0.15)(900) \ \frac{\text{kg} \cdot \text{m}^2}{\text{s}^2}$$

$$= 67.5 \ \text{J} = \boxed{68 \ \text{J}}$$

3.12. The km/h unit must first be converted to m/s before finding the kinetic energy. Note also that the work done to put an object in motion is equal to the energy of motion, or kinetic energy that it has as a result of the work. The work needed to bring the object to a stop is also equal to the kinetic energy of the moving object:

Unit conversion:

$$1 \ \frac{\text{km}}{\text{h}} = 0.2778 \ \frac{\frac{\text{m}}{\text{s}}}{\frac{\text{km}}{\text{h}}} = \left(90.0 \ \frac{\text{km}}{\text{h}}\right)\left(0.2778 \ \frac{\frac{\text{m}}{\text{s}}}{\frac{\text{km}}{\text{h}}}\right) = 25.0 \ \text{m/s}$$

(a) $KE = \dfrac{1}{2}mv^2 = \dfrac{1}{2}(1{,}000.0 \ \text{kg})\left(25.0 \ \dfrac{\text{m}}{\text{s}}\right)^2$

$$= \frac{1}{2}(1{,}000.0 \ \text{kg})\left(625 \ \frac{\text{m}^2}{\text{s}^2}\right)$$

$$= \frac{1}{2}(1{,}000.0)(625) \ \frac{\text{kg} \cdot \text{m}^2}{\text{s}^2}$$

$$= 312.5 \ \text{kJ} = \boxed{313 \ \text{kJ}}$$

(b) $\qquad W = Fd = KE = \boxed{313 \ \text{kJ}}$

(c) $\qquad KE = W = Fd = \boxed{313 \ \text{kJ}}$

3.13. $\qquad KE = \dfrac{1}{2}mv^2$

$$= \frac{1}{2}(60.0 \ \text{kg})\left(2.0 \ \frac{\text{m}}{\text{s}}\right)^2$$

$$= \frac{1}{2}(60.0 \ \text{kg})\left(4.0 \ \frac{\text{m}^2}{\text{s}^2}\right)$$

$$= 30.0 \times 4.0 \ \text{kg} \times \left(\frac{\text{m}^2}{\text{s}^2}\right)$$

$$= \boxed{120 \ \text{J}}$$

$$KE = \frac{1}{2}mv^2$$

$$= \frac{1}{2}(60.0 \ \text{kg})\left(4.0 \ \frac{\text{m}}{\text{s}}\right)^2$$

$$= \frac{1}{2}(60.0 \text{ kg})\left(16\frac{\text{m}^2}{\text{s}^2}\right)$$

$$= 30.0 \times 16 \text{ kg} \times \left(\frac{\text{m}^2}{\text{s}^2}\right)$$

$$= \boxed{480 \text{ J}}$$

Thus, doubling the speed results in a fourfold increase in kinetic energy.

3.14.
$$KE = \frac{1}{2}mv^2$$

$$= \frac{1}{2}(70.0 \text{ kg})(6.00 \text{ m/s})^2$$

$$= (35.0 \text{ kg})(36.0 \text{ m}^2/\text{s}^2)$$

$$= 35.0 \times 36.0 \text{ kg} \times \frac{\text{m}^2}{\text{s}^2}$$

$$= \boxed{1,260 \text{ J}}$$

$$KE = \frac{1}{2}mv^2$$

$$= \frac{1}{2}(140.0 \text{ kg})(6.00 \text{ m/s})^2$$

$$= (70.0 \text{ kg})(36.0 \text{ m}^2/\text{s}^2)$$

$$= 70.0 \times 36.0 \text{ kg} \times \frac{\text{m}^2}{\text{s}^2}$$

$$= \boxed{2,520 \text{ J}}$$

Thus, doubling the mass results in a doubling of the kinetic energy.

3.15. (a) The force needed is equal to the weight of the student. The English unit of a pound is a force unit, so

$$W = Fd$$

$$= (170.0 \text{ lb})(25.0 \text{ ft})$$

$$= \boxed{4,250 \text{ ft·lb}}$$

(b) Work (W) is defined as a force (F) moved through a distance (d), or $W = Fd$. Power (P) is defined as work (W) per unit of time (t), or $P = W/t$. Therefore,

$$P = \frac{Fd}{t}$$

$$= \frac{(170.0 \text{ lb})(25.0 \text{ ft})}{10.0 \text{ s}}$$

$$= \frac{(170.0)(25.0)}{10.0}\frac{\text{ft·lb}}{\text{s}}$$

$$= 425 \frac{\text{ft·lb}}{\text{s}}$$

One hp is defined as $550\frac{\text{ft·lb}}{\text{s}}$ and

$$\frac{425 \text{ ft·lb/s}}{550 \frac{\text{ft·lb/s}}{\text{hp}}} = \boxed{0.77 \text{ hp}}$$

Note that the student's power rating (425 ft·lb/s) is less than the power rating defined as 1 horsepower (550 ft·lb/s). Thus, the student's horsepower must be *less* than 1 horsepower. A simple analysis such as this will let you know if you inverted the ratio or not.

3.16. (a) The force (F) needed to lift the elevator is equal to the weight of the elevator. Since the work (W) is equal to Fd and power (P) is equal to W/t, then

$$P = \frac{Fd}{t} \qquad \therefore \qquad t = \frac{Fd}{P}$$

$$= \frac{[2,000.0 \text{ lb}][20.0 \text{ ft}]}{\left(550\frac{\text{ft·lb}}{\text{s}}\right)[20.0 \text{ hp}]}$$

$$= \frac{40,000 \text{ ft·lb}}{11,000 \text{ ft·lb}} \times \frac{1}{\text{hp}} \times \text{hp}$$
$$\quad\quad\quad\quad\frac{}{\text{s}}$$

$$= \frac{40,000 \text{ ft·lb}}{11,000 \text{ } 1} \times \frac{\text{s}}{\text{ft·lb}}$$

$$= 3.64 \frac{\text{ft·lb·s}}{\text{ft·lb}}$$

$$= \boxed{3.6 \text{ s}}$$

(b)
$$\bar{v} = \frac{d}{t}$$

$$= \frac{20.0 \text{ ft}}{3.6 \text{ s}}$$

$$= \boxed{5.6 \text{ ft/s}}$$

3.17. Since $PE_{\text{lost}} = KE_{\text{gained}}$ then $mgh = \frac{1}{2}mv^2$. Solving for v,

$$v = \sqrt{2gh} = \sqrt{(2)(32.0 \text{ ft/s}^2)(9.8 \text{ ft})}$$

$$= \sqrt{(2)(32.0)(9.8) \text{ ft}^2/\text{s}^2}$$

$$= \sqrt{627 \text{ ft}^2/\text{s}^2}$$

$$= 25 \text{ ft/s}$$

3.18.
$$KE = \frac{1}{2}mv^2 \qquad \therefore \qquad v = \sqrt{\frac{2KE}{m}}$$

$$= \sqrt{\frac{(2)\left(200,000\frac{\text{kg·m}^2}{\text{s}^2}\right)}{1,000.0 \text{ kg}}}$$

$$= \frac{400{,}000}{1{,}000.0} \frac{\text{kg·m}^2}{\text{s}^2} \times \frac{1}{\text{kg}}$$

$$= \sqrt{\frac{400{,}000}{1{,}000.0} \frac{\text{kg·m}^2}{\text{kg·s}^2}}$$

$$= \sqrt{400 \ \text{m}^2/\text{s}^2}$$

$$= \boxed{20 \ \text{m/s}}$$

3.19. The maximum velocity occurs at the lowest point with a gain of kinetic energy equivalent to the loss of potential energy in falling 3.0 in (which is 0.25 ft), so

$$KE_{\text{gained}} = PE_{\text{lost}}$$

$$\frac{1}{2} mv^2 = mgh$$

$$v = \sqrt{2gh}$$

$$= \sqrt{(2)(32 \ \text{ft/s}^2)(0.25 \ \text{ft})}$$

$$= \sqrt{(2)(32)(0.25)\text{ft/s}^2 \times \text{ft}}$$

$$= \sqrt{16 \ \text{ft}^2/\text{s}^2}$$

$$= \boxed{4.0 \ \text{ft/s}}$$

3.20. **(a)** $W = Fd$ and the force F that is needed to lift the load upward is mg, so $W = mgh$. Power is W/t, so

$$P = \frac{mgh}{t}$$

$$= \frac{(250.0 \ \text{kg})(9.8 \ \text{m/s}^2)(80.0 \ \text{m})}{39.2 \ \text{s}}$$

$$= \frac{(250.0)(9.8)(80.0)}{39.2} \frac{\text{kg}}{1} \times \frac{\text{m}}{\text{s}^2} \times \frac{\text{m}}{1} \times \frac{1}{\text{s}}$$

$$= \frac{196{,}000}{39.2} \frac{\text{kg·m}^2}{\text{s}^2} \times \frac{1}{\text{s}}$$

$$= 5{,}000 \frac{\text{J}}{\text{s}}$$

$$= \boxed{5.0 \ \text{kW}}$$

(b) There are 746 watts per horsepower, so

$$\frac{5{,}000 \ \text{W}}{746 \ \dfrac{\text{W}}{\text{hp}}} = \frac{5{,}000}{746} \frac{\text{W}}{1} \times \frac{\text{hp}}{\text{W}}$$

$$= 6.70 \frac{\text{W·hp}}{\text{W}}$$

$$= \boxed{6.70 \ \text{hp}}$$

EXERCISES **Chapter 4**

4.1. Listing the known and unknown quantities:

body temperature $\quad T_\text{F} = 98.6°$

$$T_\text{C} = ?$$

These are the quantities found in the equation for conversion of Fahrenheit to Celsius, $T_\text{C} = \dfrac{5}{9}(T_\text{F} - 32°)$, where T_F is the temperature in Fahrenheit and T_C is the temperature in Celsius. This equation describes a relationship between the two temperature scales and is used to convert a Fahrenheit temperature to Celsius. The equation is already solved for the Celsius temperature, T_C. Thus,

$$T_\text{C} = \frac{5}{9}(T_\text{F} - 32°)$$

$$= \frac{5}{9}(98.6° - 32°)$$

$$= \frac{333°}{9}$$

$$= \boxed{37° \ \text{C}}$$

4.2. $Q = mc\Delta T$

$$= (221 \ \text{g})\left(0.093 \ \frac{\text{cal}}{\text{gC°}}\right)(38.0°\text{C} - 20.0°\text{C})$$

$$= (221)(0.093)(18.0) \ \text{g} \times \frac{\text{cal}}{\text{gC°}} \times °\text{C}$$

$$= 370 \ \frac{\text{g·cal·°C}}{\text{gC°}}$$

$$= \boxed{370 \ \text{cal}}$$

4.3. First, you need to know the energy of the moving bike and rider. Since the speed is given as 36.0 km/h, convert to m/s by multiplying times 0.2778 m/s per km/h:

$$\left(36.0 \ \frac{\text{km}}{\text{h}}\right)\left(0.2778 \ \frac{\text{m/s}}{\text{km/h}}\right)$$

$$= (36.0)(0.2778) \frac{\text{km}}{\text{h}} \times \frac{\text{h}}{\text{km}} \times \frac{\text{m}}{\text{s}}$$

$$= 10.0 \ \text{m/s}$$

Then,

$$KE = \frac{1}{2} mv^2$$

$$= \frac{1}{2}(100.0 \ \text{kg})(10.0 \ \text{m/s})^2$$

$$= \frac{1}{2}(100.0 \ \text{kg})(100 \ \text{m}^2/\text{s}^2)$$

$$= \frac{1}{2}(100.0)(100) \frac{\text{kg·m}^2}{\text{s}^2}$$

$$= 5{,}000 \ \text{J}$$

Second, this energy is converted to the calorie heat unit through the mechanical equivalent of heat relationship, that 1.0 kcal = 4,184 J, or that 1.0 cal = 4.184 J. Thus,

$$\frac{5,000 \text{ J}}{4,184 \text{ J/kcal}}$$

$$1.195 \frac{\text{J}}{1} \times \frac{\text{kcal}}{\text{J}}$$

$$\boxed{1.20 \text{ kcal}}$$

4.4. First, you need to find the energy of the falling bag. Since the potential energy lost equals the kinetic energy gained, the energy of the bag just as it hits the ground can be found from

$$PE = mgh$$

$$= (15.53 \text{ kg})(9.8 \text{ m/s}^2)(5.50 \text{ m})$$

$$= (15.53)(9.8)(5.50) \frac{\text{kg·m}}{\text{s}^2} \times \text{m}$$

$$= 837 \text{ J}$$

In calories, this energy is equivalent to

$$\frac{837 \text{ J}}{4,184 \text{ J/kcal}} = 0.200 \text{ kcal}$$

Second, the temperature change can be calculated from the equation giving the relationship between a quantity of heat (Q), mass (m), specific heat of the substance (c), and the change of temperature:

$$Q = mc\Delta T \quad \therefore \quad \Delta T = \frac{Q}{mc}$$

$$= \frac{0.200 \text{ kcal}}{(15.53 \text{ kg})\left(0.200 \frac{\text{kcal}}{\text{kgC}^\circ}\right)}$$

$$= \frac{0.200}{(15.53)(0.200)} \frac{\text{kcal}}{1} \times \frac{1}{\text{kg}} \times \frac{\text{kgC}^\circ}{\text{kcal}}$$

$$= 0.0644 \frac{\text{kcal·kgC}^\circ}{\text{kcal·kg}}$$

$$= \boxed{6.44 \times 10^{-2} \,^\circ\text{C}}$$

4.5. The Calorie used by dietitians is a kilocalorie; thus 250.0 Cal is 250.0 kcal. The mechanical energy equivalent is 1 kcal = 4,184 J, so (250.0 kcal)(4,184 J/kcal) = 1,046,250 J.

Since $W = Fd$ and the force needed is equal to the weight (mg) of the person, $W = mgh = (75.0 \text{ kg})(9.8 \text{ m/s}^2)$ (10.0 m) = 7,350 J for each stairway climb.

A total of 1,046,250 J of energy from the French fries would require 1,046,250 J/7,350 J per climb, or 142.3 trips up the stairs.

4.6. For unit consistency,

$$T_\text{C} = \frac{5}{9}(T_\text{F} - 32^\circ) = \frac{5}{9}(68^\circ - 32^\circ) = \frac{5}{9}(36^\circ) = 20^\circ\text{C}$$

$$= \frac{5}{9}(32^\circ - 32^\circ) = \frac{5}{9}(0^\circ) = 0^\circ\text{C}$$

Glass Bowl:

$$Q = mc\Delta T$$

$$= (0.5 \text{ kg})\left(0.2 \frac{\text{kcal}}{\text{kg}^\circ\text{C}}\right)(20^\circ\text{C})$$

$$= (0.5)(0.2)(20) \frac{\text{kg}}{1} \times \frac{\text{kcal}}{\text{kgC}^\circ} \times \frac{^\circ\text{C}}{1}$$

$$= \boxed{2 \text{ kcal}}$$

Iron Pan:

$$Q = mc\Delta T$$

$$= (0.5 \text{ kg})\left(0.11 \frac{\text{kcal}}{\text{kgC}^\circ}\right)(20^\circ\text{C})$$

$$= (0.5)(0.11)(20) \quad \text{kg} \times \frac{\text{kcal}}{\text{kgC}^\circ} \times \,^\circ\text{C}$$

$$= \boxed{1 \text{ kcal}}$$

4.7. Note that a specific heat expressed in cal/g has the same numerical value as a specific heat expressed in kcal/kg because you can cancel the k units. You could convert 896 cal to 0.896 kcal, but one of the two conversion methods is needed for consistency with other units in the problem.

$$Q = mc\Delta T \quad \therefore \quad m = \frac{Q}{c\Delta T}$$

$$= \frac{896 \text{ cal}}{\left(0.056 \frac{\text{cal}}{\text{gC}^\circ}\right)(80.0^\circ\text{C})}$$

$$= \frac{896}{(0.056)(80.0)} \frac{\text{cal}}{1} \times \frac{\text{gC}^\circ}{\text{cal}} \times \frac{1}{\text{C}}$$

$$= 200 \text{ g}$$

$$= \boxed{0.20 \text{ kg}}$$

4.8. Since a watt is defined as a joule/s, finding the total energy in joules will tell the time:

$$Q = mc\Delta T$$

$$= (250.0 \text{ g})\left(1.00 \frac{\text{cal}}{\text{gC}^\circ}\right)(60.0^\circ\text{C})$$

$$= (250.0)(1.00)(60.0) \text{ g} \times \frac{\text{cal}}{\text{gC}^\circ} \times \,^\circ\text{C}$$

$$= 1.50 \times 10^4 \text{ cal}$$

This energy in joules is

$$(1.50 \times 10^4 \text{ cal})\left(4.184 \frac{\text{J}}{\text{cal}}\right) = 62,800 \text{ J}$$

A 300-watt heater uses energy at a rate of $300 \frac{\text{J}}{\text{s}}$, so $\frac{62,800 \text{ J}}{300 \text{ J/s}}$

$= 209$ s is required, which is $\dfrac{209\text{ s}}{60\,\frac{\text{s}}{\text{min}}} = 3.48$ min, or

$$\boxed{\text{about } 3\frac{1}{2} \text{ min}}$$

4.9.
$$Q = mc\Delta T \quad \therefore \quad c = \frac{Q}{m\Delta T}$$

$$= \frac{60.0 \text{ cal}}{(100.0\text{ g})(20.0°C)}$$

$$= \frac{60.0}{(100.0)(20.0)}\frac{\text{cal}}{\text{gC°}}$$

$$= \boxed{0.0300\,\frac{\text{cal}}{\text{gC°}}}$$

4.10. Since the problem specified a solid changing to a liquid without a temperature change, you should recognize that this is a question about a phase change only. The phase change from solid to liquid (or liquid to solid) is concerned with the latent heat of fusion. For water, the latent heat of fusion is given as 80.0 cal/g, and

$m = 250.0$ g $\qquad Q = mL_f$

$L_{f\,(\text{water})} = 80.0$ cal/g $\qquad = (250.0\text{ g})\left(80.0\,\frac{\text{cal}}{\text{g}}\right)$

$Q = ?$

$$= 250.0 \times 80.0\,\frac{\text{g·cal}}{\text{g}}$$

$$= 20{,}000 \text{ cal} = \boxed{20.0 \text{ kcal}}$$

4.11. To change water at 80.0°C to steam at 100.0°C requires two separate quantities of heat that can be called Q_1 and Q_2. The quantity Q_1 is the amount of heat needed to warm the water from 80.0°C to the boiling point, which is 100.0°C at sea level pressure ($\Delta T = 20.0°C$). The relationship between the variable involved is $Q = mc\Delta T$. The quantity Q_2 is the amount of heat needed to take 100.0° water through the phase change to steam (water vapor) at 100.0°C. The phase change from a liquid to a gas (or gas to liquid) is concerned with the latent heat of vaporization. For water, the latent heat of vaporization is given as 540.0 cal/g.

$m = 250.0$ g $\qquad Q_1 = mc\Delta T$

$L_{v\,(\text{water})} = 540.0$ cal/g $\qquad = (250.0\text{ g})\left(1.00\,\frac{\text{cal}}{\text{gC°}}\right)(20.0°C)$

$Q = ?$

$$= (250.0)(1.00)(20.0)\text{ g} \times \frac{\text{cal}}{\text{gC°}} \times °C$$

$$= 5{,}000\,\frac{\text{g·cal·°C}}{\text{gC°}}$$

$$= 5{,}000 \text{ cal}$$

$$= 5.00 \text{ kcal}$$

$Q_2 = mL_v$

$$= (250.0\text{ g})\left(540.0\,\frac{\text{cal}}{\text{g}}\right)$$

$$= 250.0 \times 540.0\,\frac{\text{g·cal}}{\text{g}}$$

$$= 135{,}000 \text{ cal}$$

$$= 135.0 \text{ kcal}$$

$Q_{\text{Total}} = Q_1 + Q_2$

$$= 5.00 \text{ kcal} + 135.0 \text{ kcal}$$

$$= \boxed{140.0 \text{ kcal}}$$

4.12. To change 20.0°C water to steam at 125.0°C requires three separate quantities of heat. First, the quantity Q_1 is the amount of heat needed to warm the water from 20.0°C to 100.0°C ($\Delta T = 80.0°C$). The quantity Q_2 is the amount of heat needed to take 100.0°C water to steam at 100.0°C. Finally, the quantity Q_3 is the amount of heat needed to warm the steam from 100.0° to 125.0°C. According to Table 4.4, the c for steam is 0.480 cal/g°C.

$m = 100.0$ g $\qquad Q_1 = mc\Delta T$

$\Delta T_{\text{water}} = 80.0°C \qquad = (100.0\text{ g})\left(1.00\,\frac{\text{cal}}{\text{gC°}}\right)(80.0°C)$

$\Delta T_{\text{steam}} = 25.0°C$

$L_{v(\text{water})} = 540.0$ cal/g $\qquad = (100.0)(1.00)(80.0)\text{ g} \times \frac{\text{cal}}{\text{gC°}} \times °C$

$c_{\text{steam}} = 0.480$ cal/gC°

$$= 8{,}000\,\frac{\text{g·cal·°C}}{\text{gC°}}$$

$$= 8{,}000 \text{ cal}$$

$$= 8.00 \text{ kcal}$$

$Q_2 = mL_v$

$$= (100.0\text{ g})\left(540.0\,\frac{\text{cal}}{\text{g}}\right)$$

$$= 100.0 \times 540.0\,\frac{\text{g·cal}}{\text{g}}$$

$$= 54{,}000 \text{ cal}$$

$$= 54.00 \text{ kcal}$$

$Q_3 = mc\Delta T$

$$= (100.0\text{ g})\left(0.480\,\frac{\text{cal}}{\text{gC°}}\right)(25.0°C)$$

$$= (100.0)(0.480)(25.0)\text{ g} \times \frac{\text{cal}}{\text{gC°}} \times °C$$

$$= 1{,}200\,\frac{\text{g·cal·°C}}{\text{gC°}}$$

$$= 1{,}200 \text{ cal}$$

$$= 1.20 \text{ kcal}$$

$Q_{\text{total}} = Q_1 + Q_2 + Q_3$

$$= 8.00 \text{ kcal} + 54.00 \text{ kcal} + 1.20 \text{ kcal}$$

$$= \boxed{63.20 \text{ kcal}}$$

4.13. (a) **Step 1:** Cool water from 18.0°C to 0°C.

$$Q_1 = mc\Delta T$$

$$= (400.0 \text{ g})\left(1.00 \frac{\text{cal}}{\text{gC°}}\right)(18.0°C)$$

$$= (400.0)(1.00)(18.0) \text{ g} \times \frac{\text{cal}}{\text{gC°}} \times °C$$

$$= 7{,}200 \frac{\text{g·cal·}°\!\!\!\!\diagup C}{\text{g}\diagup\!\!\!\!C°}$$

$$= 7{,}200 \text{ cal}$$

$$= 7.20 \text{ kcal}$$

Step 2: Find the energy needed for the phase change of water at 0°C to ice at 0°C.

$$Q_2 = mL_f$$

$$= (400.0 \text{ g})\left(80.0 \frac{\text{cal}}{\text{g}}\right)$$

$$= 400.0 \times 80.0 \frac{\text{g·cal}}{\diagup\!\!\!\!g}$$

$$= 32{,}000 \text{ cal}$$

$$= 32.0 \text{ kcal}$$

Step 3: Cool the ice from 0°C to ice at −5.00°C.

$$Q_3 = mc\Delta T$$

$$= (400.0 \text{ g})\left(0.500 \frac{\text{cal}}{\text{gC°}}\right)(5.00°C)$$

$$= 400.0 \times 0.500 \times 5.00 \text{ g} \times \frac{\text{cal}}{\text{gC°}} \times °C$$

$$= 1{,}000 \frac{\text{g·cal·}°\!\!\!\!\diagup C}{\text{g}\diagup\!\!\!\!C°}$$

$$= 1{,}000 \text{ cal}$$

$$= 1.00 \text{ kcal}$$

$$Q_{total} = Q_1 + Q_2 + Q_3$$

$$= 7.20 \text{ kcal} + 32.0 \text{ kcal} + 1.00 \text{ kcal}$$

$$= \boxed{40.2 \text{ kcal}}$$

(b) $Q = mL_v \quad \therefore \quad m = \dfrac{Q}{L_v}$

$$= \frac{40{,}200 \text{ cal}}{40.0 \dfrac{\text{cal}}{\text{g}}}$$

$$= \frac{40{,}200}{40.0} \frac{\text{cal}}{1} \times \frac{\text{g}}{\text{cal}}$$

$$= 1{,}005 \frac{\text{cal·g}}{\text{cal}}$$

$$= \boxed{1.01 \times 10^3 \text{ g}}$$

4.14.

$$W = J(Q_H - Q_L)$$

$$= 4{,}184 \frac{\text{J}}{\text{kcal}} (0.3000 \text{ kcal} - 0.2000 \text{ kcal})$$

$$= 4{,}184 \frac{\text{J}}{\text{kcal}} (0.1000 \text{ kcal})$$

$$= 4{,}184 \times 0.1000 \frac{\text{J·kcal}}{\text{kcal}}$$

$$= \boxed{418.4 \text{ J}}$$

4.15.

$$W = J(Q_H - Q_L)$$

$$= 4{,}184 \frac{\text{J}}{\text{kcal}} (55.0 \text{ kcal} - 40.0 \text{ kcal})$$

$$= 4{,}184 \frac{\text{J}}{\text{kcal}} (15.0 \text{ kcal})$$

$$= 4{,}184 \times 15.0 \frac{\text{J·kcal}}{\text{kcal}}$$

$$= 62{,}760 \text{ J}$$

$$= \boxed{62.8 \text{ kJ}}$$

EXERCISES **Chapter 5**

5.1. (a) $f = 3$ Hz $\qquad T = \dfrac{1}{f}$

$\lambda = 2$ cm

$T = ? \qquad\qquad = \dfrac{1}{3\dfrac{1}{s}}$

$$= \frac{1}{3}\frac{s}{1}$$

$$= \boxed{0.3 \text{ s}}$$

(b) $f = 3$ Hz $\qquad v = \lambda f$

$\lambda = 2$ cm $\qquad\quad = (2 \text{ cm})\left(3\dfrac{1}{s}\right)$

$T = ?$

$$= 2 \times 3 \text{ cm} \times \frac{1}{s}$$

$$= \boxed{6 \frac{\text{cm}}{\text{s}}}$$

5.2. **Step 1:**

$t = 20.0°C \qquad v_{T_{P(m/s)}} = v_0 + \left(\dfrac{0.600 \text{ m/s}}{°C}\right)(T_P)$

$f = 20{,}000$ Hz

$v = ? \qquad\qquad = 331\dfrac{\text{m}}{\text{s}} + \left(\dfrac{0.600 \text{ m/s}}{°C}\right)(20.0°C)$

$\lambda = ?$

$$= 331\frac{\text{m}}{\text{s}} + 12.0\frac{\text{m}}{\text{s}}$$

$$= 343\frac{\text{m}}{\text{s}}$$

Step 2:

$$v = \lambda f \quad \therefore \quad \lambda = \frac{v}{f}$$

$$= \frac{343 \, \frac{m}{s}}{20{,}000 \, \frac{1}{s}}$$

$$= \frac{343}{20{,}000} \, \frac{m}{s} \times \frac{s}{1}$$

$$= 0.01715 \, m$$

$$= \boxed{0.02 \, m}$$

5.3. $f_1 = 440 \, \text{Hz}$ $f_b = f_2 - f_1$

$f_2 = 446 \, \text{Hz}$ $= (446 \, \text{Hz}) - (440 \, \text{Hz})$

$f_b = ?$ $= \boxed{6 \, \text{Hz}}$

Note that the smaller frequency is subtracted from the larger one to avoid negative beats.

5.4.   **Step 1:** Assume room temperature (20.0°C) to obtain the velocity:

$f = 2.00 \times 10^7 \, \text{Hz}$ $v_{T_{P(m/s)}} = v_0 + \left(\frac{0.600 \, \frac{m}{s}}{°C} \right)(T_P)$

$\lambda = ?$

$v = ?$ $= 331 \, \frac{m}{s} + \left(\frac{0.600 \, \frac{m}{s}}{°C} \right)(20.0°C)$

$$= 331 \, \frac{m}{s} + 12.0 \, \frac{m}{s}$$

$$= 343 \, \frac{m}{s}$$

Step 2:

$$v = \lambda f \quad \therefore \quad \lambda = \frac{v}{f}$$

$$= \frac{343 \, \frac{m}{s}}{2.00 \times 10^7 \, \frac{1}{s}}$$

$$= \frac{343}{2.00 \times 10^7} \, \frac{m}{s} \times \frac{s}{1}$$

$$= 1.715 \times 10^{-5} \, m$$

$$= \boxed{1.7 \times 10^{-5} \, m}$$

5.5.   **Step 1:** Assume room temperature (20.0°C) to obtain the velocity (yes, you can have room temperature outside a room):

$d = 150 \, m$

$t = ?$

$v = ?$

$$v_{T_{P(m/s)}} = v_0 + \left(\frac{0.600 \, \frac{m}{s}}{°C} \right)(T_P)$$

$$= 331 \, \frac{m}{s} + \left(\frac{0.600 \, \frac{m}{s}}{°C} \right)(20.0°C)$$

$$= 331 \, \frac{m}{s} + 12.0 \, \frac{m}{s}$$

$$= 343 \, \frac{m}{s}$$

Step 2:

$$v = \frac{d}{t} \quad \therefore \quad t = \frac{d}{v}$$

$$= \frac{150.0 \, m}{343 \, \frac{m}{s}}$$

$$= \frac{150.0}{343} \, \frac{m}{1} \times \frac{s}{m}$$

$$= 0.4373177 \, s$$

$$= \boxed{0.437 \, s}$$

5.6.   **Step 1:** Find the velocity of sound in ft/s at a temperature of 20.0°C

$t = 0.500 \, s$

$T = 20.0°C$

$d = ?$

$v = ?$ $v_{T_{P(ft/s)}} = v_0 + \left(\frac{2.00 \, \frac{ft}{s}}{°C} \right)(T_P)$

$$= 1{,}087 \, \frac{ft}{s} + \left(\frac{2.00 \, \frac{ft}{s}}{°C} \right)(20.0°C)$$

$$= 1{,}087 \, \frac{ft}{s} + 40.0 \, \frac{ft}{s}$$

$$= 1{,}127 \, \frac{ft}{s}$$

Step 2:

$$v = \frac{d}{t} \quad \therefore \quad d = vt$$

$$= \left(1{,}127 \, \frac{ft}{s} \right)(0.500 \, s)$$

$$= 1{,}127 \times 0.500 \, \frac{ft}{s} \times s$$

$$= 563.5 \, ft$$

$$= 564 \, ft$$

Step 3: The distance to the building is half the distance the sound traveled, so

$$\frac{564}{2} = \boxed{282 \, ft}$$

5.7. **Step 1:**

$t = 1.75$ s $\qquad v = \dfrac{d}{t} \quad \therefore \quad d = vt$

$v = 1,530$ m/s

$d = ?$

$\qquad = \left(1,530\,\dfrac{m}{s}\right)(1.75\text{ s})$

$\qquad = 1,530 \times 1.75\,\dfrac{m}{s} \times s$

$\qquad = 2,677.5$ m

Step 2: The sonar signal traveled from the ship to the bottom, then back to the ship, so the distance to the bottom is half of the distance traveled:

$\dfrac{2,677.5\text{ m}}{2} = 1,338.75$ m

$\qquad = \boxed{1,340\text{ m}}$

5.8.

$f = 660$ Hz $\qquad v = \lambda f$

$\lambda = 9.0$ m

$v = ?$

$\qquad = (9.0\text{ m})\left(660\,\dfrac{1}{s}\right)$

$\qquad = 9.0 \times 660\text{ m} \times \dfrac{1}{s}$

$\qquad = 5,940\,\dfrac{m}{s}$

$\qquad = \boxed{5,900\,\dfrac{m}{s}}$

5.9. **Step 1:**

$t = 2.5$ s $\qquad v_{T_{P(m/s)}} = v_0 + \left(\dfrac{0.600\,\frac{m}{s}}{°C}\right)(T_P)$

$T = 20.0°C$

$d = ?$

$v = ?$

$\qquad = 331\,\dfrac{m}{s} + \left(\dfrac{0.600\,\frac{m}{s}}{°C}\right)(20.0°C)$

$\qquad = 331\,\dfrac{m}{s} + 12.0\,\dfrac{m}{s}$

$\qquad = 343\,\dfrac{m}{s}$

Step 2:

$v = \dfrac{d}{t} \quad \therefore \quad d = vt$

$\qquad = \left(343\,\dfrac{m}{s}\right)(2.5\text{ s})$

$\qquad = 343 \times 2.5\,\dfrac{m}{s} \times s$

$\qquad = 857.5$ m

$\qquad = \boxed{860\text{ m}}$

5.10. According to table 5.1, sound moves through air at 0°C with a velocity of 331 m/s and through steel with a velocity of 5,940 m/s. Therefore,

(a) $v = 331$ m/s $\qquad v = \dfrac{d}{t} \quad \therefore \quad d = vt$

$t = 8.00$ s

$d = ?$

$\qquad = \left(331\,\dfrac{m}{s}\right)(8.00\text{ s})$

$\qquad = 331 \times 8.00\,\dfrac{m}{s} \times s$

$\qquad = 2,648$ m

$\qquad = \boxed{2.65\text{ km}}$

(b) $v = 5,940$ m/s $\qquad v = \dfrac{d}{t} \quad \therefore \quad d = vt$

$t = 8.00$ s

$d = ?$

$\qquad = \left(5,940\,\dfrac{m}{s}\right)(8.00\text{ s})$

$\qquad = 5,940 \times 8.00\,\dfrac{m}{s} \times s$

$\qquad = 47,520$ m

$\qquad = \boxed{47.5\text{ km}}$

5.11. $\qquad v = f\lambda$

$\qquad = \left(10\,\dfrac{1}{s}\right)(0.50\text{ m})$

$\qquad = 5\,\dfrac{m}{s}$

5.12. The distance between two *consecutive* condensations (or rarefactions) is one wavelength, so $\lambda = 3.00$ m and

$\qquad v = f\lambda$

$\qquad = \left(112.0\,\dfrac{1}{s}\right)(3.00\text{ m})$

$\qquad = 336\,\dfrac{m}{s}$

5.13. **(a)** One complete wave every 4.0 s means that $T = 4.0$ s.

(b) $\qquad f = \dfrac{1}{T}$

$\qquad = \dfrac{1}{4.0\text{ s}}$

$\qquad = \dfrac{1}{4.0}\,\dfrac{1}{s}$

$\qquad = 0.25\,\dfrac{1}{s}$

$\qquad = \boxed{0.25\text{ Hz}}$

5.14. The distance from one condensation to the next is one wavelength, so

$$v = f\lambda \quad \therefore \quad \lambda = \frac{v}{f}$$

$$= \frac{330 \,\frac{m}{s}}{260 \,\frac{1}{s}}$$

$$= \frac{330}{260} \,\frac{m}{s} \times \frac{s}{1}$$

$$= \boxed{1.3 \text{ m}}$$

5.15. **(a)** $v = f\lambda = \left(256 \,\frac{1}{s}\right)(1.34 \text{ m})$ $\quad = \boxed{343 \text{ m/s}}$

(b) $= \left(440.0 \,\frac{1}{s}\right)(0.780 \text{ m})$ $\quad = \boxed{343 \text{ m/s}}$

(c) $= \left(750.0 \,\frac{1}{s}\right)(0.457 \text{ m})$ $\quad = \boxed{343 \text{ m/s}}$

(d) $= \left(2,500.0 \,\frac{1}{s}\right)(0.1372 \text{ m})$ $\quad = \boxed{343 \text{ m/s}}$

5.16. The speed of sound at 0.0°C is 1,087 ft/s, and

(a) $v_{T_F} = v_0 + \left[\frac{2.00 \text{ ft/s}}{°C}\right][T_P]$

$$= 1,087 \text{ ft/s} + \left[\frac{2.00 \text{ ft/s}}{°C}\right][0.0°C]$$

$$= 1,087 + (2.00)(0.0) \text{ ft/s} + \frac{\text{ft/s}}{°C} \times °C$$

$$= 1,087 \text{ ft/s} + 0.0 \text{ ft/s}$$

$$= \boxed{1,087 \text{ ft/s}}$$

(b) $v_{20°} = 1,087 \text{ ft/s} + \left[\frac{2.00 \text{ ft/s}}{°C}\right][20.0°C]$

$$= 1,087 \text{ ft/s} + 40.0 \text{ ft/s}$$

$$= \boxed{1,127 \text{ ft/s}}$$

(c) $v_{40°} = 1,087 \text{ ft/s} + \left[\frac{2.00 \text{ ft/s}}{°C}\right][40.0°C]$

$$= 1,087 \text{ ft/s} + 80.0 \text{ ft/s}$$

$$= \boxed{1,167 \text{ ft/s}}$$

(d) $v_{80°} = 1,087 \text{ ft/s} + \left[\frac{2.00 \text{ ft/s}}{°C}\right][80.0°C]$

$$= 1,087 \text{ ft/s} + 160 \text{ ft/s}$$

$$= \boxed{1,247 \text{ ft/s}}$$

5.17. For consistency with the units of the equation given, 43.7°F is first converted to 6.50°C. The velocity of sound in this air is:

$$v_{T_F} = v_0 + \left[\frac{2.00 \text{ ft/s}}{°C}\right][T_P]$$

$$= 1,087 \text{ ft/s} + \left[\frac{2.00 \text{ ft/s}}{°C}\right][6.50°C]$$

$$= 1,087 \text{ ft/s} + 13.0 \text{ ft/s}$$

$$= 1,100 \text{ ft/s}$$

The distance that a sound with this velocity travels in the given time is

$$v = \frac{d}{t} \quad \therefore \quad d = vt$$

$$= (1,100 \text{ ft/s})(4.80 \text{ s})$$

$$= (1,100)(4.80) \,\frac{\text{ft} \cdot s}{s}$$

$$= 5,280 \text{ ft}$$

$$\frac{5,280 \text{ ft}}{2}$$

$$= 2,640 \text{ ft}$$

$$= \boxed{2,600 \text{ ft}}$$

Since the sound traveled from the rifle to the cliff and then back, the cliff must be about one-half mile away.

5.18. This problem requires three steps, (1) conversion of the °F temperature value to °C, (2) calculating the velocity of sound in air at this temperature, and (3) calculating the distance from the calculated velocity and the given time:

$$v_{T_F} = v_0 + \left[\frac{2.00 \text{ ft/s}}{°C}\right][T_P]$$

$$= 1,087 \text{ ft/s} + \left[\frac{2.00 \text{ ft/s}}{°C}\right][26.67°C]$$

$$= 1,087 \text{ ft/s} + 53.0 \text{ ft/s} = 1,140 \text{ ft/s}$$

$$v = \frac{d}{t} \quad \therefore \quad d = vt$$

$$= (1,140 \text{ ft/s})(4.63 \text{ s})$$

$$= \boxed{5,280 \text{ ft (one mile)}}$$

5.19. **(a)** $v = f\lambda \quad \therefore \quad \lambda = \frac{v}{f}$

$$= \frac{1,125 \,\frac{\text{ft}}{s}}{440 \,\frac{1}{s}}$$

$$= \frac{1,125}{440} \,\frac{\text{ft}}{s} \times \frac{s}{1}$$

$$= 2.56 \,\frac{\text{ft} \cdot s}{s}$$

$$= \boxed{2.6 \text{ ft}}$$

(b) $v = f\lambda \quad \therefore \quad \lambda = \dfrac{v}{f}$

$$= \dfrac{5{,}020}{440}\,\dfrac{\text{ft}}{\text{s}} \times \dfrac{\text{s}}{1}$$

$$= 11.4\text{ ft} = \boxed{11\text{ ft}}$$

EXERCISES Chapter 6

6.1. First, recall that a negative charge means an excess of electrons. Second, the relationship between the total charge (q), the number of electrons (n), and the charge of a single electron (e) is $q = ne$. The fundamental charge of a single ($n = 1$) electron (e) is 1.60×10^{-19} C. Thus

$$q = ne \quad \therefore \quad n = \dfrac{q}{e}$$

$$= \dfrac{1.00 \times 10^{-14}\text{ C}}{1.60 \times 10^{-19}\,\dfrac{\text{C}}{\text{electron}}}$$

$$= \dfrac{1.00 \times 10^{-14}}{1.60 \times 10^{-19}}\,\dfrac{\text{C}}{1} \times \dfrac{\text{electron}}{\text{C}}$$

$$= 6.25 \times 10^4\,\dfrac{\text{C·electron}}{\text{C}}$$

$$= \boxed{6.25 \times 10^4\text{ electron}}$$

6.2. **(a)** Both balloons have negative charges so the force is repulsive, pushing the balloons away from each other.

(b) The magnitude of the force can be found from Coulomb's law:

$$F = \dfrac{kq_1q_2}{d^2}$$

$$= \dfrac{(9.00 \times 10^9\text{ N·m}^2/\text{C}^2)(3.00 \times 10^{-14}\text{ C})(2.00 \times 10^{-12}\text{ C})}{(2.00 \times 10^{-2}\text{m})^2}$$

$$= \dfrac{(9.00 \times 10^9)(3.00 \times 10^{-14})(2.00 \times 10^{-12})\,\dfrac{\text{N·m}^2}{\text{C}^2} \times \text{C} \times \text{C}}{4.00 \times 10^{-4}\quad \text{m}^2}$$

$$= \dfrac{5.40 \times 10^{-16}}{4.00 \times 10^{-4}}\,\dfrac{\text{N·m}^2}{\text{C}^2} \times \text{C}^2 \times \dfrac{1}{\text{m}^2}$$

$$= \boxed{1.35 \times 10^{-12}\text{ N}}$$

6.3. $\dfrac{\text{potential}}{\text{difference}} = \dfrac{\text{work}}{\text{charge}}$

or

$$V = \dfrac{W}{q}$$

$$= \dfrac{7.50\text{ J}}{5.00\text{ C}}$$

$$= 1.50\,\dfrac{\text{J}}{\text{C}}$$

$$= \boxed{1.50\text{ V}}$$

6.4. $\dfrac{\text{electric}}{\text{current}} = \dfrac{\text{charge}}{\text{time}}$

or

$$I = \dfrac{q}{t}$$

$$= \dfrac{6.00\text{ C}}{2.00\text{ s}}$$

$$= 3.00\,\dfrac{\text{C}}{\text{s}}$$

$$= \boxed{3.00\text{ A}}$$

6.5. A current of 1.00 amp is defined as 1.00 coulomb/s. Since the fundamental charge of the electron is 1.60×10^{-19} C/electron,

$$\dfrac{1.00\,\dfrac{\text{C}}{\text{s}}}{1.60 \times 10^{-19}\,\dfrac{\text{C}}{\text{electron}}}$$

$$= 6.25 \times 10^{18}\,\dfrac{\text{C}}{\text{s}} \times \dfrac{\text{electron}}{\text{C}}$$

$$= \boxed{6.25 \times 10^{18}\,\dfrac{\text{electrons}}{\text{s}}}$$

6.6. $R = \dfrac{V}{I}$

$$= \dfrac{120.0\text{ V}}{4.00\text{ A}}$$

$$= 30.0\,\dfrac{\text{V}}{\text{A}}$$

$$= \boxed{30.0\ \Omega}$$

6.7. $R = \dfrac{V}{I} \quad \therefore \quad I = \dfrac{V}{R}$

$$= \dfrac{120.0\text{ V}}{60.0\,\dfrac{\text{V}}{\text{A}}}$$

$$= \dfrac{120.0}{60.0}\,\text{V} \times \dfrac{\text{A}}{\text{V}}$$

$$= \boxed{2.00\text{ A}}$$

6.8. **(a)** $R = \dfrac{V}{I} \quad \therefore \quad V = IR$

$$= (1.20\text{ A})\left(10.0\,\dfrac{\text{V}}{\text{A}}\right)$$

$$= \boxed{12.0\text{ V}}$$

(b) Power = (current)(potential difference)

or

$$P = IV$$

$$= \left(1.20\,\dfrac{\text{C}}{\text{s}}\right)\left(12.0\,\dfrac{\text{J}}{\text{C}}\right)$$

$$= (1.20)(12.0)\frac{\cancel{C}}{s} \times \frac{J}{\cancel{C}}$$

$$= 14.4 \frac{J}{s}$$

$$= \boxed{14.4 \text{ W}}$$

6.9. Note that there are two separate electrical units that are rates: (1) the amp (coulomb/s), and (2) the watt (joule/s). The question asked for a rate of using energy. Energy is measured in joules, so you are looking for the power of the radio in watts. To find watts $(P = IV)$, you will need to calculate the current (I) since it is not given. The current can be obtained from the relationship of Ohm's law:

$$I = \frac{V}{R}$$

$$= \frac{3.00 \text{ V}}{15.0 \frac{V}{A}}$$

$$= 0.200 \text{ A}$$

$$P = IV$$

$$= (0.200 \text{ C/s})(3.00 \text{ J/C})$$

$$= \boxed{0.600 \text{ W}}$$

6.10. $\text{cost} = \dfrac{(\text{watts})(\text{time})(\text{rate})}{1,000 \frac{W}{kW}}$

$$= \frac{(1,200 \text{ W})(0.25 \text{ h})\left(\frac{\$0.10}{kWh}\right)}{1,000 \frac{W}{kW}}$$

$$= \frac{(1,200)(0.25)(0.10)}{1,000} \frac{W}{1} \times \frac{h}{1} \times \frac{\$}{kWh} \times \frac{kW}{W}$$

$$= \boxed{\$0.03} \text{ (3 cents)}$$

6.11. The relationship between power (P), current (I), and volts (V) will provide a solution. Since the relationship considers power in watts the first step is to convert horsepower to watts. One horsepower is equivalent to 746 watts, so:

$$(746 \text{ W/hp})(2.00 \text{ hp}) = 1,492 \text{ W}$$

$$P = IV \quad \therefore \quad I = \frac{P}{V}$$

$$= \frac{1,492 \frac{J}{s}}{12.0 \frac{J}{C}}$$

$$= \frac{1,492}{12.0} \frac{\cancel{J}}{s} \times \frac{C}{\cancel{J}}$$

$$= 124.3 \frac{C}{s}$$

$$= \boxed{124 \text{ A}}$$

6.12. (a) The rate of using energy is joule/s, or the watt. Since 1.00 hp = 746 W,

inside motor: $(746 \text{ W/hp})(1/3 \text{ hp}) = 249 \text{ W}$

outside motor: $(746 \text{ W/hp})(1/3 \text{ hp}) = 249 \text{ W}$

compressor motor: $(746 \text{ W/hp})(3.70 \text{ hp}) = 2,760 \text{ W}$

$249 \text{ W} + 249 \text{ W} + 2,760 \text{ W} = \boxed{3,258 \text{ W}}$

(b) $\dfrac{(3,258 \text{ W})(1.00 \text{ h})\left(\dfrac{\$0.10}{kWh}\right)}{1,000 \dfrac{W}{kW}} = \$0.33 \text{ per hour}$

(c) $(\$0.33/\text{h})(12 \text{ h/day})(30 \text{ day/mo}) = \boxed{\$118.80}$

6.13. The solution is to find how much current each device draws and then to see if the total current is less or greater than the breaker rating:

$$\text{Toaster: } I = \frac{V}{R} = \frac{120 \text{ V}}{15 \text{ V/A}} = 8.0 \text{ A}$$

$$\text{Motor: } (0.20 \text{ hp})(746 \text{ W/hp}) = 150 \text{ W}$$

$$I = \frac{P}{V} = \frac{150 \text{ J/s}}{102 \text{ J/C}} = 1.3 \text{ A}$$

$$\text{Three 100 W bulbs: } 3 \times 100 \text{ W} = 300 \text{ W}$$

$$I = \frac{P}{V} = \frac{300 \text{ J/s}}{120 \text{ J/C}} = 2.5 \text{ A}$$

$$\text{Iron: } I = \frac{P}{V} = \frac{600 \text{ J/s}}{120 \text{ J/C}} = 5.0 \text{ A}$$

The sum of the currents is 8.0 A + 1.3 A + 2.5 A + 5.0 A = 16.8 A, so the total current is greater than 15.0 amp and the circuit breaker will trip.

6.14. (a) $V_p = 1,200 \text{ V}$

$N_p = 1 \text{ loop}$

$N_s = 200 \text{ loops}$

$V_s = ?$

$$\frac{V_p}{N_p} = \frac{V_s}{N_s} \quad \therefore \quad V_s = \frac{V_p N_s}{N_p}$$

$$V_s = \frac{(1,200 \text{ V})(200 \text{ loop})}{1 \text{ loop}}$$

$$= \boxed{240,000 \text{ V}}$$

(b) $I_p = 40 \text{ A} \qquad V_p I_p = V_s I_s \quad \therefore \quad I_s = \dfrac{V_p I_p}{V_s}$

$I_s = ? \qquad I_s = \dfrac{1,200 \text{ V} \times 40 \text{ A}}{240,000 \text{ V}}$

$$= \frac{1,200 \times 40}{240,000} \frac{V \cdot A}{V}$$

$$= \boxed{0.2 \text{ A}}$$

6.15. **(a)** $V_s = 12$ V

$I_s = 0.5$ A

$V_P = 120$ V

$\dfrac{N_P}{N_s} = ?$

$\dfrac{V_P}{N_P} = \dfrac{V_s}{N_s}$ $\therefore$ $\dfrac{N_P}{N_s} = \dfrac{V_P}{V_s}$

$\dfrac{N_P}{N_s} = \dfrac{120\ \text{V}}{12\ \text{V}} = \dfrac{10}{1}$

or

$\boxed{10 \text{ primary to 1 secondary}}$

(b) $I_P = ?$ $V_P I_P = V_s I_s$ $\therefore$ $I_P = \dfrac{V_s I_s}{V_P}$

$I_P = \dfrac{(12\ \text{V})(0.5\ \text{A})}{120\ \text{V}}$

$= \dfrac{12 \times 0.5}{120} \dfrac{\text{V·A}}{\text{V}}$

$= \boxed{0.05\ \text{A}}$

(c) $P_s = ?$ $P_s = I_s V_s$

$= (0.5\ \text{A})(12\ \text{V})$

$= 0.5 \times 12 \dfrac{\text{C}}{\text{s}} \times \dfrac{\text{J}}{\text{C}}$

$= 6 \dfrac{\text{J}}{\text{s}}$

$= \boxed{6\ \text{W}}$

6.16. **(a)** $V_P = 120$ V

$N_P = 50$ loops

$N_s = 150$ loops

$I_P = 5.0$ A

$V_s = ?$

$\dfrac{V_P}{N_P} = \dfrac{V_s}{N_s}$ $\therefore$ $V_s = \dfrac{V_P N_s}{N_P}$

$V_s = \dfrac{120\ \text{V} \times 150\ \text{loops}}{50\ \text{loops}}$

$= \dfrac{120 \times 150}{50} \dfrac{\text{V·loops}}{\text{loops}}$

$= \boxed{360\ \text{V}}$

(b) $I_s = ?$ $V_P I_P = V_s I_s$ $\therefore$ $I_s = \dfrac{V_P I_P}{V_s}$

$I_s = \dfrac{(120\ \text{V})(5.0\ \text{A})}{360\ \text{V}}$

$= \dfrac{120 \times 5.0}{360} \dfrac{\text{V·A}}{\text{V}}$

$= \boxed{1.7\ \text{A}}$

(c) $P_s = ?$ $P_s = I_s V_s$

$= \left(1.7 \dfrac{\text{C}}{\text{s}}\right)\left(360 \dfrac{\text{J}}{\text{C}}\right)$

$= 1.7 \times 360 \dfrac{\text{C}}{\text{s}} \times \dfrac{\text{J}}{\text{C}}$

$= 612 \dfrac{\text{J}}{\text{s}}$

$= \boxed{600\ \text{W}}$

EXERCISES **Chapter 7**

7.1. The relationship between the speed of light in a transparent material (v), the speed of light in a vacuum ($c = 3.00 \times 10^8$ m/s) and the index of refraction (n) is $n = c/v$. According to Table 7.1, the index of refraction for water is $n = 1.33$ and for ice is $n = 1.31$.

(a) $c = 3.00 \times 10^8$ m/s

$n = 1.33$

$v = ?$

$n = \dfrac{c}{v}$ $\therefore$ $v = \dfrac{c}{n}$

$v = \dfrac{3.00 \times 10^8\ \text{m/s}}{1.33}$

$= \boxed{2.26 \times 10^8\ \text{m/s}}$

(b) $c = 3.00 \times 10^8$ m/s

$n = 1.31$

$v = ?$

$v = \dfrac{3.00 \times 10^8\ \text{m/s}}{1.31}$

$= \boxed{2.29 \times 10^8\ \text{m/s}}$

7.2. $d = 1.50 \times 10^8$ km

$= 1.50 \times 10^{11}$ m

$c = 3.00 \times 10^8$ m/s

$t = ?$

$v = \dfrac{d}{t}$ $\therefore$ $t = \dfrac{d}{v}$

$t = \dfrac{1.50 \times 10^{11}\ \text{m}}{3.00 \times 10^8 \dfrac{\text{m}}{\text{s}}}$

$= \dfrac{1.50 \times 10^{11}}{3.00 \times 10^8}\ \text{m} \times \dfrac{\text{s}}{\text{m}}$

$= 5.00 \times 10^2 \dfrac{\text{m·s}}{\text{m}}$

$= \dfrac{5.00 \times 10^2\ \text{s}}{60.0 \dfrac{\text{s}}{\text{min}}}$

$= \dfrac{5.00 \times 10^2}{60.0}\ \text{s} \times \dfrac{\text{min}}{\text{s}}$

$= \boxed{8.33\ \text{min}}$

7.3. $d = 6.00 \times 10^9$ km

$= 6.00 \times 10^{12}$ m

$c = 3.00 \times 10^8$ m/s

$t = ?$

$$v = \frac{d}{t} \quad \therefore \quad t = \frac{d}{v}$$

$$t = \frac{6.00 \times 10^{12}\,\text{m}}{3.00 \times 10^{8}\,\dfrac{\text{m}}{\text{s}}}$$

$$= \frac{6.00 \times 10^{12}}{3.00 \times 10^{8}}\,\text{m} \times \frac{\text{s}}{\text{m}}$$

$$= 2.00 \times 10^{4}\,\text{s}$$

$$= \frac{2.00 \times 10^{4}\,\text{s}}{3{,}600\,\dfrac{\text{s}}{\text{h}}}$$

$$= \frac{2.00 \times 10^{4}}{3.600 \times 10^{3}}\,\text{s} \times \frac{\text{h}}{\text{s}}$$

$$= \boxed{5.56\,\text{h}}$$

7.4. From equation 7.1, note that both angles are measured from the normal and that the angle of incidence (θ_i) equals the angle of reflection (θ_r), or

$$\theta_i = \theta_r \quad \therefore \quad \boxed{\theta_i = 10°}$$

7.5.
$$v = 2.20 \times 10^{8}\,\text{m/s}$$

$$c = 3.00 \times 10^{8}\,\text{m/s}$$

$$n = ?$$

$$n = \frac{c}{v}$$

$$= \frac{3.00 \times 10^{8}\,\dfrac{\text{m}}{\text{s}}}{2.20 \times 10^{8}\,\dfrac{\text{m}}{\text{s}}}$$

$$= 1.36$$

According to Table 7.1, the substance with an index of refraction of 1.36 is $\boxed{\text{ethyl alcohol.}}$

7.6. **(a)** From equation 7.3:

$$\lambda = 6.00 \times 10^{-7}\,\text{m} \qquad c = \lambda f \quad \therefore \quad f = \frac{c}{\lambda}$$

$$c = 3.00 \times 10^{8}\,\text{m/s}$$

$$f = ?$$

$$f = \frac{3.00 \times 10^{8}\,\dfrac{\text{m}}{\text{s}}}{6.00 \times 10^{-7}\,\text{m}}$$

$$= \frac{3.00 \times 10^{8}}{6.00 \times 10^{-7}}\,\frac{\text{m}}{\text{s}} \times \frac{1}{\text{m}}$$

$$= 5.00 \times 10^{14}\,\frac{1}{\text{s}}$$

$$= \boxed{5.00 \times 10^{14}\,\text{Hz}}$$

(b) From equation 7.5:

$$f = 5.00 \times 10^{14}\,\text{Hz}$$

$$h = 6.63 \times 10^{-34}\,\text{J·s}$$

$$E = ?$$

$$E = hf$$

$$= (6.63 \times 10^{-34}\,\text{J·s})\left(5.00 \times 10^{14}\,\frac{1}{\text{s}}\right)$$

$$= (6.63 \times 10^{-34})(5.00 \times 10^{14})\,\text{J·s} \times \frac{1}{\text{s}}$$

$$= \boxed{3.32 \times 10^{-19}\,\text{J}}$$

7.7. First, you can find the energy of one photon of the peak intensity wavelength (5.60×10^{-7} m) by using equation 7.3 to find the frequency, then equation 7.5 to find the energy:

Step 1: $c = \lambda f \quad \therefore \quad f = \dfrac{c}{\lambda}$

$$= \frac{3.00 \times 10^{8}\,\dfrac{\text{m}}{\text{s}}}{5.60 \times 10^{-7}\,\text{m}}$$

$$= 5.36 \times 10^{14}\,\text{Hz}$$

Step 2: $E = hf$

$$= (6.63 \times 10^{-34}\,\text{J·s})(5.36 \times 10^{14}\,\text{Hz})$$

$$= 3.55 \times 10^{-19}\,\text{J}$$

Step 3: Since one photon carries an energy of 3.55×10^{-19} J and the overall intensity is 1,000.0 W, each square meter must receive an average of

$$\frac{1{,}000.0\,\dfrac{\text{J}}{\text{s}}}{3.55 \times 10^{-19}\,\dfrac{\text{J}}{\text{photon}}}$$

$$\frac{1.000 \times 10^{3}}{3.55 \times 10^{-19}}\,\frac{\text{J}}{\text{s}} \times \frac{\text{photon}}{\text{J}}$$

$$\boxed{2.82 \times 10^{21}\,\frac{\text{photon}}{\text{s}}}$$

7.8. **(a)** $f = 4.90 \times 10^{14}\,\text{Hz} \qquad c = \lambda f \quad \therefore \quad \lambda = \dfrac{c}{f}$

$$c = 3.00 \times 10^{8}\,\text{m/s}$$

$$\lambda = ?$$

$$\lambda = \frac{3.00 \times 10^{8}\,\dfrac{\text{m}}{\text{s}}}{4.90 \times 10^{14}\,\dfrac{1}{\text{s}}}$$

$$= \frac{3.00 \times 10^{8}}{4.90 \times 10^{14}}\,\frac{\text{m}}{\text{s}} \times \frac{\text{s}}{1}$$

$$= \boxed{6.12 \times 10^{-7}\,\text{m}}$$

(b) According to Table 7.2, this is the frequency and wavelength of orange light.

7.9. $f = 5.00 \times 10^{20}$ Hz

$h = 6.63 \times 10^{-34}$ J·s

$E = ?$

$$E = hf$$

$$= (6.63 \times 10^{-34} \text{ J·s})\left(5.00 \times 10^{20} \, \frac{1}{s}\right)$$

$$= (6.63 \times 10^{-34})(5.00 \times 10^{20}) \text{ J·s} \times \frac{1}{s}$$

$$= \boxed{3.32 \times 10^{-13} \text{ J}}$$

7.10. $\lambda = 1.00$ mm

$= 0.001$ m

$f = ?$

$c = 3.00 \times 10^8$ m/s

$h = 6.63 \times 10^{-34}$ J·s

$E = ?$

Step 1: $c = \lambda f \quad \therefore \quad f = \dfrac{v}{\lambda}$

$$f = \frac{3.00 \times 10^8 \, \frac{m}{s}}{1.00 \times 10^{-3} \text{ m}}$$

$$= \frac{3.00 \times 10^8}{1.00 \times 10^{-3}} \, \frac{m}{s} \times \frac{1}{m}$$

$$= 3.00 \times 10^{11} \text{Hz}$$

Step 2: $E = hf$

$$= (6.63 \times 10^{-34} \text{ J·s})\left(3.00 \times 10^{11} \, \frac{1}{s}\right)$$

$$= (6.63 \times 10^{-34})(3.00 \times 10^{11}) \text{ J·s} \times \frac{1}{s}$$

$$= \boxed{1.99 \times 10^{-22} \text{ J}}$$

7.11. The index of refraction is found from $n = \dfrac{c}{v}$, where n is the index of refraction of a transparent material, c is the speed of light in a vaccum, and v is the speed of light in the material. The index of refraction of glass is found in Table 7.1 ($n = 1.50$).

$n = 1.50$

$c = 3.00 \times 10^8$ m/s

$v = ?$

$n = \dfrac{c}{v} \quad \therefore \quad v = \dfrac{c}{n}$

$$= \frac{3.00 \times 10^8 \, \frac{m}{s}}{1.50}$$

$$= \boxed{2.00 \times 10^8 \, \frac{m}{s}}$$

7.12. Listing the known and unknown quantities:

Wavelength	$\lambda = 5.60 \times 10^{-7}$ m
Speed of light	$c = 3.00 \times 10^{-8}$ m/s
Frequency	$f = ?$

The relationship between the wavelength (λ), frequency (f), and speed of light in a vacuum (c), is found in equation 7.3, $c = \lambda f$.

$$c = \lambda f \quad \therefore \quad f = \frac{c}{\lambda}$$

$$= \frac{3.00 \times 10^8 \, \frac{m}{s}}{5.60 \times 10^{-7} \text{ m}}$$

$$= \frac{3.00 \times 10^8}{5.60 \times 10^{-7}} \, \frac{m}{s} \times \frac{1}{m}$$

$$= 5.40 \times 10^{14} \, \frac{1}{s}$$

$$= \boxed{5.40 \times 10^{14} \text{ Hz}}$$

7.13. Listing the known and unknown quantities:

Frequency	$f = 5.00 \times 10^{14}$ Hz
Planck's constant	$h = 6.63 \times 10^{-34}$ J·s
Energy	$E = ?$

The relationship between the frequency (f) and energy (E) of a photon is found in equation 7.4, $E = hf$.

$$E = hf$$

$$= (6.63 \times 10^{-34} \text{ J·s})\left(5.00 \times 10^{14} \, \frac{1}{s}\right)$$

$$= (6.63 \times 10^{-34})(5.00 \times 10^{14}) \text{ J·s} \times \frac{1}{s}$$

$$= 3.32 \times 10^{-19} \, \frac{\text{J·s}}{s}$$

$$= \boxed{3.32 \times 10^{-19} \text{ J}}$$

7.14. Listing the known and unknown quantities:

Frequency	$f = 6.50 \times 10^{14}$ Hz
Planck's constant	$h = 6.63 \times 10^{-34}$ J·s
Energy	$E = ?$

The relationship between the frequency (f) and energy (E) of a photon is found in equation 7.4, $E = hf$.

$$E = hf$$

$$= (6.63 \times 10^{-34} \text{ J·s})\left(6.50 \times 10^{14} \, \frac{1}{s}\right)$$

$$= (6.63 \times 10^{-34})(6.50 \times 10^{14}) \text{ J·s} \times \frac{1}{s}$$

$$= 4.31 \times 10^{-19} \frac{\text{J·s}}{\text{s}}$$

$$= \boxed{4.31 \times 10^{-19} \text{ J}}$$

7.15   Listing the known and unknown quantities:

Wavelength $\lambda = 5.60 \times 10^{-7}$ m

Planck's constant $h = 6.63 \times 10^{-34}$ J·s

Speed of light $c = 3.00 \times 10^{8}$ m/s

Frequency $f = ?$

First, you can find the energy of one photon of the peak intensity wavelength (5.60×10^{-7} m) by using the relationship between the wavelength (λ), frequency (f), and speed of light in a vacuum (c), $c = \lambda f$, then use the relationship between the frequency (f) and energy (E) of a photon, $E = hf$, to find the energy:

Step 1: $c = \lambda f$ $\quad \therefore \quad f = \dfrac{c}{\lambda}$

$$= \frac{3.00 \times 10^{8} \frac{\text{m}}{\text{s}}}{5.60 \times 10^{-7} \text{ m}}$$

$$= \frac{3.00 \times 10^{8}}{5.60 \times 10^{-7}} \frac{\text{m}}{\text{s}} \times \frac{1}{\text{m}}$$

$$= 5.36 \times 10^{14} \frac{1}{\text{s}}$$

$$= 5.36 \times 10^{14} \text{ Hz}$$

Step 2: $E = hf$

$$= (6.63 \times 10^{-34} \text{ J·s})\left(5.36 \times 10^{14} \frac{1}{\text{s}}\right)$$

$$= (6.63 \times 10^{-34})(5.36 \times 10^{14}) \text{ J·s} \times \frac{1}{\text{s}}$$

$$= 3.55 \times 10^{-19} \frac{\text{J·s}}{\text{s}}$$

$$= 3.55 \times 10^{-19} \text{ J}$$

Step 3: Since one photon carries an energy of 3.55×10^{-19} J and the overall intensity is 1,000.0 W, for each square meter there must be an average of

$$\frac{1,000.0 \frac{\text{J}}{\text{s}}}{3.55 \times 10^{-19} \frac{\text{J}}{\text{photon}}}$$

$$\frac{1.000 \times 10^{-3}}{3.55 \times 10^{-19}} \frac{\text{J}}{\text{s}} \times \frac{\text{photon}}{\text{J}}$$

$$\boxed{2.82 \times 10^{21} \frac{\text{photon}}{\text{s}}}$$

EXERCISES Chapter 8

8.1. $m = 1.68 \times 10^{-27}$ kg

$v = 3.22 \times 10^{3}$ m/s

$h = 6.63 \times 10^{-34}$ J·s

$\lambda = ?$

$$\lambda = \frac{h}{mv}$$

$$= \frac{6.63 \times 10^{-34} \text{ J·s}}{(1.68 \times 10^{-27} \text{ kg})\left(3.22 \times 10^{3} \frac{\text{m}}{\text{s}}\right)}$$

$$= \frac{6.63 \times 10^{-34}}{(1.68 \times 10^{-27})(3.22 \times 10^{3})} \frac{\text{J·s}}{\text{kg} \times \frac{\text{m}}{\text{s}}}$$

$$= \frac{6.63 \times 10^{-34}}{5.41 \times 10^{-24}} \frac{\frac{\text{kg·m}^2}{\text{s·s}} \times \text{s}}{\text{kg} \times \frac{\text{m}}{\text{s}}}$$

$$= 1.23 \times 10^{-10} \frac{\text{kg·m·m}}{\text{s}} \times \frac{1}{\text{kg}} \times \frac{\text{s}}{\text{m}}$$

$$= \boxed{1.23 \times 10^{-10} \text{ m}}$$

8.2.   **(a)** $n = 6$

$E_L = -13.6$ eV

$E_6 = ?$

$$E_n = \frac{E_L}{n^2}$$

$$E_6 = \frac{-13.6 \text{ eV}}{6^2}$$

$$= \frac{-13.6 \text{ eV}}{36}$$

$$= \boxed{-1.378 \text{ eV}}$$

(b) $= (-0.378 \text{ eV})\left(1.60 \times 10^{-19} \frac{\text{J}}{\text{eV}}\right)$

$$= (-0.378)(1.60 \times 10^{-19}) \text{ eV} \times \frac{\text{J}}{\text{eV}}$$

$$= \boxed{-6.05 \times 10^{-20} \text{ J}}$$

8.3.   **(a)** Energy is related to the frequency and Planck's constant in equation 8.1, $E = hf$. From equation 8.4,

$$hf = E_H - E_L \quad \therefore \quad E = E_H - E_L$$

For $n = 6$, $E_H = 6.05 \times 10^{-20}$ J

For $n = 2$, $E_L = 5.44 \times 10^{-19}$ J

$$E = ? \text{ J}$$

$$E = E_H - E_L$$
$$= (-6.05 \times 10^{-20}\,\text{J}) - (-5.44 \times 10^{-19}\,\text{J})$$
$$= \boxed{4.84 \times 10^{-19}\,\text{J}}$$

(b) $E_H = -0.377\,\text{eV}^*$

$\quad\quad E_L = -3.40\,\text{eV}^*$

$\quad\quad E = ?\,\text{eV}$

$\quad\quad E = E_H - E_L$

$\quad\quad\quad = (-0.377\,\text{eV}) - (-3.40\,\text{eV})$

$\quad\quad\quad = \boxed{3.02\,\text{eV}}$

$\quad\quad$*From figure 8.11

8.4. For $n = 6$, $E_H = -6.05 \times 10^{-20}\,\text{J}$

$\quad\quad$ For $n = 2$, $E_L = -5.44 \times 10^{-19}\,\text{J}$

$\quad\quad\quad h = 6.63 \times 10^{-34}\,\text{J·s}$

$\quad\quad\quad f = ?$

$$hf = E_H - E_L \quad \therefore \quad f = \frac{E_H - E_L}{h}$$

$$f = \frac{(-6.05 \times 10^{-20}\,\text{J}) - (-5.44 \times 10^{-19}\,\text{J})}{6.63 \times 10^{-34}\,\text{J·s}}$$

$$= \frac{4.84 \times 10^{-19}\,\cancel{\text{J}}}{6.63 \times 10^{-34}\,\cancel{\text{J}}\text{·s}}$$

$$= 7.29 \times 10^{14}\,\frac{1}{\text{s}}$$

$$= \boxed{7.29 \times 10^{14}\,\text{Hz}}$$

8.5. $(n = 1) = -13.6\,\text{eV} \quad\quad E_n = \dfrac{E_1}{n^2}$

$\quad\quad\quad E = ? \quad\quad\quad\quad\quad = \dfrac{-13.6\,\text{eV}}{1^2}$

$\quad\quad\quad\quad\quad\quad\quad\quad\quad\quad\quad = -13.6\,\text{eV}$

Since the energy of the electron is $-13.6\,\text{eV}$, it will require $13.6\,\text{eV}$ (or $2.17 \times 10^{-18}\,\text{J}$) to remove the electron.

8.6. $q/m = -1.76 \times 10^{11}\,\text{C/kg}$

$\quad\quad\quad q = -1.60 \times 10^{-19}\,\text{C}$

$\quad\quad\quad m = ?$

$$\text{mass} = \frac{\text{charge}}{\text{charge/mass}}$$

$$= \frac{-1.60 \times 10^{-19}\,\text{C}}{-1.76 \times 10^{11}\,\dfrac{\text{C}}{\text{kg}}}$$

$$= \frac{-1.60 \times 10^{-19}}{-1.76 \times 10^{11}}\,\cancel{\text{C}} \times \frac{\text{kg}}{\cancel{\text{C}}}$$

$$= \boxed{9.09 \times 10^{-31}\,\text{kg}}$$

8.7. $\lambda = -1.67 \times 10^{-10}\,\text{m}$

$\quad\quad\quad m = 9.11 \times 10^{-31}\,\text{kg}$

$\quad\quad\quad v = ?$

$$\lambda = \frac{h}{mv} \quad\quad \therefore \quad\quad v = \frac{h}{m\lambda}$$

$$v = \frac{6.63 \times 10^{-34}\,\text{J·s}}{(9.11 \times 10^{-31}\,\text{kg})(1.67 \times 10^{-10}\,\text{m})}$$

$$= \frac{6.63 \times 10^{-34}}{(9.11 \times 10^{-31})(1.67 \times 10^{-10})}\,\frac{\text{J·s}}{\text{kg·m}}$$

$$= \frac{6.63 \times 10^{-34}}{1.52 \times 10^{-40}}\,\frac{\dfrac{\text{kg·m}^2}{\text{s·}\cancel{\text{s}}} \times \cancel{\text{s}}}{\text{kg·m}}$$

$$= 4.36 \times 10^6\,\frac{\cancel{\text{kg}}\text{·m·}\cancel{\text{m}}}{\text{s}} \times \frac{1}{\cancel{\text{kg}}} \times \frac{1}{\cancel{\text{m}}}$$

$$= \boxed{4.36 \times 10^6\,\frac{\text{m}}{\text{s}}}$$

8.8. **(a)** Boron: $1s^2 2s^2 2p^1$
   (b) Aluminum: $1s^2 2s^2 2p^6 3s^2 3p^1$
   (c) Potassium: $1s^2 2s^2 2p^6 3s^2 3p^6 4s^1$

8.9. **(a)** Boron is atomic number 5 and there are 5 electrons.
   (b) Aluminum is atomic number 13 and there are 13 electrons.
   (c) Potassium is atomic number 19 and there are 19 electrons.

8.10. **(a)** Argon: $1s^2 2s^2 2p^6 3s^2 3p^6$
   (b) Zinc: $1s^2 2s^2 2p^6 3s^2 3p^6 4s^2 3d^{10}$
   (c) Bromine: $1s^2 2s^2 2p^6 3s^2 3p^6 4s^2 3d^{10} 4p^5$

8.11. Atomic weight is the weighted average of the isotopes as they occur in nature. Thus

$\quad\quad$ Lithium-6: 6.01512 u $\times$ 0.0742 = 0.446 u
$\quad\quad$ Lithium-7: 7.016 u $\times$ 0.9258 = 6.4054 u

Lithium-6 contributes 0.446 u of the weighted average and lithium-7 contributes 6.4954 u. The atomic weight of lithium is therefore

$\quad\quad\quad\quad$ 0.446 u

$\quad\quad\quad\quad$ +6.4954 u

$\quad\quad\quad\quad$ 6.941 u

8.12. Recall that the subscript is the atomic number, which identifies the number of protons. In a neutral atom, the number of protons equals the number of electrons, so the atomic number tells you the number of electrons, too. The superscript is the mass number, which identifies the number of neutrons and the number of protons in the nucleus. The number of neutrons is therefore the mass number minus the atomic number.

	Protons	Neutrons	Electrons
(a)	6	6	6
(b)	1	0	1
(c)	18	22	18
(d)	1	1	1
(e)	79	118	79
(f)	92	143	92

8.13.

		Period	Family
(a)	Radon (Rn)	6	VIIIA
(b)	Sodium (Na)	3	IA
(c)	Copper (Cu)	4	IB
(d)	Neon (Ne)	2	VIIIA
(e)	Iodine (I)	5	VIIA
(f)	Lead (Pb)	6	IVA

8.14. Recall that the number of outer-shell electrons is the same as the family number for the representative elements:

(a) Li: 1   (d) Cl: 7
(b) N: 5   (e) Ra: 2
(c) F: 7   (f) Be: 2

8.15. The same information that was used in question 5 can be used to draw the dot notation (see Figure 8.20):

(a) $\overset{\cdot}{\underset{\cdot}{B}}\cdot$ (c) $\overset{}{Ca}\cdot$ (e) $\cdot\overset{\cdot\cdot}{\underset{}{O}}:$

(b) $:\overset{\cdot\cdot}{Br}:$ (d) $K\cdot$ (f) $\cdot\overset{\cdot\cdot}{\underset{}{S}}:$

8.16. The charge is found by identifying how many electrons are lost or gained in achieving the noble gas structure:

(a) Boron 3+
(b) Bromine 1−
(c) Calcium 2+
(d) Potassium 1+
(e) Oxygen 2−
(f) Nitrogen 3−

8.17. Metals have one, two, or three outer electrons and are located in the left two-thirds of the periodic table. Semiconductors are adjacent to the line that separates the metals and nonmetals. Look at the periodic table on the inside back cover and you will see:

(a) Krypton—nonmetal
(b) Cesium—metal
(c) Silicon—semiconductor
(d) Sulfur—nonmetal
(e) Molybdenum—metal
(f) Plutonium—metal

8.18. (a) Bromine gained an electron to acquire a 1− charge, so it must be in family VIIA (the members of this family have seven electrons and need one more to acquire the noble gas structure).
(b) Potassium must have lost one electron, so it is in IA.
(c) Aluminum lost three electrons, so it is in IIIA.
(d) Sulfur gained two electrons, so it is in VIA.
(e) Barium lost two electrons, so it is in IIA.
(f) Oxygen gained two electrons, so it is in VIA.

8.19. (a) $^{16}_{8}O$ (c) $^{3}_{1}H$
(b) $^{23}_{11}Na$ (d) $^{35}_{17}Cl$

9.1.

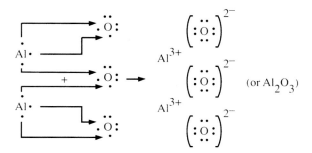

9.2. (a) Sulfur is in family VIA, so sulfur has six valence electrons and will need two more to achieve a stable outer structure like the noble gases. Two more outer shell electrons will give the sulfur atom a charge of 2−. Copper^{2+} will balance the 2− charge of sulfur, so the name is copper(II) sulfide. Note the "-ide" ending for compounds that have only two different elements.

(b) Oxygen is in family VIA, so oxygen has six valence electrons and will have a charge of 2−. Using the crossover technique in reverse, you can see that the charge on the oxygen is 2−, and the charge on the iron is 3−. Therefore the name is iron(III) oxide.

(c) From information in (a) and (b), you know that oxygen has a charge of 2−. The chromium ion must have the same charge to make a neutral compound as it must be, so the name is chromium(II) oxide. Again, note the "-ide" ending for a compound with two different elements.

(d) Sulfur has a charge of 2−, so the lead ion must have the same positive charge to make a neutral compound. The name is lead(II) sulfide.

9.3. The name of some common polyatomic ions are in Table 9.3. Using this table as a reference, the names are

(a) hydroxide
(b) sulfite
(c) hypochlorite
(d) nitrate
(e) carbonate
(f) perchlorate

9.4. The Roman numeral tells you the charge on the variable-charge elements. The charges for the polyatomic ions are found in Table 9.3. The charges for metallic elements can be found in Tables 9.1 and 9.2. Using these resources and the crossover technique, the formulas are as follows:

(a) $Fe(OH)_3$
(b) $Pb_3(PO_4)_2$
(c) $ZnCO_3$
(d) NH_4NO_3
(e) $KHCO_3$
(f) K_2SO_3

9.5. Table 9.7 has information about the meaning of prefixes and stem names used in naming covalent compounds. (a), for example, asks for the formula of carbon tetrachloride. Carbon has no prefixes, so there is one carbon atom, and it comes first in the formula because it comes first in the name. The "tetra-" prefix means four, so there are four chlorine atoms. The name ends in "-ide," so you know there are only two elements in the compound. The symbols can be obtained from the list of elements on the inside back cover of this text. Using all this information from the name, you can think out the formula for carbon tetrachloride. The same process is used for the other compounds and formulas:

(a) CCl_4
(b) H_2O
(c) MnO_2
(d) SO_3
(e) N_2O_5
(f) As_2S_5

9.6. Again using information from Table 9.7, this question requires you to reverse the thinking procedure you learned in question 5.

(a) carbon monoxide
(b) carbon dioxide
(c) carbon disulfide
(d) dinitrogen monoxide
(e) tetraphosphorus trisulfide
(f) dinitrogen trioxide

9.7. The types of bonds formed are predicted by using the electronegativity scale in Table 9.5 and finding the absolute difference. On this basis:

(a) Difference = 1.7, which means ionic bond
(b) Difference = 0, which means covalent
(c) Difference = 0, which means covalent
(d) Difference = 0.4, which means covalent
(e) Difference = 3.0, which means ionic
(f) Difference = 1.6, which means polar covalent and almost ionic

---

10.1. (a) $MgCl_2$ is an ionic compound, so the formula has to be empirical.

(b) C_2H_2 is a covalent compound, so the formula might be molecular. Since it is not the simplest whole number ratio (which would be CH), then the formula is molecular.

(c) BaF_2 is ionic; the formula is empirical.

(d) C_8H_{18} is not the simplest whole number ratio of a covalent compound, so the formula is molecular.

(e) CH_4 is covalent, but the formula might or might not be molecular (?).

(f) S_8 is a nonmetal bonded to a nonmetal (itself); this is a molecular formula.

10.2. (a) $CuSO_4$

$$1 \text{ of Cu} = 1 \times 63.5 \text{ u} = 63.5 \text{ u}$$
$$1 \text{ of S} = 1 \times 32.1 \text{ u} = 32.1 \text{ u}$$
$$4 \text{ of O} = 4 \times 16.0 \text{ u} = \underline{64.0 \text{ u}}$$
$$159.6 \text{ u}$$

(b) CS_2

$$1 \text{ of C} = 1 \times 12.0 \text{ u} = 12.0 \text{ u}$$
$$2 \text{ of S} = 2 \times 32.0 \text{ u} = \underline{64.0 \text{ u}}$$
$$76.0 \text{ u}$$

(c) $CaSO_4$

$$1 \text{ of Ca} = 1 \times 40.1 \text{ u} = 40.1 \text{ u}$$
$$1 \text{ of S} = 1 \times 32.0 \text{ u} = 32.0 \text{ u}$$
$$4 \text{ of O} = 4 \times 16.0 \text{ u} = \underline{64.0 \text{ u}}$$
$$136.1 \text{ u}$$

(d) Na_2CO_3

$$2 \text{ of Na} = 2 \times 23.0 \text{ u} = 46.0 \text{ u}$$
$$1 \text{ of C} = 1 \times 12.0 \text{ u} = 12.0 \text{ u}$$
$$3 \text{ of O} = 3 \times 16.0 \text{ u} = \underline{48.0 \text{ u}}$$
$$106.0 \text{ u}$$

10.3. (a) FeS_2

$$\text{For Fe: } \frac{(55.9 \text{ u Fe})(1)}{119.9 \text{ u } FeS_2} \times 100\% \; FeS_2 = 46.6\% \text{ Fe}$$

$$\text{For S: } \frac{(32.0 \text{ u S})(2)}{119.9 \text{ u } FeS_2} \times 100\% \; FeS_2 = 53.4\% \text{ S}$$

or $(100\% \text{ FeS}_2) - (46.6\% \text{ Fe}) = 53.4\% \text{ S}$

(b) H_3BO_3

$$\text{For H: } \frac{(1.0 \text{ u H})(3)}{61.8 \text{ u } H_3BO_3} \times 100\% \; H_3BO_3 = 4.85\% \text{ H}$$

$$\text{For B: } \frac{(10.8 \text{ u B})(1)}{61.8 \text{ u } H_3BO_3} \times 100\% \; H_3BO_3 = 17.5\% \text{ B}$$

$$\text{For O: } \frac{(16 \text{ u O})(3)}{61.8 \text{ u } H_3BO_3} \times 100\% \; H_3BO_3 = 77.7\% \text{ O}$$

(c) $NaHCO_3$

For Na: $\dfrac{(23.0 \text{ u Na})(1)}{84.0 \text{ u } \overline{NaHCO_3}} \times 100\% \, \overline{NaHCO_3} = 27.4\%$ Na

For H: $\dfrac{(1.0 \text{ u H})(1)}{84.0 \text{ u } \overline{NaHCO_3}} \times 100\% \, \overline{NaHCO_3} = 1.2\%$ H

For C: $\dfrac{(12.0 \text{ u C})(1)}{84.0 \text{ u } \overline{NaHCO_3}} \times 100\% \, \overline{NaHCO_3} = 14.3\%$ C

For O: $\dfrac{(16.0 \text{ u O})(3)}{84.0 \text{ u } \overline{NaHCO_3}} \times 100\% \, \overline{NaHCO_3} = 57.1\%$ O

(d) $C_9H_8O_4$

For C: $\dfrac{(12.0 \text{ u C})(9)}{180.0 \text{ u } \overline{C_9H_8O_4}} \times 100\% \, \overline{C_9H_8O_4} = 60.0\%$ C

For H: $\dfrac{(1.0 \text{ u H})(8)}{180.0 \text{ u } \overline{C_9H_8O_4}} \times 100\% \, \overline{C_9H_8O_4} = 4.4\%$ H

For O: $\dfrac{(16.0 \text{ u O})(4)}{180.0 \text{ u } \overline{C_9H_8O_4}} \times 100\% \, \overline{C_9H_8O_4} = 35.6\%$ O

10.4. **(a)** $2\,SO_2 + O_2 \rightarrow 2\,SO_3$

(b) $4\,P + 5\,O_2 \rightarrow 2\,P_2O_5$

(c) $2\,Al + 6\,HCl \rightarrow 2\,AlCl_3 + 3\,H_2$

(d) $2\,NaOH + H_2SO_4 \rightarrow Na_2SO_4 + 2\,H_2O$

(e) $Fe_2O_3 + 3\,CO \rightarrow 2\,Fe + 3\,CO_2$

(f) $3\,Mg(OH)_2 + 2\,H_3PO_4 \rightarrow Mg_3(PO_4)_2 + 6\,H_2O$

10.5. **(a)** General form of $XY + AZ \rightarrow XZ + AY$ with precipitate formed: Ion exchange reaction.

(b) General form of $X + Y \rightarrow XY$: Combination reaction.

(c) General form of $XY \rightarrow X + Y + \ldots$: Decomposition reaction.

(d) General form of $X + Y \rightarrow XY$: Combination reaction.

(e) General form of $XY + A \rightarrow AY + X$: Replacement reaction.

(f) General form of $X + Y \rightarrow XY$: Combination reaction.

10.6. **(a)** $C_5H_{12(g)} + 8\,O_{2(g)} \rightarrow 5\,CO_{2(g)} + 6\,H_2O_{(g)}$

(b) $HCl_{(aq)} + NaOH_{(aq)} \rightarrow NaCl_{(aq)} + H_2O_{(l)}$

(c) $2\,Al_{(s)} + Fe_2O_{3(s)} \rightarrow Al_2O_{3(s)} + 2\,Fe_{(l)}$

(d) $Fe_{(s)} + CuSO_{4(aq)} \rightarrow FeSO_{4(aq)} + Cu_{(s)}$

(e) $MgCl_{(aq)} + Fe(NO_3)_{2(aq)} \rightarrow$ No reaction (all possible compounds are soluble and no gas or water was formed).

(f) $C_6H_{10}O_{5(s)} + 6\,O_{2(g)} \rightarrow 6\,CO_{2(g)} + 5\,H_2O_{(g)}$

10.7. **(a)** $2\,KClO_3 \xrightarrow{\Delta} 2\,KCl_{(s)} + 3\,O_2 \uparrow$

(b) $2\,Al_2O_{3(l)} \xrightarrow{\text{elec}} 4\,Al_{(s)} + 3\,O_2 \uparrow$

(c) $CaCO_{3(s)} \xrightarrow{\Delta} CaO_{(s)} + CO_2 \uparrow$

10.8. **(a)** $2\,Na_{(s)} + 2\,H_2O_{(l)} \rightarrow 2\,NaOH_{(aq)} + H_2 \uparrow$

(b) $Au_{(s)} + HCl_{(aq)} \rightarrow$ No reaction (gold is below hydrogen in the activity series).

(c) $Al_{(s)} + FeCl_{3(aq)} \rightarrow AlCl_{3(aq)} + Fe_{(s)}$

(d) $Zn_{(s)} + CuCl_{2(aq)} \rightarrow ZnCl_{2(aq)} + Cu_{(s)}$

10.9. **(a)** $NaOH_{(aq)} + HNO_{3(aq)} \rightarrow NaNO_{3(aq)} + H_2O_{(l)}$

(b) $CaCl_{2(aq)} + KNO_{3(aq)} \rightarrow$ No reaction

(c) $3\,Ba(NO_3)_{2(aq)} + 2\,Na_3PO_{4(aq)} \rightarrow 6\,NaNO_{3(aq)} + Ba_3(PO_4)_2 \downarrow$

(d) $2\,KOH_{(aq)} + ZnSO_{4(aq)} \rightarrow K_2SO_{4(aq)} + Zn(OH)_2 \downarrow$

10.10. One mole of oxygen combines with 2 moles of acetylene, so 0.5 mole of oxygen would be needed for 1 mole of acetylene. Therefore, 1 L of C_2H_2 requires 0.5 L of O_2.

EXERCISES **Chapter 11**

11.1. $m_{\text{solute}} = 1.75$ g

$m_{\text{solution}} = 50.0$ g

% weight = ?

% solute $= \dfrac{m_{\text{solute}}}{m_{\text{solution}}} \times 100\%$ solution

$= \dfrac{1.75 \text{ g NaCl}}{50.0 \text{ g solution}} \times 100\%$ solution

$= \boxed{3.50\% \text{ NaCl}}$

11.2. $m_{\text{solution}} = 103.5$ g

$m_{\text{solute}} = 3.50$ g

% weight = ?

% solute $= \dfrac{m_{\text{solute}}}{m_{\text{solution}}} \times 100\%$ solution

$= \dfrac{3.50 \text{ g NaCl}}{103.5 \text{ g solution}} \times 100\%$ solution

$= \boxed{3.38\% \text{ NaCl}}$

11.3. Since ppm is defined as the weight unit of solute in 1,000,000 weight units of solution, the percent by weight can be calculated just like any other percent. The weight of the dissolved sodium and chlorine ions is the part, and the weight of the solution is the whole, so

% $= \dfrac{\text{part}}{\text{whole}} \times 100\%$

$= \dfrac{30{,}113 \text{ g NaCl ions}}{1{,}000{,}000 \text{ g seawater}} \times 100\%$ seawater

$= \boxed{3.00\% \text{ NaCl ions}}$

11.4. $m_{\text{solution}} = 250$ g

% solute = 3.0%

$m_{\text{solute}} = ?$

% solute $= \dfrac{m_{\text{solute}}}{m_{\text{solution}}} \times 100\%$ solution

$\therefore$

$m_{\text{solute}} = \dfrac{(m_{\text{solution}})(\% \text{ solute})}{100\% \text{ solution}}$

D-29   **Appendix D** Solutions for Group A Parallel Exercises   **631**

$$= \frac{(250 \text{ g})(3.0\%)}{100\%}$$

$$= \boxed{7.5 \text{ g}}$$

11.5. % solution $= 12\%$ solution

$$V_{\text{solution}} = 200 \text{ mL}$$

$$V_{\text{solute}} = ?$$

$$\% \text{ solution} = \frac{V_{\text{solute}}}{V_{\text{solution}}} \times 100\% \text{ solution}$$

$$\therefore$$

$$V_{\text{solute}} = \frac{(\% \text{ solution})(V_{\text{solution}})}{100\% \text{ solution}}$$

$$= \frac{(12\% \text{ solution})(200 \text{ mL})}{100\% \text{ solution}}$$

$$= \boxed{24 \text{ mL alcohol}}$$

11.6. % solution $= 40\%$

$$V_{\text{solution}} = 50 \text{ mL}$$

$$V_{\text{solute}} = ?$$

$$\% \text{ solution} = \frac{V_{\text{solute}}}{V_{\text{solution}}} \times 100\% \text{ solution}$$

$$\therefore$$

$$V_{\text{solute}} = \frac{(\% \text{ solution})(V_{\text{solution}})}{100\% \text{ solution}}$$

$$= \frac{(40\% \text{ solution})(50 \text{ mL})}{100\% \text{ solution}}$$

$$= \boxed{20 \text{ mL alcohol}}$$

11.7. **(a)** $\% \text{ concentration} = \dfrac{\text{ppm}}{1 \times 10^4}$

$$= \frac{5}{1 \times 10^4}$$

$$= \boxed{0.0005\% \text{ DDT}}$$

(b) $\% \text{ part} = \dfrac{\text{part}}{\text{whole}} \times 100\% \text{ whole}$

$$\therefore$$

$$\text{whole} = \frac{(100\%)(\text{part})}{\% \text{ part}}$$

$$= \frac{(100\%)(17.0 \text{ g})}{0.0005\%} = \boxed{3{,}400{,}000 \text{ g or } 3{,}400 \text{ kg}}$$

11.8.

(a) $\overbrace{HC_2H_3O_{2(aq)}}^{\text{acid}}$ + $\boxed{H_2O_{(l)}}_{\text{base}}$ $\rightarrow H_3O^+_{(aq)} + C_2H_3O_2^-_{(aq)}$

(b) $\boxed{C_6H_6NH_{2(l)}}_{\text{base}}$ + $\overbrace{H_2O_{(l)}}^{\text{acid}}$ $\rightarrow C_6H_6NH_3^+_{(aq)} + OH^-_{(aq)}$

(c) $\overbrace{HClO_{4(aq)}}^{\text{acid}}$ + $\boxed{HC_2H_3O_{2(aq)}}_{\text{base}}$ $\rightarrow H_2C_2H_3O_2^+_{(aq)} + ClO_4^-_{(aq)}$

(d) $\boxed{H_2O_{(l)}}_{\text{base}}$ + $\overbrace{H_2O_{(l)}}^{\text{acid}}$ $\rightarrow H_3O^+_{(aq)} + OH^-_{(aq)}$

12.1.

(a)

(b)

(c) 2,2-dimethylpropane

12.2. *n*-hexane

3-methylpentane

2-methylpentane

2,2-dimethylbutane

```
        H   CH₃ H   H
        |    |   |   |
   H — C — C — C — C — H
        |    |   |   |
        H   CH₃ H   H
```

12.3.

(a)
```
        H   H  CH₃ CH₃ H   H   H   H
        |   |   |   |   |   |   |   |
   H — C — C — C — C — C — C — C — C — H
        |   |   |   |   |   |   |   |
        H   H  CH₃ H   H   H   H   H
```

(b)
```
        H   CH₃ H   H   H
        |    |   |   |   |
   H — C = C — C — C — C — H
                 |   |   |
                 H   H   H
```

(c)
```
        H   H           CH₃ H   H
        |   |            |   |   |
   H — C — C — C ≡ C — C — C — C — H
        |   |            |   |   |
        H   H           CH₃ H   H
```

12.4. (a) 2-chloro-4-methylpentane

(b) 2-methyl-1-pentene

(c) 3-ethyl-4-methyl-2-pentene

12.5. The 2,2,3-trimethylbutane is more highly branched, so it will have the higher octane rating.

2,2,3-trimethylbutane
```
        H   CH₃ CH₃ H
        |    |   |   |
   H — C — C — C — C — H
        |    |   |   |
        H   CH₃ H   H
```

2,2,-dimethylpentane
```
        H   CH₃ H   H   H
        |    |   |   |   |
   H — C — C — C — C — C — H
        |    |   |   |   |
        H   CH₃ H   H   H
```

12.6 (a) alcohol
(b) amide
(c) ether
(d) ester
(e) organic acid

13.1. (a) cobalt-60: 27 protons, 33 neutrons

(b) potassium-40: 19 protons, 21 neutrons

(c) neon-24: 10 protons, 14 neutrons

(d) lead-208: 82 protons, 126 neutrons

13.2. (a) $^{60}_{27}Co$ (c) $^{24}_{10}Ne$

(b) $^{40}_{19}K$ (d) $^{204}_{82}Pb$

13.3. (a) cobalt-60: Radioactive because odd numbers of protons (27) and odd numbers of neutrons (33) are usually unstable.

(b) potassium-40: Radioactive, again having an odd number of protons (19) and an odd number of neutrons (21).

(c) neon-24: Stable, because even numbers of protons and neutrons are usually stable.

(d) lead-208: Stable, because even numbers of protons and neutrons *and* because 82 is a particularly stable number of nucleons.

13.4. (a) $^{56}_{26}Fe \rightarrow \,^{0}_{-1}e + \,^{56}_{27}Co$

(b) $^{7}_{4}Be \rightarrow \,^{0}_{-1}e + \,^{7}_{5}B$

(c) $^{64}_{29}Cu \rightarrow \,^{0}_{-1}e + \,^{64}_{30}Zn$

(d) $^{24}_{11}Na \rightarrow \,^{0}_{-1}e + \,^{24}_{12}Mg$

(e) $^{214}_{82}Pb \rightarrow \,^{0}_{-1}e + \,^{214}_{83}Bi$

(f) $^{32}_{15}P \rightarrow \,^{0}_{-1}e + \,^{32}_{16}S$

13.5. (a) $^{235}_{92}U \rightarrow \,^{4}_{2}He + \,^{231}_{90}Th$

(b) $^{226}_{88}Ra \rightarrow \,^{4}_{2}He + \,^{222}_{86}Rn$

(c) $^{239}_{94}Pu \rightarrow \,^{4}_{2}He + \,^{235}_{92}U$

(d) $^{214}_{83}Bi \rightarrow \,^{4}_{2}He + \,^{210}_{81}Tl$

(e) $^{230}_{90}Th \rightarrow \,^{4}_{2}He + \,^{226}_{88}Ra$

(f) $^{210}_{84}Po \rightarrow \,^{4}_{2}He + \,^{206}_{82}Pb$

13.6. Thirty-two days is four half-lives. After the first half-life (8 days), 1/2 oz will remain. After the second half-life (8 + 8, or 16 days), 1/4 oz will remain. After the third half-life (8 + 8 + 8, or 24 days), 1/8 oz will remain. After the fourth half-life (8 + 8 + 8 + 8, or 32 days), 1/16 oz will remain, or 6.3×10^{-2} oz.

Glossary

A

absolute humidity a measure of the actual amount of water vapor in the air at a given time—for example, in grams per cubic meter

absolute magnitude a classification scheme to compensate for the distance differences to stars; calculations of the brightness that stars would appear to have if they were all at a defined, standard distance of 10 parsecs

absolute scale temperature scale set so that zero is at the theoretical lowest temperature possible, which would occur when all random motion of molecules has ceased

absolute zero the theoretical lowest temperature possible, which occurs when all random motion of molecules has ceased

abyssal plain the practically level plain of the ocean floor

acceleration a change in velocity per change in time; by definition, this change in velocity can result from a change in speed, a change in direction, or a combination of changes in speed and direction

accretion disk fat bulging disk of gas and dust from the remains of the gas cloud that forms around a protostar

achondrites homogeneously textured stony meteorites

acid any substance that is a proton donor when dissolved in water; generally considered a solution of hydronium ions in water that can neutralize a base, forming a salt and water

acid-base indicator a vegetable dye used to distinguish acid and base solutions by a color change

air mass a large, more or less uniform body of air with nearly the same temperature and moisture conditions throughout

air mass weather the weather experienced within a given air mass; characterized by slow, gradual changes from day to day

alcohol an organic compound with a general formula of ROH, where R is one of

the hydrocarbon groups—for example, methyl or ethyl

aldehyde an organic molecule with the general formula RCHO, where R is one of the hydrocarbon groups—for example, methyl or ethyl

alkali metals members of family IA of the periodic table, having common properties of shiny, low-density metals that can be cut with a knife and that react violently with water to form an alkaline solution

alkaline earth metals members of family IIA of the periodic table, having common properties of soft, reactive metals that are less reactive than alkali metals

alkanes hydrocarbons with single covalent bonds between the carbon atoms

alkenes hydrocarbons with a double covalent carbon-carbon bond

alkyne hydrocarbon with a carbon-carbon triple bond

allotropic forms elements that can have several different structures with different physical properties—for example, graphite and diamond are two allotropic forms of carbon

alpha particle the nucleus of a helium atom (two protons and two neutrons) emitted as radiation from a decaying heavy nucleus; also known as an alpha ray

alpine glaciers glaciers that form at high elevations in mountainous regions

alternating current an electric current that first moves one direction, then the opposite direction with a regular frequency

amino acids organic functional groups that form polypeptides and proteins

amp unit of electric current; equivalent to C/s

ampere full name of the unit amp

amplitude the extent of displacement from the equilibrium condition; the size of a wave from the rest (equilibrium) position

angle of incidence angle of an incident (arriving) ray or particle to a surface; measured from a line perpendicular to the surface (the normal)

angle of reflection angle of a reflected ray or particle from a surface; measured from a line perpendicular to the surface (the normal)

angular momentum quantum number in the quantum mechanics model of the atom, one of four descriptions of the energy state of an electron wave; this quantum number describes the energy sublevels of electrons within the main energy levels of an atom

angular unconformity a boundary in rock where the bedding planes above and below the time interruption unconformity are not parallel, meaning probable tilting or folding followed by a significant period of erosion, which in turn was followed by a period of deposition

annular eclipse occurs when the penumbra reaches the surface of the earth; as seen from the earth, the sun forms a bright ring around the disk of the new moon

Antarctic circle parallel identifying the limit toward the equator where the sun appears above the horizon all day for six months during the summer; located at 66.5°S latitude

anticline an arch-shaped fold in layered bedrock

anticyclone a high-pressure center with winds flowing away from the center; associated with clear, fair weather

antinode region of maximum amplitude between adjacent nodes in a standing wave

apogee the point at which the Moon's elliptical orbit takes the Moon farthest from Earth

apparent local noon the instant when the Sun crosses the celestial meridian at any particular longitude

apparent local solar time the time found from the position of the sun in the sky; the shadow of the gnomon on a sundial

apparent magnitude a classification scheme for different levels of brightness of stars that you see; brightness values range from 1 to 6 with the number 1 (first magnitude) assigned to the brightest star and the number 6 (sixth magnitude) assigned to the faintest star that can be seen

apparent solar day the interval between two consecutive crossings of the celestial meridian by the sun

aquifer a layer of sand, gravel, or other highly permeable material beneath the surface that is saturated with water and is capable of producing water in a well or spring

Arctic circle parallel identifying the limit toward the equator where the sun appears above the horizon all day for one day up to six months during the summer; located at 66.5°N latitude

arid dry climate classification; receives less than 25 cm (10 in) precipitation per year

aromatic hydrocarbon organic compound with at least one benzene ring structure; cyclic hydrocarbons and their derivatives

artesian term describing the condition where confining pressure forces groundwater from a well to rise above the aquifer

asbestos The common name for any one of several incombustible fibrous minerals that will not melt or ignite and can be woven into a fireproof cloth or used directly in fireproof insulation; about six different commercial varieties of asbestos are used, one of which has been linked to cancer under heavy exposure

asteroids small rocky bodies left over from the formation of the solar system; most are accumulated in a zone between the orbits of Mars and Jupiter

asthenosphere a plastic, mobile layer of the earth's structure that extends around the earth below the lithosphere; ranges in thickness from a depth of 130 km to 160 km

astronomical unit the radius of Earth's orbit is defined as one astronomical unit (A.U.)

atmospheric stability the condition of the atmosphere related to the temperature of the air at increasing altitude compared to the temperature of a rising parcel of air at increasing altitude

atom the smallest unit of an element that can exist alone or in combination with other elements

atomic mass unit relative mass unit (u) of an isotope based on the standard of the carbon-12 isotope, which is defined as a mass of exactly 12.00 u; one atomic mass unit (1 u) is 1/12 the mass of a carbon-12 atom

atomic number the number of protons in the nucleus of an atom

atomic weight weighted average of the masses of stable isotopes of an element as they occur in nature, based on the abundance of each isotope of the element and the atomic mass of the isotope compared to C-12

autumnal equinox one of two times a year that daylight and night are of equal length; occurs on or about September 23 and identifies the beginning of the fall season

avalanche a mass movement of a wide variety of materials such as rocks, snow, trees, soils, and so forth in a single chaotic flow; also called debris avalanche

Avogadro's number the number of C-12 atoms in exactly 12.00 g of C; 6.02×10^{23} atoms or other chemical units; the number of chemical units in one mole of a substance

axis the imaginary line about which a planet or other object rotates

B

background radiation ionizing radiation (alpha, beta, gamma, etc.) from natural sources; between 100 and 500 millirems/yr of exposure to natural radioactivity from the environment

Balmer series a set of four line spectra, narrow lines of color emitted by hydrogen atom electrons as they drop from excited states to the ground state

band of stability a region of a graph of the number of neutrons versus the number of protons in nuclei; nuclei that have the neutron-to-proton ratios located in this band do not undergo radioactive decay

barometer an instrument that measures atmospheric pressure, used in weather forecasting and in determining elevation above sea level

base any substance that is a proton acceptor when dissolved in water; generally considered a solution that forms hydroxide ions in water that can neutralize an acid, forming a salt and water

basin a large, bowl-shaped fold in the land into which streams drain; also a small enclosed or partly enclosed body of water

batholith a large volume of magma that has cooled and solidified below the surface, forming a large mass of intrusive rock

beat rhythmic increases and decreases of volume from constructive and destructive interference between two sound waves of slightly different frequencies

beta particle high-energy electron emitted as ionizing radiation from a decaying nucleus; also known as a beta ray

big bang theory current model of galactic evolution in which the universe was created from an intense and brilliant explosion from a primeval fireball

binding energy the energy required to break a nucleus into its constituent protons and neutrons; also the energy equivalent released when a nucleus is formed

black hole the theoretical remaining core of a supernova that is so dense that even light cannot escape

blackbody radiation electromagnetic radiation emitted by an ideal material (the blackbody) that perfectly absorbs and perfectly emits radiation

body wave a seismic wave that travels through the earth's interior, spreading outward from a disturbance in all directions

Bohr model model of the structure of the atom that attempted to correct the deficiencies of the solar system model and account for the Balmer series

boiling point the temperature at which a phase change of liquid to gas takes place through boiling; the same temperature as the condensation point

boundary the division between two regions of differing physical properties

Bowen's reaction series crystallization series that occurs as a result of the different freezing point temperatures of various minerals present in magma

breaker a wave whose front has become so steep that the top part has broken forward of the wave, breaking into foam, especially against a shoreline

British thermal unit the amount of energy or heat needed to increase the temperature of 1 pound of water 1 degree Fahrenheit (abbreviated Btu)

C

calorie the amount of energy (or heat) needed to increase the temperature of 1 gram of water 1 degree Celsius

Calorie the dieter's "calorie"; equivalent to 1 kilocalorie

carbohydrates organic compounds that include sugars, starches, and cellulose; carbohydrates are used by plants and animals for structure, protection, and food

carbon film a type of fossil formed when the volatile and gaseous constituents of a buried organic structure are distilled away, leaving a carbon film as a record

carbonation in chemical weathering   a reaction that occurs naturally between carbonic acid (H_2CO_3) and rock minerals

cast   sediments deposited by groundwater in a mold, taking the shape and external features of the organism that was removed to form the mold, then gradually changing to sedimentary rock

cathode rays   negatively charged particles (electrons) that are emitted from a negative terminal in an evacuated glass tube

celestial equator   line of the equator of the earth directly above the earth; the equator of the earth projected on the celestial sphere

celestial meridian   an imaginary line in the sky directly above you that runs north through the north celestial pole, south through the south celestial pole, and back around the other side to make a big circle around the earth

celestial sphere   a coordinate system of lines used to locate objects in the sky by imagining a huge turning sphere surrounding the earth with the stars and other objects attached to the sphere; the latitude and longitude lines of the earth's surface are projected to the celestial sphere

cellulose   a polysaccharide abundant in plants that forms the fibers in cell walls that preserve the structure of plant materials

Celsius scale   referent scale that defines numerical values for measuring hotness or coldness, defined as degrees of temperature; based on the reference points of the freezing point of water and the boiling point of water at sea-level pressure, with 100 degrees between the two points

cementation   process by which spaces between buried sediment particles under compaction are filled with binding chemical deposits, binding the particles into a rigid, cohesive mass of a sedimentary rock

Cenozoic   one of four geologic eras; the time of recent life, meaning the fossils of this era are identical to the life found on the earth today

centigrade   alternate name for the Celsius scale

centrifugal force   an apparent outward force on an object following a circular path that is a consequence of the third law of motion

centripetal force   the force required to pull an object out of its natural straight-line path and into a circular path; centripetal means "center seeking"

Cepheid variable   a bright variable star that can be used to measure distance

chain reaction   a self-sustaining reaction where some of the products are able to produce more reactions of the same kind; in a nuclear chain reaction neutrons are the products that produce more nuclear reactions in a self-sustaining series

chemical bond   an attractive force that holds atoms together in a compound

chemical change   a change in which the identity of matter is altered and new substances are formed

chemical energy   a form of energy involved in chemical reactions associated with changes in internal potential energy; a kind of potential energy that is stored and later released during a chemical reaction

chemical equation   concise way of describing what happens in a chemical reaction

chemical equilibrium   occurs when two opposing reactions happen at the same time and at the same rate

chemical reaction   a change in matter where different chemical substances are created by forming or breaking chemical bonds

chemical sediments   ions from rock materials that have been removed from solution, for example, carbonate ions removed by crystallization or organisms to form calcium carbonate chemical sediments

chemical weathering   the breakdown of minerals in rocks by chemical reactions with water, gases of the atmosphere, or solutions

chemistry   the science concerned with the study of the composition, structure, and properties of substances and the transformations they undergo

Chinook   a warm wind that has been warmed by compression; also called Santa Ana

chondrites   subdivision of stony meteorites containing small, spherical lumps of silicate minerals or glass

chondrules   small, spherical lumps of silicate minerals or glass found in some meteorites

cinder cone volcano   a volcanic cone that formed from cinders, sharp-edged rock fragments that cooled from frothy blobs of lava as they were thrown into the air

cirque   a bowl-like depression in the side of a mountain, usually at the upper end of a mountain valley, formed by glacial erosion

clastic sediments   weathered rock fragments that are in various states of being

broken down from solid bedrock; boulders, gravel, and silt

climate   the general pattern of weather that occurs in a region over a number of years

coalescence process (meteorology)   the process by which large raindrops form from the merging and uniting of millions of tiny water droplets

cold front   the front that is formed as a cold air mass moves into warmer air

combination chemical reaction   a synthesis reaction in which two or more substances combine to form a single compound

comets   celestial objects originating from the outer edges of the solar system that move about the sun in highly elliptical orbits; solar heating and pressure from the solar wind form a tail on the comet that points away from the Sun

compaction   the process of pressure from a depth of overlying sediments squeezing the deeper sediments together and squeezing water out

composite volcano   a volcanic cone that formed from a buildup of alternating layers of cinders, ash, and lava flows

compound   a pure chemical substance that can be decomposed by a chemical change into simpler substances with a fixed mass ratio

compressive stress   a force that tends to compress the surface as the earth's plates move into each other

concentration   an arbitrary description of the relative amounts of solute and solvent in a solution; a larger amount of solute makes a concentrated solution, and a small amount of solute makes a dilute concentration

condensation (sound)   a compression of gas molecules; a pulse of increased density and pressure that moves through the air at the speed of sound

condensation (water vapor)   where more vapor or gas molecules are returning to the liquid state than are evaporating

condensation nuclei   tiny particles such as tiny dust, smoke, soot, and salt crystals that are suspended in the air on which water condenses

condensation point   the temperature at which a gas or vapor changes back to a liquid

conduction   the transfer of heat from a region of higher temperature to a region of lower temperature by increased kinetic energy moving from molecule to molecule

constructive interference   the condition in which two waves arriving at the same place, at the same time and in phase, add amplitudes to create a new wave

continental air mass dry air masses that form over large land areas

continental climate a climate influenced by air masses from large land areas; hot summers and cold winters

continental drift a concept that continents shift positions on the earth's surface, moving across the surface rather than being fixed, stationary landmasses

continental glaciers glaciers that cover a large area of a continent, e.g., Greenland and the Antarctic

continental shelf a feature of the ocean floor; the flooded margins of the continents that form a zone of relatively shallow water adjacent to the continents

continental slope a feature of the ocean floor; a steep slope forming the transition between the continental shelf and the deep ocean basin

control rods rods inserted between fuel rods in a nuclear reactor to absorb neutrons and thus control the rate of the nuclear chain reaction

controlled experiment an experiment that allows for a comparison of two events that are identical in all but one respect

convection transfer of heat from a region of higher temperature to a region of lower temperature by the displacement of high-energy molecules—for example, the displacement of warmer, less dense air (higher kinetic energy) by cooler, more dense air (lower kinetic energy)

convection cell complete convective circulation pattern; also, slowly turning regions in the plastic asthenosphere that might drive the motion of plate tectonics

convection zone (of a star) part of the interior of a star according to a model; the region directly above the radiation zone where gases are heated by the radiation zone below and move upward by convection to the surface, where they emit energy in the form of visible light, ultraviolet radiation, and infrared radiation

conventional current opposite to electron current—that is, considers an electric current to consist of a drift of positive charges that flow from the positive terminal to the negative terminal of a battery

convergent boundaries boundaries that occur between two plates moving toward each other

core (of the earth) the center part of the earth, which consists of a solid inner part and liquid outer part, making up about 15 percent of the earth's total volume and about one-third of its mass

core (of a star) dense, very hot region of a star where nuclear fusion reactions release gamma and X-ray radiation

Coriolis effect the apparent deflection due to the rotation of the earth; it is to the right in the Northern Hemisphere

correlation the determination of the equivalence in geologic age by comparing the rocks in two separate locations

coulomb unit used to measure quantity of electric charge; equivalent to the charge resulting from the transfer of 6.24 billion billion particles such as the electron

Coulomb's law relationship between charge, distance, and magnitude of the electrical force between two bodies

covalent bond a chemical bond formed by the sharing of a pair of electrons

covalent compound chemical compound held together by a covalent bond or bonds

creep the slow downhill movement of soil down a steep slope

crest the high mound of water that is part of a wave; also refers to the condensation, or high-pressure part, of a sound wave

critical angle limit to the angle of incidence when all light rays are reflected internally

critical mass mass of fissionable material needed to sustain a chain reaction

crude oil petroleum pumped from the ground that has not yet been refined into usable products

crust the outermost part of the earth's interior structure; the thin, solid layer of rock that rests on top of the Mohorovicic discontinuity

curie unit of nuclear activity defined as 3.70×10^{10} nuclear disintegrations per second

cycle a complete vibration

cyclone a low-pressure center where the winds move into the low-pressure center and are forced upward; a low-pressure center with clouds, precipitation, and stormy conditions

D

data measurement information used to describe something

data points points that may be plotted on a graph to represent simultaneous measurements of two related variables

daylight saving time setting clocks ahead one hour during the summer to more effectively utilize the longer days of summer, then setting the clocks back in the fall

decibel scale a nonlinear scale of loudness based on the ratio of the intensity level of a sound to the intensity at the threshold of hearing

decomposition chemical reaction a chemical reaction in which a compound is broken down into the elements that make up the compound, into simpler compounds, or into elements and simpler compounds

deep-focus earthquakes earthquakes that occur in the lower part of the upper mantle, between 350 and 700 km below the surface of the earth

deflation the widespread removal of base materials from the surface by the wind

degassing process whereby gases and water vapor were released from rocks heated to melting during the early stages of the formation of a planet

delta a somewhat triangular deposit at the mouth of a river formed where a stream flowing into a body of water slowed and lost its sediment-carrying ability

density the compactness of matter described by a ratio of mass (or weight) per unit volume

density current an ocean current that flows because of density differences in seawater

destructive interference the condition in which two waves arriving at the same point at the same time out of phase add amplitudes to create zero total disturbance

dew condensation of water vapor into droplets of liquid on surfaces

dew point temperature the temperature at which condensation begins

diastrophism all-inclusive term that means any and all possible movements of the earth's plates, including drift, isostatic adjustment, and any other process that deforms or changes the earth's surface by movement

diffuse reflection light rays reflected in many random directions, as opposed to the parallel rays reflected from a perfectly smooth surface such as a mirror

dike a tabular-shaped intrusive rock that formed when magma moved into joints or faults that cut across other rock bodies

direct current an electrical current that always moves in one direction

direct proportion when two variables increase or decrease together in the same ratio (at the same rate)

disaccharides two monosaccharides joined together with the loss of a water molecule; examples of disaccharides are sucrose (table sugar), lactose, and maltose

dispersion the effect of spreading colors of light into a spectrum with a material that has an index of refraction that varies with wavelength

divergent boundaries boundaries that occur between two plates moving away from each other

divide line separating two adjacent watersheds

dome a large, upwardly bulging, symmetrical fold that resembles a dome

Doppler effect an apparent shift in the frequency of sound or light due to relative motion between the source of the sound or light and the observer

double bond covalent bond formed when two pairs of electrons are shared by two atoms

dune a hill, low mound, or ridge of windblown sand or other sediments

E

earthflow a mass movement of a variety of materials such as soil, rocks, and water with a thick, fluid-like flow

earthquake a quaking, shaking, vibrating, or upheaval of the earth's surface

earthquake epicenter point on the earth's surface directly above an earthquake focus

earthquake focus place where seismic waves originate beneath the surface of the earth

echo a reflected sound that can be distinguished from the original sound, which usually arrives 0.1 s or more after the original sound

eclipse when the shadow of a celestial body falls on the surface of another celestial body

El Niño changes in atmospheric pressure systems, ocean currents, water temperatures, and wind patterns that seem to be linked to worldwide changes in the weather

elastic rebound the sudden snap of stressed rock into new positions; the recovery from elastic strain that results in an earthquake

elastic strain an adjustment to stress in which materials recover their original shape after a stress is released

electric circuit consists of a voltage source that maintains an electrical potential, a continuous conducting path for a current to follow, and a device where work is done by the electrical potential; a switch in the circuit

is used to complete or interrupt the conducting path

electric current the flow of electric charge

electric field force field produced by an electrical charge

electric field lines a map of an electric field representing the direction of the force that a positive test charge would experience; the direction of an electric field shown by lines of force

electric generator a mechanical device that uses wire loops rotating in a magnetic field to produce electromagnetic induction in order to generate electricity

electric potential energy potential energy due to the position of a charge near other charges

electrical conductors materials that have electrons that are free to move throughout the material; for example, metals

electrical energy a form of energy from electromagnetic interactions; one of five forms of energy—mechanical, chemical, radiant, electrical, and nuclear

electrical force a fundamental force that results from the interaction of electrical charge and is billions and billions of times stronger than the gravitational force; sometimes called the "electromagnetic force" because of the strong association between electricity and magnetism

electrical insulators electrical nonconductors, or materials that obstruct the flow of electric current

electrical nonconductors materials that have electrons that are not moved easily within the material—for example, rubber; electrical nonconductors are also called electrical insulators

electrical resistance the property of opposing or reducing electric current

electrolyte water solution of ionic substances that conducts an electric current

electromagnet a magnet formed by a solenoid that can be turned on and off by turning the current on and off

electromagnetic force one of four fundamental forces; the force of attraction or repulsion between two charged particles

electromagnetic induction process in which current is induced by moving a loop of wire in a magnetic field or by changing the magnetic field

electron subatomic particle that has the smallest negative charge possible, usually found in an orbital of an atom, but gained or lost when atoms become ions

electron configuration the arrangement of electrons in orbitals and suborbitals about the nucleus of an atom

electron current opposite to conventional current; that is, considers electric current to consist of a drift of negative charges that flows from the negative terminal to the positive terminal of a battery

electron dot notation notation made by writing the chemical symbol of an element with dots around the symbol to indicate the number of outer shell electrons

electron pair a pair of electrons with different spin quantum numbers that may occupy an orbital

electron volt the energy gained by an electron moving across a potential difference of one volt; equivalent to 1.60×10^{-19} J

electronegativity the comparative ability of atoms of an element to attract bonding electrons

electrostatic charge an accumulated electric charge on an object from a surplus or deficiency of electrons; also called "static electricity"

element a pure chemical substance that cannot be broken down into anything simpler by chemical or physical means; there are over 100 known elements, the fundamental materials of which all matter is made

empirical formula identifies the elements present in a compound and describes the simplest whole number ratio of atoms of these elements with subscripts

energy the ability to do work

English system a system of measurement that originally used sizes of parts of the human body as referents

entropy the measure of disorder in thermodynamics

eons major blocks of time in the earth's geologic history

epochs subdivisions of geologic periods

equation a statement that describes a relationship in which quantities on one side of the equal sign are identical to quantities on the other side

equation of time the cumulative variation between the apparent local solar time and the mean solar time

equinoxes Latin meaning "equal nights"; time when daylight and night are of equal length, which occurs during the spring equinox and the autumnal equinox

eras the major blocks of time in the earth's geologic history; the Cenozoic, Mesozoic, Paleozoic, and Precambrian

erosion the process of physically removing weathered materials; for example, rock fragments are physically picked up by an erosion agent such as a stream or a glacier

esters class of organic compounds with the general structure of RCOOR′, where R is one of the hydrocarbon groups—for example, methyl or ethyl; esters make up fats, oils, and waxes and some give fruit and flowers their taste and odor

ether class of organic compounds with the general formula ROR′, where R is one of the hydrocarbon groups—for example, methyl or ethyl; mostly used as industrial and laboratory solvents

excited states as applied to an atom, describes the energy state of an atom that has electrons in a state above the minimum energy state for that atom; as applied to a nucleus, describes the energy state of a nucleus that has particles in a state above the minimum energy state for that nuclear configuration

exfoliation the fracturing and breaking away of curved, sheetlike plates from bare rock surfaces via physical or chemical weathering, resulting in dome-shaped hills and rounded boulders

exosphere the outermost layer of the atmosphere where gas molecules merge with the diffuse vacuum of space

experiment a re-creation of an event in a way that enables a scientist to gain valid and reliable empirical evidence

external energy the total potential and kinetic energy of an everyday-sized object

extrusive igneous rocks fine-grained igneous rocks formed as lava cools rapidly on the surface

F

Fahrenheit scale referent scale that defines numerical values for measuring hotness or coldness, defined as degrees of temperature; based on the reference points of the freezing point of water and the boiling point of water at sea-level pressure, with 180 degrees between the two points

family vertical columns of the periodic table consisting of elements that have similar properties

fats organic compounds of esters formed from glycerol and three long-chain carboxylic acids that are also called triglycerides; called fats in animals and oils in plants

fault a break in the continuity of a rock formation along which relative movement has occurred between the rocks on either side

fault plane the surface along which relative movement has occurred between the rocks on either side; the surface of the break in continuity of a rock formation

ferromagnesian silicates silicates that contain iron and magnesium; examples include the dark-colored minerals olivine, augite, hornblende, and biotite

first law of motion every object remains at rest or in a state of uniform straight-line motion unless acted on by an unbalanced force

first law of thermodynamics a statement of the law of conservation of energy in the relationship between internal energy, work, and heat

first quarter the moon phase between the new phase and the full phase when the moon is perpendicular to a line drawn through the earth and the sun; one-half of the lighted moon can be seen from the earth, so this phase is called the first quarter

floodplain the wide, level floor of a valley built by a stream; the river valley where a stream floods when it spills out of its channel

fluids matter that has the ability to flow or be poured; the individual molecules of a fluid are able to move, rolling over or by one another

focus the place beneath the surface where the waves of an earthquake originate

folds bends in layered bedrock as a result of stress or stresses that occurred when the rock layers were in a ductile condition, probably under considerable confining pressure from deep burial

foliation the alignment of flat crystal flakes of a rock into parallel sheets

force a push or pull capable of changing the state of motion of an object; a force has magnitude (strength) as well as direction

force field a model describing action at a distance by giving the magnitude and direction of force on a unit particle; considers a charge or a mass to alter the space surrounding it and a second charge or mass to interact with the altered space with a force

formula describes what elements are in a compound and in what proportions

formula weight the sum of the atomic weights of all the atoms in a chemical formula

fossil any evidence of former prehistoric life

fossil fuels organic fuels that contain the stored radiant energy of the sun converted to chemical energy by plants or animals

that lived millions of years ago; coal, petroleum, and natural gas are the common fossil fuels

Foucault pendulum a heavy mass swinging from a long wire that can be used to provide evidence about the rotation of the earth

fracture strain an adjustment to stress in which materials crack or break as a result of the stress

free fall when objects fall toward the earth with no forces acting upward; air resistance is neglected when considering an object to be in free fall

freezing point the temperature at which a phase change of liquid to solid takes place; the same temperature as the melting point for a given substance

frequency the number of cycles of a vibration or of a wave occurring in one second, measured in units of cycles per second (hertz)

freshwater water that is not saline and is fit for human consumption

front the boundary, or thin transition zone, between air masses of different temperatures

frost ice crystals formed by water vapor condensing directly from the vapor phase; frozen water vapor that forms on objects

frost wedging the process of freezing and thawing water in small rock pores and cracks that become larger and larger, eventually forcing pieces of rock to break off

fuel rod long zirconium alloy tubes containing fissionable material for use in a nuclear reactor

full moon the moon phase when the earth is between the sun and the moon and the entire side of the moon facing the earth is illuminated by sunlight

functional group the atom or group of atoms in an organic molecule that is responsible for the chemical properties of a particular class or group of organic chemicals

fundamental charge smallest common charge known; the magnitude of the charge of an electron and a proton, which is 1.60×10^{-19} coulomb

fundamental frequency the lowest frequency (longest wavelength) that can set up standing waves in an air column or on a string

fundamental properties a property that cannot be defined in simpler terms other than to describe how it is measured; the fundamental properties are length, mass, time, and charge

G

g symbol representing the acceleration of an object in free fall due to the force of gravity; its magnitude is 9.8 m/s² (32 ft/s²)

galactic clusters gravitationally bound subgroups of as many as 1,000 stars that move together within the Milky Way galaxy

galaxy group of billions and billions of stars that form the basic unit of the universe; for example, Earth is part of the solar system, which is located in the Milky Way galaxy

gamma ray very-short-wavelength electromagnetic radiation emitted by decaying nuclei

gases a phase of matter composed of molecules that are relatively far apart moving freely in a constant, random motion and having weak cohesive forces acting between them, resulting in the characteristic indefinite shape and indefinite volume of a gas

gasohol solution of ethanol and gasoline

Geiger counter a device that indirectly measures ionizing radiation (beta and/or gamma) by detecting "avalanches" of electrons that are able to move because of the ions produced by the passage of ionizing radiation

geologic time scale a "calendar" of geologic history based on the appearance and disappearance of particular fossils in the sedimentary rock record

geomagnetic time scale time scale established from the number and duration of magnetic field reversals during the past 6 million years

giant planets the large outer planets Jupiter, Saturn, Uranus, and Neptune that all have similar densities and compositions

glacier a large mass of ice on land that is formed from compacted snow and slowly moves under its own weight

globular clusters symmetrical and tightly packed clusters of as many as a million stars that move together as subgroups within the Milky Way galaxy

glycerol an alcohol with three hydroxyl groups per molecule; for example, glycerin (1,2,3-propanetriol)

glycogen a highly branched polysaccharide synthesized by the human body and stored in the muscles and liver; serves as a direct reserve source of energy

glycol an alcohol with two hydroxyl groups per molecule; for example, ethylene glycol that is used as an antifreeze

gram-atomic weight the mass in grams of one mole of an element that is numerically equal to its atomic weight

gram-formula weight the mass in grams of one mole of a compound that is numerically equal to its formula weight

gram-molecular weight the gram-formula weight of a molecular compound

granite light-colored, coarse-grained igneous rock common on continents; igneous rocks formed by blends of quartz and feldspars, with small amounts of micas, hornblende, and other minerals

greenhouse effect the process of increasing the temperature of the lower parts of the atmosphere through redirecting energy back toward the surface; the absorption and reemission of infrared radiation by carbon dioxide, water vapor, and a few other gases in the atmosphere

ground state energy state of an atom with electrons at the lowest energy state possible for that atom

groundwater water from a saturated zone beneath the surface; water from beneath the surface that supplies wells and springs

gyre the great circular systems of moving water in each ocean

H

hail a frozen form of precipitation, sometimes with alternating layers of clear and opaque, cloudy ice

hair hygrometer a device that measures relative humidity from changes in the length of hair

half-life the time required for one-half of the unstable nuclei in a radioactive substance to decay into a new element

halogen member of family VIIA of the periodic table, having common properties of very reactive nonmetallic elements common in salt compounds

hard water water that contains relatively high concentrations of dissolved salts of calcium and magnesium

heat total internal energy of molecules, which is increased by gaining energy from a temperature difference (conduction, convection, radiation) or by gaining energy from a form conversion (mechanic, chemical, radiant, electrical, nuclear)

heat of formation energy released in a chemical reaction

Heisenberg uncertainty principle you cannot measure both the exact momentum and the exact position of a subatomic particle at the same time—the more exactly one of the two is known, the less certain you are of the value of the other

hertz unit of frequency; equivalent to one cycle per second

Hertzsprung-Russell diagram diagram to classify stars with a temperature-luminosity graph

high short for high-pressure center (anticyclone), which is associated with clear, fair weather

high latitudes latitudes close to the poles; those that sometimes receive no solar radiation at noon

high-pressure center another term for anticyclone

horsepower measurement of power defined as a power rating of 550 ft·lb/s

hot spots sites on the earth's surface where plumes of hot rock materials rise from deep within the mantle

humid moist climate classification; receives more than 50 cm (20 in) precipitation per year

humidity the amount of water vapor in the air; see *relative humidity*

hurricane a tropical cyclone with heavy rains and winds exceeding 120 km/h

hydration the attraction of water molecules for ions; a reaction that occurs between water and minerals that make up rocks

hydrocarbon an organic compound consisting of only the two elements hydrogen and carbon

hydrocarbon derivatives organic compounds that can be thought of as forming when one or more hydrogen atoms on a hydrocarbon have been replaced by an element or a group of elements other than hydrogen

hydrogen bond a weak to moderate bond between the hydrogen end (+) of a polar molecule and the negative end (−) of a second polar molecule

hydrologic cycle water vapor cycling into and out of the atmosphere through continuous evaporation of liquid water from the surface and precipitation of water back to the surface

hydronium ion a molecule of water with an attached hydrogen ion, H_3O^+

hypothesis a tentative explanation of a phenomenon that is compatible with the data and provides a framework for understanding and describing that phenomenon

I

ice-crystal process a precipitation-forming process that brings water droplets of a cloud together through the formation of ice crystals

ice-forming nuclei small, solid particles suspended in air; ice can form on the suspended particles

igneous rocks rocks that formed from magma, which is a hot, molten mass of melted rock materials

impulse a change of motion is brought about by an impulse; the product of the size of an applied force and the time the force is applied

incandescent matter emitting visible light as a result of high temperature; for example, a lightbulb, a flame from any burning source, and the sun are all incandescent sources because of high temperature

incident ray line representing the direction of motion of incoming light approaching a boundary

inclination of Earth axis tilt of Earth's axis measured from the plane of the ecliptic (23.5°); considered to be the same throughout the year

index fossils distinctive fossils of organisms that lived only a brief time; used to compare the age of rocks exposed in two different locations

index of refraction the ratio of the speed of light in a vacuum to the speed of light in a material

inertia a property of matter describing the tendency of an object to resist a change in its state of motion; an object will remain in unchanging motion or at rest in the absence of an unbalanced force

infrasonic sound waves having too low a frequency to be heard by the human ear; sound having a frequency of less than 20 Hz

inorganic chemistry the study of all compounds and elements in which carbon is not the principal element

insulators materials that are poor conductors of heat—for example, heat flows slowly through materials with air pockets because the molecules making up air are far apart; also, materials that are poor conductors of electricity, for example, glass or wood

intensity a measure of the energy carried by a wave

interference phenomenon of light whereby the relative phase difference between two light waves produces light or dark spots, a result of light's wavelike nature

intermediate-focus earthquakes earthquakes that occur in the upper part of the mantle, between 70 and 350 km below the surface of the earth

intermolecular forces forces of interaction between molecules

internal energy sum of all the potential energy and all the kinetic energy of all the molecules of an object

international date line the 180° meridian is arbitrarily called the international date line; used to compensate for cumulative time zone changes by adding or subtracting a day when the line is crossed

intertropical convergence zone a part of the lower troposphere in a belt from 10°N to 10°S of the equator where air is heated, expands, and becomes less dense and rises around the belt

intrusive igneous rocks coarse-grained igneous rocks formed as magma cools slowly deep below the surface

inverse proportion the relationship in which the value of one variable increases while the value of the second variable decreases at the same rate (in the same ratio)

inversion a condition of the troposphere when temperature increases with height rather than decreasing with height; a cap of cold air over warmer air that results in increased air pollution

ion an atom or a particle that has a net charge because of the gain or loss of electrons; polyatomic ions are groups of bonded atoms that have a net charge

ion exchange reaction a reaction that takes place when the ions of one compound interact with the ions of another, forming a solid that comes out of solution, a gas, or water

ionic bond chemical bond of electrostatic attraction between negative and positive ions

ionic compounds chemical compounds that are held together by ionic bonds—that is, bonds of electrostatic attraction between negative and positive ions

ionization process of forming ions from molecules

ionization counter a device that measures ionizing radiation (alpha, beta, gamma, etc.) by indirectly counting the ions produced by the radiation

ionized an atom or a particle that has a net charge because it has gained or lost electrons

ionosphere refers to that part of the atmosphere—parts of the thermosphere and upper mesosphere—where free electrons and ions reflect radio waves around the earth and where the northern lights occur

iron meteorites meteorite classification group whose members are composed mainly of iron

island arcs curving chains of volcanic islands that occur over belts of deep-seated earthquakes; for example, the Japanese and Indonesian islands

isomers chemical compounds with the same molecular formula but different molecular structure; compounds that are made from the same numbers of the same elements but have different molecular arrangements

isotope atoms of an element with identical chemical properties but with different masses; isotopes are atoms of the same element with different numbers of neutrons

J

jet stream a powerful, winding belt of wind near the top of the troposphere that tends to extend all the way around the earth, moving generally from the west in both hemispheres at speeds of 160 km/h or more

joint a break in the continuity of a rock formation without a relative movement of the rock on either side of the break

joule metric unit used to measure work and energy; can also be used to measure heat; equivalent to newton-meter

K

Kelvin scale a temperature scale that does not have arbitrarily assigned referent points, and zero means nothing; the zero point on the Kelvin scale (also called absolute scale) is the lowest limit of temperature, where all random kinetic energy of molecules ceases

ketone an organic compound with the general formula RCOR′, where R is one of the hydrocarbon groups; for example, methyl or ethyl

kilocalorie the amount of energy required to increase the temperature of 1 kilogram of water 1 degree Celsius: equivalent to 1,000 calories

kilogram the fundamental unit of mass in the metric system of measurement

kinetic energy the energy of motion; can be measured from the work done to put an object in motion, from the mass and velocity of the object while in motion, or from the amount of work the object can do because of its motion

kinetic molecular theory the collection of assumptions that all matter is made up of tiny atoms and molecules that interact physically, that explain the various states of matter, and that have an average kinetic energy that defines the temperature of a substance

Kuiper Belt a disk-shaped region of small icy bodies some 30 to 100 AU from the Sun; the source of short-period comets

L

L-wave seismic waves that move on the solid surface of the earth much as water waves move across the surface of a body of water

laccolith an intrusive rock feature that formed when magma flowed into the plane of contact between sedimentary rock layers, then raised the overlying rock into a blister-like uplift

lake a large inland body of standing water

landforms the features of the surface of Earth such as mountains, valleys, and plains

landslide general term for rapid movement of any type or mass of materials

last quarter the moon phase between the full phase and the new phase when the Moon is perpendicular to a line drawn through Earth and the Sun; one-half of the lighted Moon can be seen from Earth, so this phase is called the last quarter

latent heat refers to the heat "hidden" in phase changes

latent heat of fusion the heat absorbed when 1 gram of a substance changes from the solid to the liquid phase, or the heat released by 1 gram of a substance when changing from the liquid phase to the solid phase

latent heat of vaporization the heat absorbed when 1 gram of a substance changes from the liquid phase to the gaseous phase, or the heat released when 1 gram of gas changes from the gaseous phase to the liquid phase

laterites highly leached soils of tropical climates; usually red with high iron and aluminum oxide content

latitude the angular distance from the equator to a point on a parallel that tells you how far north or south of the equator the point is located

lava magma, or molten rock, that is forced to the surface from a volcano or a crack in the earth's surface

law of conservation of energy energy is never created or destroyed; it can only be converted from one form to another as the total energy remains constant

law of conservation of mass same as law of conservation of matter; mass, including single atoms, is neither created nor destroyed in a chemical reaction

law of conservation of matter matter is neither created nor destroyed in a chemical reaction

law of conservation of momentum the total momentum of a group of interacting objects remains constant in the absence of external forces

light ray model using lines to show the direction of motion of light to describe the travels of light

light-year the distance that light travels through empty space in one year, approximately 9.5×10^{12} km $(5.86 \times 10^{12}$ mi)

line spectrum narrow lines of color in an otherwise dark spectrum; these lines can be used as "fingerprints" to identify gases

linear scale a scale, generally on a graph, where equal intervals represent equal changes in the value of a variable

lines of force lines drawn to make an electric field strength map, with each line originating on a positive charge and ending on a negative charge; each line represents a path on which a charge would experience a constant force and lines closer together mean a stronger electric field

liquids a phase of matter composed of molecules that have interactions stronger than those found in a gas but not strong enough to keep the molecules near the equilibrium positions of a solid, resulting in the characteristic definite volume but indefinite shape of a liquid

liter a metric system unit of volume usually used for liquids

lithosphere solid layer of the earth's structure that is above the asthenosphere and includes the entire crust, the Moho, and the upper part of the mantle

loess very fine dust or silt that has been deposited by the wind over a large area

longitude angular distance of a point east or west from the prime meridian on a parallel

longitudinal wave a mechanical disturbance that causes particles to move closer together and farther apart in the same direction that the wave is traveling

longshore current a current that moves parallel to the shore, pushed along by waves that move accumulated water from breakers

loudness a subjective interpretation of a sound that is related to the energy of the

vibrating source, related to the condition of the transmitting medium, and related to the distance involved

low latitudes latitudes close to the equator; those that sometimes receive vertical solar radiation at noon

luminosity the total amount of energy radiated into space each second from the surface of a star

luminous an object or objects that produce visible light; for example, the sun, stars, lightbulbs, and burning materials are all luminous

lunar eclipse occurs when the moon is full and the sun, moon, and earth are lined up so the shadow of earth falls on the moon

lunar highlands light-colored mountainous regions of the moon

M

macromolecule very large molecule, with a molecular weight of thousands or millions of atomic mass units, that is made up of a combination of many smaller, similar molecules

magma a mass of molten rock material either below or on the earth's crust from which igneous rock is formed by cooling and hardening

magnetic domain tiny physical regions in permanent magnets, approximately 0.01 to 1 mm, that have magnetically aligned atoms, giving the domain an overall polarity

magnetic field model used to describe how magnetic forces on moving charges act at a distance

magnetic poles the ends, or sides, of a magnet about which the force of magnetic attraction seems to be concentrated

magnetic quantum number from the quantum mechanics model of the atom, one of four descriptions of the energy state of an electron wave; this quantum number describes the energy of an electron orbital as the orbital is oriented in space by an external magnetic field, a kind of energy sub-sublevel

magnetic reversal the flipping of polarity of the earth's magnetic field as the north magnetic pole and the south magnetic pole exchange positions

main sequence stars normal, mature stars that use their nuclear fuel at a steady rate; stars on the Hertzsprung-Russell diagram in a narrow band that runs from the top left to the lower right

manipulated variable in an experiment, a quantity that can be controlled or manipulated; also known as the independent variable

mantle   middle part of the earth's interior; a 2,870 km (about 1,780 mile) thick shell between the core and the crust

maria   smooth, dark areas on the moon

marine climate   a climate influenced by air masses from over an ocean, with mild winters and cool summers compared to areas farther inland

maritime air mass   a moist air mass that forms over the ocean

mass   a measure of inertia, which means a resistance to a change of motion

mass defect   the difference between the sum of the masses of the individual nucleons forming a nucleus and the actual mass of that nucleus

mass movement   erosion caused by the direct action of gravity

mass number   the sum of the number of protons and neutrons in a nucleus defines the mass number of an atom; used to identify isotopes; for example, uranium-238

matter   anything that occupies space and has mass

matter waves   any moving object has wave properties, but at ordinary velocities these properties are observed only for objects with a tiny mass; term for the wavelike properties of subatomic particles

mean solar day   is 24 hours long and is averaged from the mean solar time

mean solar time   a uniform time averaged from the apparent solar time

meanders   winding, circuitous turns or bends of a stream

measurement   the process of comparing a property of an object to a well-defined and agreed-upon referent

mechanical energy   the form of energy associated with machines, objects in motion, and objects having potential energy that results from gravity

mechanical weathering   the physical breaking up of rocks without any changes in their chemical composition

melting point   the temperature at which a phase change of solid to liquid takes place; the same temperature as the freezing point for a given substance

Mercalli scale   expresses the relative intensity of an earthquake in terms of effects on people and buildings using Roman numerals that range from I to XII

meridians   north-south running arcs that intersect at both poles and are perpendicular to the parallels

mesosphere   the term means "middle layer"—the solid, dense layer of the earth's structure below the asthenosphere but above the core; also the layer of the atmosphere below the thermosphere and above the stratosphere

Mesozoic   one of four geologic eras; the time of middle life, meaning some of the fossils for this time period are similar to the life found on the earth today, but many are different from anything living today

metal   matter having the physical properties of conductivity, malleability, ductility, and luster

metamorphic rocks   previously existing rocks that have been changed into a distinctly different rock by heat, pressure, or hot solutions

meteor   the streak of light and smoke that appears in the sky when a meteoroid is made incandescent by friction with the earth's atmosphere

meteor shower   event when many meteorites fall in a short period of time

meteorite   the solid iron or stony material of a meteoroid that survives passage through the earth's atmosphere and reaches the surface

meteoroids   remnants of comets and asteroids in space

meteorology   the science of understanding and predicting weather

meter   the fundamental metric unit of length

metric system   a system of referent units based on invariable referents of nature that have been defined as standards

microclimate   a local, small-scale pattern of climate; for example, the north side of a house has a different microclimate than the south side

middle latitudes   latitudes equally far from the poles and equator; between the high and low latitudes

mineral   a naturally occurring, inorganic solid element or chemical compound with a crystalline structure

miscible fluids   fluids that can mix in any proportion

mixture   matter made of unlike parts that have a variable composition and can be separated into their component parts by physical means

model   a mental or physical representation of something that cannot be observed directly that is usually used as an aid to understanding

moderator   a substance in a nuclear reactor that slows fast neutrons so the neutrons can participate in nuclear reactions

Mohorovicic discontinuity   boundary between the crust and mantle that is marked by a sharp increase in the velocity of seismic waves as they pass from the crust to the mantle

molarity   a measure of the concentration of a solution—number of moles of a solute dissolved in one liter of solution

mold   the preservation of the shape of an organism by the dissolution of the remains of a buried organism, leaving an empty space where the remains were

mole   an amount of a substance that contains Avogadro's number of atoms, ions, molecules, or any other chemical unit; a mole is thus 6.02×10^{23} atoms, ions, or other chemical units

molecular formula   a chemical formula that identifies the actual numbers of atoms in a molecule

molecular weight   the formula weight of a molecular substance

molecule   from the chemical point of view: a particle composed of two or more atoms held together by an attractive force called a chemical bond; from the kinetic theory point of view: smallest particle of a compound or gaseous element that can exist and still retain the characteristic properties of a substance

momentum   the product of the mass of an object times its velocity

monadnocks   hills of resistant rock that are found on peneplains

monosaccharides   simple sugars that are mostly 6-carbon molecules such as glucose and fructose

moraines   deposits of bulldozed rocks and other mounded materials left behind by a melted glacier

mountain   a natural elevation of the earth's crust that rises above the surrounding surface

mudflow   a mass movement of a slurry of debris and water with the consistency of a thick milkshake

N

natural frequency   the frequency of vibration of an elastic object that depends on the size, composition, and shape of the object

neap tide   period of less-pronounced high and low tides: occurs when the sun and moon are at right angles to one another

nebula   a diffuse mass of interstellar clouds of hydrogen gas or dust

negative electric charge   one of the two types of electric charge; repels other negative charges and attracts positive charges

negative ion   atom or particle that has a surplus, or imbalance, of electrons and, thus, a negative charge

net force   the resulting force after all vector forces have been added; if a net force is zero, all the forces have canceled each other and there is not an unbalanced force

neutralized   acid or base properties have been lost through a chemical reaction

neutron   neutral subatomic particle usually found in the nucleus of an atom

neutron star   very small superdense remains of a supernova with a center core of pure neutrons

new crust zone   zone of a divergent boundary where new crust is formed by magma upwelling at the boundary

new moon   the moon phase when the Moon is between Earth and the Sun and the entire side of the Moon facing Earth is dark

newton   a unit of force defined as kg·m/s²; that is, a 1 newton force is needed to accelerate a 1 kg mass 1 m/s²

noble gas   members of family VIII of the periodic table, having common properties of colorless, odorless, chemically inert gases; also known as rare gases or inert gases

node   regions on a standing wave that do not oscillate

noise   sounds made up of groups of waves of random frequency and intensity

nonelectrolytes   water solutions that do not conduct an electric current; covalent compounds that form molecular solutions and cannot conduct an electric current

nonferromagnesian silicates   silicates that do not contain iron or magnesium ions; examples include the minerals of muscovite (white mica), the feldspars, and quartz

nonmetal   an element that is brittle (when a solid), does not have a metallic luster, is a poor conductor of heat and electricity, and is not malleable or ductile

nonsilicates   minerals that do not have the silicon-oxygen tetrahedra in their crystal structure

noon   the event of time when the sun moves across the celestial meridian

normal   a line perpendicular to the surface of a boundary

normal fault   a fault where the hanging wall has moved downward with respect to the foot wall

north celestial pole   a point directly above the north pole of the earth; the point above the north pole on the celestial sphere

north pole   the north pole of a magnet or lodestone is "north seeking," meaning that the pole of a magnet points northward when the magnet is free to turn

nova   a star that explodes or suddenly erupts and increases in brightness

nuclear energy   the form of energy from reactions involving the nucleus, the innermost part of an atom

nuclear fission   nuclear reaction of splitting a massive nucleus into more stable, less-massive nuclei with an accompanying release of energy

nuclear force   one of four fundamental forces, a strong force of attraction that operates over very short distances between subatomic particles; this force overcomes the electric repulsion of protons in a nucleus and binds the nucleus together

nuclear fusion   nuclear reaction of low-mass nuclei fusing together to form more stable and more massive nuclei with an accompanying release of energy

nuclear reactor   steel vessel in which a controlled chain reaction of fissionable materials releases energy

nucleons   name used to refer to both the protons and neutrons in the nucleus of an atom

nucleus   tiny, relatively massive and positively charged center of an atom containing protons and neutrons; the small, dense center of an atom

numerical constant   a constant without units; a number

oblate spheroid   the shape of the earth—a somewhat squashed spherical shape

observed lapse rate   the rate of change in temperature compared to change in altitude

occluded front   a front that has been lifted completely off the ground into the atmosphere, forming a cyclonic storm

ocean   the single, continuous body of salt water on the surface of the earth

ocean basin   the deep bottom of the ocean floor, which starts beyond the continental slope

ocean currents   streams of water within the ocean that stay in about the same path as

they move over large distances; steady and continuous onward movement of a channel of water in the ocean

ocean wave   a moving disturbance that travels across the surface of the ocean

oceanic ridges   long, high, continuous, suboceanic mountain chains; for example, the Mid-Atlantic Ridge in the center of the Atlantic Ocean Basin

oceanic trenches   long, narrow, deep troughs with steep sides that run parallel to the edges of continents

octet rule   a generalization that helps keep track of the valence electrons in most representative elements; atoms of the representative elements (A families) attempt to acquire an outer orbital with eight electrons through chemical reactions

ohm   unit of resistance; equivalent to volts/amps

Ohm's law   the electric potential difference is directly proportional to the product of the current times the resistance

oil field   petroleum accumulated and trapped in extensive porous rock structure or structures

oils   organic compounds of esters formed from glycerol and three long-chain carboxylic acids that are also called triglycerides; called fats in animals and oils in plants

Oort cloud   a spherical "cloud" of small, icy bodies from 30,000 AU out to a light-year from the sun; the source of long-period comets

opaque   materials that do not allow the transmission of any light

orbital   the region of space around the nucleus of an atom where an electron is likely to be found

ore mineral   mineral deposits with an economic value

organic acids   acids derived from organisms; organic compounds with a general formula of RCOOH, where R is one of the hydrocarbon groups; for example, methyl or ethyl

organic chemistry   the study of compounds in which carbon is the principal element

orientation of the earth's axis   direction that the earth's axis points; considered to be the same throughout the year

origin   the only point on a graph where both the x and y variables have a value of zero at the same time

overtones   higher resonant frequencies that occur at the same time as the fundamental frequency, giving a musical instrument its characteristic sound quality

oxbow lake a small body of water, or lake, that formed when two bends of a stream came together and cut off a meander

oxidation the process of a substance losing electrons during a chemical reaction; a reaction between oxygen and the minerals making up rocks

oxidation-reduction reaction a chemical reaction in which electrons are transferred from one atom to another; sometimes called "redox" for short

oxidizing agents substances that take electrons from other substances

ozone shield concentration of ozone in the upper portions of the stratosphere that absorbs potentially damaging ultraviolet radiation, preventing it from reaching the surface of the earth

P

P-wave a pressure, or compressional wave in which a disturbance vibrates materials back and forth in the same direction as the direction of wave movement

P-wave shadow zone a region on the earth between 103° and 142° of arc from an earthquake where no P-waves are received; believed to be explained by P-waves being refracted by the core

Paleozoic one of four geologic eras; time of ancient life, meaning the fossils from this time period are very different from anything living on the earth today

parallels reference lines on the earth used to identify where in the world you are northward or southward from the equator; east and west running circles that are parallel to the equator on a globe with the distance from the equator called the latitude

parts per billion concentration ratio of parts of solute in every one billion parts of solution (ppb); could be expressed as ppb by volume or as ppb by weight

parts per million concentration ratio of parts of solute in every one million parts of solution (ppm); could be expressed as ppm by volume or as ppm by weight

Pauli exclusion principle no two electrons in an atom can have the same four quantum numbers; thus, a maximum of two electrons can occupy a given orbital

peneplain a nearly flat landform that is the end result of the weathering and erosion of the land surface

penumbra the zone of partial darkness in a shadow

percent by volume the volume of solute in 100 volumes of solution

percent by weight the weight of solute in 100 weight units of solution

perigee when the Moon's elliptical orbit brings the Moon closest to Earth

period (geologic time) subdivisions of geologic eras

period (periodic table) horizontal rows of elements with increasing atomic numbers; runs from left to right on the element table

period (wave) the time required for one complete cycle of a wave

periodic law similar physical and chemical properties recur periodically when the elements are listed in order of increasing atomic number

permeability the ability to transmit fluids through openings, small passageways, or gaps

permineralization the process that forms a fossil by alteration of an organism's buried remains by circulating groundwater depositing calcium carbonate, silica, or pyrite

petroleum oil that comes from oil-bearing rock, a mixture of hydrocarbons that is believed to have formed from ancient accumulations of buried organic materials such as remains of algae

pH scale scale that measures the acidity of a solution with numbers below 7 representing acids, 7 representing neutral, and numbers above 7 representing bases

Phanerozoic the eon of an abundant fossil record and living organisms

phase change the action of a substance changing from one state of matter to another; a phase change always absorbs or releases internal potential energy that is not associated with a temperature change

phases of matter the different physical forms that matter can take as a result of different molecular arrangements, resulting in characteristics of the common phases of a solid, liquid, or gas

photoelectric effect the movement of electrons in some materials as a result of energy acquired from absorbed light

photon a quanta of energy in a light wave; the particle associated with light

physical change a change of the state of a substance but not the identity of the substance

pitch the frequency of a sound wave

Planck's constant proportionality constant in the relationship between the energy of vibrating molecules and their frequency of vibration; a value of 6.63×10^{-34} Js

plane of the ecliptic the plane of Earth's orbit

plasma a phase of matter; a very hot gas consisting of electrons and atoms that have been stripped of their electrons because of high kinetic energies

plastic strain an adjustment to stress in which materials become molded or bent out of shape under stress and do not return to their original shape after the stress is released

plate tectonics the theory that the earth's crust is made of rigid plates that float on the upper mantle

plunging folds synclines and anticlines that are not parallel to the surface of the earth

polar air mass cold air mass that forms in cold regions

polar climate zone climate zone of the high latitudes; average monthly temperatures stay below 10°C (50°F), even during the warmest month of the year

polar covalent bond a covalent bond in which there is an unequal sharing of bonding electrons

polarized light whose constituent transverse waves are all vibrating in the same plane; also known as plane-polarized light

Polaroid a film that transmits only polarized light

polyatomic ion ion made up of many atoms

polymers huge, chainlike molecules made of hundreds or thousands of smaller repeating molecular units called monomers

polysaccharides polymers consisting of monosaccharide units joined together in straight or branched chains; starches, glycogen, or cellulose

pond a small body of standing water, smaller than a lake

porosity the ratio of pore space to the total volume of a rock or soil sample, expressed as a percentage; freely admitting the passage of fluids through pores or small spaces between parts of the rock or soil

positive electric charge one of the two types of electric charge; repels other positive charges and attracts negative charges

positive ion atom or particle that has a net positive charge due to an electron or electrons being torn away

potential energy energy due to position; energy associated with changes in position (e.g., gravitational potential energy) or changes in shape (e.g., compressed or stretched spring)

power the rate at which energy is transferred or the rate at which work is performed; defined as work per unit of time

Prearchean the earliest of the eons before life

Precambrian one of four geologic eras; the time before the time of ancient life, meaning the rocks for this time period contain very few fossils

precession the slow wobble of the axis of the earth similar to the wobble of a spinning top

precipitation water that falls to the surface of the earth in the solid or liquid form

pressure defined as force per unit area; for example, pounds per square inch (lb/in^2)

primary coil part of a transformer; a coil of wire that is connected to a source of alternating current

primary loop part of the energy-converting system of a nuclear power plant; the closed pipe system that carries heated water from the nuclear reactor to a steam generator

prime meridian the referent meridian (0°) that passes through the Greenwich Observatory in England

principal quantum number from the quantum mechanics model of the atom, one of four descriptions of the energy state of an electron wave; this quantum number describes the main energy level of an electron in terms of its most probable distance from the nucleus

principle of crosscutting relationships a frame of reference based on the understanding that any geologic feature that cuts across or is intruded into a rock mass must be younger than the rock mass

principle of faunal succession a frame of reference based on the understanding that life forms have changed through time as old life forms disappear from the fossil record and new ones appear, but the same form is never exactly duplicated independently at two different times in history

principle of original horizontality a frame of reference based on the understanding that on a large scale sediments are deposited in flat-lying layers, so any layers of sedimentary rocks that are not horizontal have been subjected to forces that have deformed the earth's surface

principle of superposition a frame of reference based on the understanding that an undisturbed sequence of horizontal rock layers is arranged in chronological order with the oldest layers at the bottom, and each consecutive layer will be younger than the one below it

principle of uniformity a frame of reference of slow, uniform changes in the earth's history; the processes changing rocks today are the processes that changed them in the past, or "the present is the key to the past"

proof a measure of ethanol concentration of an alcoholic beverage; proof is double the concentration by volume; for example, 50 percent by volume is 100 proof

properties qualities or attributes that, taken together, are usually unique to an object; for example, color, texture, and size

proportionality constant a constant applied to a proportionality statement that transforms the statement into an equation

proteins macromolecular polymers made of smaller molecules of amino acids, with molecular weight from about six thousand to fifty million; proteins are amino acid polymers with roles in biological structures or functions; without such a function, they are known as polypeptides

Proterozoic the eon before the Phanerozoic, meaning "beginning life"

protogalaxy collection of gas, dust, and young stars in the process of forming a galaxy

proton subatomic particle that has the smallest possible positive charge, usually found in the nucleus of an atom

protoplanet nebular model a model of the formation of the solar system that states that the planets formed from gas and dust left over from the formation of the sun

protostar an accumulation of gases that will become a star

psychrometer a two-thermometer device used to measure the relative humidity

Ptolemaic system geocentric model of the structure of the solar system that uses epicycles to explain retrograde motion

pulsars the source of regular, equally spaced pulsating radio signals believed to be the result of the magnetic field of a rotating neutron star

pure substance materials that are the same throughout and have a fixed definite composition

pure tone sound made by very regular intensities and very regular frequencies from regular repeating vibrations

Q

quad one quadrillion Btu (10^{15} Btu); used to describe very large amounts of energy

quanta fixed amounts; usually referring to fixed amounts of energy absorbed or emitted by matter ("quanta" is plural, and "quantum" is singular)

quantities measured properties; includes the numerical value of the measurement and the unit used in the measurement

quantum mechanics model of the atom based on the wave nature of subatomic particles, the mechanics of electron waves; also called wave mechanics

quantum numbers numbers that describe energy states of an electron; in the Bohr model of the atom, the orbit quantum numbers could be any whole number 1, 2, 3, and so on out from the nucleus; in the quantum mechanics model of the atom, four quantum numbers are used to describe the energy state of an electron wave

R

rad a measure of radiation received by a material (radiation absorbed dose)

radiant energy the form of energy that can travel through space; for example, visible light and other parts of the electromagnetic spectrum

radiation the transfer of heat from a region of higher temperature to a region of lower temperature by greater emission of radiant energy from the region of higher temperature

radiation zone part of the interior of a star according to a model; the region directly above the core where gamma and X rays from the core are absorbed and reemitted, with the radiation slowly working its way outward

radioactive decay the natural spontaneous disintegration or decomposition of a nucleus

radioactive decay constant a specific constant for a particular isotope that is the ratio of the rate of nuclear disintegration per unit of time to the total number of radioactive nuclei

radioactive decay series series of decay reactions that begins with one radioactive nucleus that decays to a second nucleus that decays to a third nucleus and so on until a stable nucleus is reached

radioactivity spontaneous emission of particles or energy from an atomic nucleus as it disintegrates

radiometric age age of rocks determined by measuring the radioactive decay of unstable elements within the crystals of certain minerals in the rocks

rarefaction a thinning or pulse of decreased density and pressure of gas molecules

ratio a relationship between two numbers, one divided by the other; the ratio of distance per time is speed

real image an image generated by a lens or mirror that can be projected onto a screen

red giant stars one of two groups of stars on the Hertzsprung-Russell diagram that have a different set of properties than the main sequence stars; bright, low temperature giant stars that are enormously bright for their temperature

redox reaction short name for oxidation-reduction reaction

reducing agent supplies electrons to the substance being reduced in a chemical reaction

referent referring to or thinking of a property in terms of another, more familiar object

reflected ray a line representing direction of motion of light reflected from a boundary

reflection the change when light, sound, or other waves bounce backward off a boundary

refraction a change in the direction of travel of light, sound, or other waves crossing a boundary

rejuvenation process of uplifting land that renews the effectiveness of weathering and erosion processes

relative dating dating the age of a rock unit or geologic event relative to some other unit or event

relative humidity ratio (times 100%) of how much water vapor is in the air to the maximum amount of water vapor that could be in the air at a given temperature

rem measure of radiation that considers the biological effects of different kinds of ionizing radiation

replacement (chemical reaction) reaction in which an atom or polyatomic ion is replaced in a compound by a different atom or polyatomic ion

replacement (fossil formation) process in which an organism's buried remains are altered by circulating groundwaters carrying elements in solution; the removal of original materials by dissolutions and the replacement of new materials an atom or molecule at a time

representative elements name given to the members of the A-group families of the periodic table; also called the main-group elements

reservoir a natural or artificial pond or lake used to store water, control floods, or generate electricity; a body of water stored for public use

resonance when the frequency of an external force matches the natural frequency and standing waves are set up

responding variable the variable that responds to changes in the manipulated variable; also known as the dependent variable because its value depends on the value of the manipulated variable

reverberation apparent increase in volume caused by reflections, usually arriving within 0.1 second after the original sound

reverse fault a fault where the hanging wall has moved upward with respect to the foot wall

revolution the motion of a planet as it orbits the sun

Richter scale expresses the intensity of an earthquake in terms of a scale with each higher number indicating 10 times more ground movement and about 30 times more energy released than the preceding number

ridges long, rugged mountain chains rising thousands of meters above the abyssal plains of the ocean basin

rift a split or fracture in a rock formation, in a land formation, or in the crust of the earth

rip current strong, brief current that runs against the surf and out to sea

rock a solid aggregation of minerals or mineral materials that have been brought together into a cohesive solid

rock cycle understanding of igneous, sedimentary, or metamorphic rock as a temporary state in an ongoing transformation of rocks into new types; the process of rocks continually changing from one type to another

rock flour rock pulverized by a glacier into powdery, silt-sized sediment

rockfall the rapid tumbling, bouncing, or free fall of rock fragments from a cliff or steep slope

rockslide a sudden, rapid movement of a coherent unit of rock along a clearly defined surface or plane

rotation the spinning of a planet on its axis

runoff water moving across the surface of the earth as opposed to soaking into the ground

S

S-wave a sideways, or shear wave in which a disturbance vibrates materials from side to side, perpendicular to the direction of wave movement

S-wave shadow zone a region of the earth more than 103° of arc away from the epicenter of an earthquake where S-waves

are not recorded; believed to be the result of the core of the earth being a liquid, or at least acting like a liquid

salinity a measure of dissolved salts in seawater, defined as the mass of salts dissolved in 1,000 g of solution

salt any ionic compound except one with hydroxide or oxide ions

San Andreas fault in California, the boundary between the North American Plate and the Pacific Plate that runs north-south for some 1,300 km (800 miles) with the Pacific Plate moving northwest and the North American Plate moving southeast

saturated air air in which an equilibrium exists between evaporation and condensation; the relative humidity is 100 percent

saturated molecule an organic molecule that has the maximum number of hydrogen atoms possible

saturated solution the apparent limit to dissolving a given solid in a specified amount of water at a given temperature; a state of equilibrium that exists between dissolving solute and solute coming out of solution

scientific law a relationship between quantities, usually described by an equation in the physical sciences; is more important and describes a wider range of phenomena than a scientific principle

scientific principle a relationship between quantities concerned with a specific, or narrow range of observations and behavior

scintillation counter a device that indirectly measures ionizing radiation (alpha, beta, gamma, etc.) by measuring the flashes of light produced when the radiation strikes a phosphor

sea a smaller part of the ocean with characteristics that distinguish it from the larger ocean

sea breeze cool, dense air from over water moving over land as part of convective circulation

seafloor spreading the process by which hot, molten rock moves up from the interior of the earth to emerge along mid-oceanic rifts, flowing out in both directions to create new rocks

seamounts steep, submerged volcanic peaks on the abyssal plain

second the standard unit of time in both the metric and English systems of measurement

second law of motion the acceleration of an object is directly proportional to the net

force acting on that object and inversely proportional to the mass of the object

second law of thermodynamics a statement that the natural process proceeds from a state of higher order to a state of greater disorder

secondary coil part of a transformer, a coil of wire in which the voltage of the original alternating current in the primary coil is stepped up or down by way of electromagnetic induction

secondary loop part of nuclear power plant; the closed-pipe system that carries steam from a steam generator to the turbines, then back to the steam generator as feedwater

sedimentary rocks rocks formed from particles or dissolved minerals from previously existing rocks

sediments accumulations of silt, sand, or gravel that settled out of the atmosphere or out of water

seismic waves vibrations that move as waves through any part of the earth, usually associated with earthquakes, volcanoes, or large explosions

seismograph an instrument that measures and records seismic wave data

semiarid climate classification between arid and humid; receives between 25 and 50 cm (10 and 20 in) precipitation per year

semiconductors elements that have properties between those of a metal and those of a nonmetal, sometimes conducting an electric current and sometimes acting like an electrical insulator depending on the conditions and their purity; also called metalloids

shallow-focus earthquakes earthquakes that occur from the surface down to 70 km deep

shear stress produced when two plates slide past one another or by one plate sliding past another plate that is not moving

shell model of the nucleus model of the nucleus that has protons and neutrons moving in energy levels or shells in the nucleus (similar to the shell structure of electrons in an atom)

shield volcano a broad, gently sloping volcanic cone constructed of solidified lava flows

shock wave a large, intense wave disturbance of very high pressure; the pressure wave created by an explosion, for example

sidereal day the interval between two consecutive crossings of the celestial meridian by a particular star

sidereal month the time interval between two consecutive crossings of the moon across any star

sidereal year the time interval required for the earth to move around its orbit so that the sun is again in the same position against the stars

silicates minerals that contain silicon-oxygen tetrahedra either isolated or joined together in a crystal structure

sill a tabular-shaped intrusive rock that formed when magma moved into the plane of contact between sedimentary rock layers

simple harmonic motion the vibratory motion that occurs when there is a restoring force opposite to and proportional to a displacement

single bond covalent bond in which a single pair of electrons is shared by two atoms

slope the ratio of changes in the y variable to changes in the x variable or how fast the y-value increases as the x-value increases

soil a mixture of unconsolidated weathered earth materials and humus, which is altered, decay-resistant organic matter

solar constant the averaged solar power received by the outermost part of the earth's atmosphere when the sunlight is perpendicular to the outer edge and the earth is at an average distance from the sun; about 1,370 watts per square meter

solenoid a cylindrical coil of wire that becomes electromagnetic when a current runs through it

solids a phase of matter with molecules that remain close to fixed equilibrium positions due to strong interactions between the molecules, resulting in the characteristic definite shape and definite volume of a solid

solstices time when the sun is at its maximum or minimum altitude in the sky, known as the summer solstice and the winter solstice

solubility dissolving ability of a given solute in a specified amount of solvent, the concentration that is reached as a saturated solution is achieved at a particular temperature

solute the component of a solution that dissolves in the other component; the solvent

solution a homogeneous mixture of ions or molecules of two or more substances

solvent the component of a solution present in the larger amount; the solute dissolves in the solvent to make a solution

sonic boom sound waves that pile up into a shock wave when a source is traveling at or faster than the speed of sound

sound quality characteristic of the sound produced by a musical instrument; determined by the presence and relative strengths of the overtones produced by the instrument

south celestial pole a point directly above the south pole of earth; the point above the south pole on the celestial sphere

south pole short for "south seeking"; the pole of a magnet that points southward when it is free to turn

specific heat each substance has its own specific heat, which is defined as the amount of energy (or heat) needed to increase the temperature of 1 gram of a substance 1 degree Celsius

speed a measure of how fast an object is moving—the rate of change of position per change in time; speed has magnitude only and does not include the direction of change

spin quantum number from the quantum mechanics model of the atom, one of four descriptions of the energy state of an electron wave; this quantum number describes the spin orientation of an electron relative to an external magnetic field

spring equinox one of two times a year that daylight and night are of equal length; occurs on or about March 21 and identifies the beginning of the spring season

spring tides unusually high and low tides that occur every two weeks because of the relative positions of the earth, moon, and sun

standard atmospheric pressure the average atmospheric pressure at sea level, which is also known as normal pressure; the standard pressure is 29.92 inches or 760.0 mm of mercury (1,013.25 millibar)

standard time zones 15° wide zones defined to have the same time throughout the zone, defined as the mean solar time at the middle of each zone

standard unit a measurement unit established as the standard upon which the value of the other referent units of the same type are based

standing waves condition where two waves of equal frequency traveling in opposite directions meet and form stationary regions of maximum displacement due to constructive interference and stationary regions of zero displacement due to destructive interference

starch group of complex carbohydrates (polysaccharides) that plants use as a stored food source and that serves as an important source of food for animals

stationary front occurs when the edge of a front is not advancing

steam generator   part of nuclear power plant; the heat exchanger that heats feedwater from the secondary loop to steam with the very hot water from the primary loop

step-down transformer   a transformer that decreases the voltage of a current

step-up transformer   a transformer that increases the voltage of a current

stony-iron meteorites   meteorites composed of silicate minerals and metallic iron

stony meteorites   meteorites composed mostly of silicate minerals that usually make up rocks on the earth

storm   a rapid and violent weather change with strong winds, heavy rain, snow, or hail

strain   adjustment to stress; a rock unit might respond to stress by changes in volume, changes in shape, or breaking

stratopause   the upper boundary of the stratosphere

stratosphere   the layer of the atmosphere above the troposphere where temperature increases with height

stream   a large or small body of running water

stress   a force that tends to compress, pull apart, or deform rock; stress on rocks in the earth's solid outer crust results as the earth's plates move into, away from, or alongside each other

strong acid   acid that ionizes completely in water, with all molecules dissociating into ions

strong base   base that is completely ionic in solution and has hydroxide ions

subduction zone   the region of a convergent boundary where the crust of one plate is forced under the crust of another plate into the interior of the earth

sublimation   the phase change of a solid directly into a vapor or gas

submarine canyons   a feature of the ocean basin; deep, steep-sided canyons that cut through the continental slopes

summer solstice   in the Northern Hemisphere, the time when the sun reaches its maximum altitude in the sky, which occurs on or about June 22 and identifies the beginning of the summer season

superconductors   some materials in which, under certain conditions, the electrical resistance approaches zero

supercooled   water in the liquid phase when the temperature is below the freezing point

supernova   a rare catastrophic explosion of a star into an extremely bright, but short-lived phenomenon

supersaturated   containing more than the normal saturation amount of a solute at a given temperature

surf   the zone where breakers occur; the water zone between the shoreline and the outermost boundary of the breakers

surface wave   a seismic wave that moves across the earth's surface, spreading across the surface as water waves spread on the surface of a pond from a disturbance

swell   regular groups of low-profile, long-wavelength waves that move continuously

syncline   a trough-shaped fold in layered bedrock

synodic month   the interval of time from new moon to new moon (or any two consecutive identical phases)

T

talus   steep, conical or apron-like accumulations of rock fragments at the base of a slope

temperate climate zone   climate zone of the middle latitudes; average monthly temperatures stay between 10°C and 18°C (50°F and 64°F) throughout the year

temperature   how hot or how cold something is; a measure of the average kinetic energy of the molecules making up a substance

tensional stress   the opposite of compressional stress; occurs when one part of a plate moves away from another part that does not move

terrestrial planets   planets Mercury, Venus, Earth, and Mars that have similar densities and compositions as compared to the outer giant planets

theory   a broad, detailed explanation that guides the development of hypotheses and interpretations of experiments in a field of study

thermometer   a device used to measure the hotness or coldness of a substance

thermosphere   thin, high, outer atmospheric layer of the earth where the molecules are far apart and have a high kinetic energy

third law of motion   whenever two objects interact, the force exerted on one object is equal in size and opposite in direction to the force exerted on the other object; forces always occur in matched pairs that are equal and opposite

thrust fault   a reverse fault with a low-angle fault plane

thunderstorm   a brief, intense electrical storm with rain, lightning, thunder, strong winds, and sometimes hail

tidal bore   a strong tidal current, sometimes resembling a wave, produced in very long, very narrow bays as the tide rises

tidal currents   a steady and continuous onward movement of water produced in narrow bays by the tides

tides   periodic rise and fall of the level of the sea from the gravitational attraction of the moon and sun

tornado   a long, narrow, funnel-shaped column of violently whirling air from a thundercloud that moves destructively over a narrow path when it touches the ground

total internal reflection   condition where all light is reflected back from a boundary between materials; occurs when light arrives at a boundary at the critical angle or beyond

total solar eclipse   eclipse that occurs when the earth, the moon, and the sun are lined up so the new moon completely covers the disk of the sun; the umbra of the moon's shadow falls on the surface of the earth

transform boundaries   in plate tectonics, boundaries that occur between two plates sliding horizontally by each other along a long, vertical fault; sudden jerks along the boundary result in the vibrations of earthquakes

transformer   a device consisting of a primary coil of wire connected to a source of alternating current and a secondary coil of wire in which electromagnetic induction increases or decreases the voltage of the source

transition elements   members of the B-group families of the periodic table

transparent   term describing materials that allow the transmission of light; for example, glass and clear water are transparent materials

transportation   the movement of eroded materials by agents such as rivers, glaciers, wind, or waves

transverse wave   a mechanical disturbance that causes particles to move perpendicular to the direction that the wave is traveling

trenches   a long, relatively narrow, steep-sided trough that occurs along the edges of the ocean basins

triglyceride   organic compound of esters formed from glycerol and three long-chain carboxylic acids; also called fats in animals and oil in plants

triple bond   covalent bond formed when three pairs of electrons are shared by two atoms

tropic of Cancer   parallel identifying the northern limit where the sun appears directly overhead; located at 23.5°N latitude

tropic of Capricorn   parallel identifying the southern limit where the sun appears directly overhead; located at 23.5°S latitude

tropical air mass   a warm air mass from warm regions

tropical climate zone   climate zone of the low latitudes; average monthly temperatures stay above 18°C (64°F), even during the coldest month of the year

tropical cyclone   a large, violent circular storm that is born over the warm, tropical ocean near the equator; also called hurricane (Atlantic and eastern Pacific) and typhoon (in western Pacific)

tropical year   the time interval between two consecutive spring equinoxes; used as standard for the common calendar year

tropopause   the upper boundary of the troposphere, identified by the altitude where the temperature stops decreasing and remains constant with increasing altitude

troposphere   layer of the atmosphere from the surface to where the temperature stops decreasing with height

trough   the low mound of water that is part of a wave; also refers to the rarefaction, or low-pressure part of a sound wave

tsunami   very large, fast, and destructive ocean wave created by an undersea earthquake, landslide, or volcanic explosion; a seismic sea wave

turbidity current   a muddy current produced by underwater landslides

typhoon   the name for hurricanes in the western Pacific

U

ultrasonic   sound waves too high in frequency to be heard by the human ear; frequencies above 20,000 Hz

umbra   the inner core of a complete shadow

unconformity   a time break in the rock record

undertow   a current beneath the surface of the water produced by the return of water from the shore to the sea

unit   in measurement, a well-defined and agreed-upon referent

universal law of gravitation   every object in the universe is attracted to every other object with a force directly proportional to the product of their masses and inversely

proportional to the square of the distance between the centers of the two masses

unpolarized light   light consisting of transverse waves vibrating in all conceivable random directions

unsaturated molecule   an organic molecule that does not contain the maximum number of hydrogen atoms; a molecule that can add more hydrogen atoms because of the presence of double or triple bonds

V

valence   the number of covalent bonds an atom can form

valence electrons   electrons of the outermost shell; the electrons that determine the chemical properties of an atom and the electrons that participate in chemical bonding

Van Allen belts   belts of radiation caused by cosmic-ray particles becoming trapped and following the earth's magnetic field lines between the poles

vapor   the gaseous state of a substance that is normally in the liquid state

variable   changing quantity usually represented by a letter or symbol

velocity   describes both the speed and direction of a moving object; a change in velocity is a change in speed, in direction of travel, or both

ventifacts   rocks sculpted by wind abrasion

vernal equinox   another name for the spring equinox, which occurs on or about March 21 and marks the beginning of the spring season

vibration   a back-and-forth motion that repeats itself

virtual image   an image where light rays appear to originate from a mirror or lens; this image cannot be projected on a screen

volcanism   volcanic activity; the movement of magma

volcano   a hill or mountain formed by the extrusion of lava or rock fragments from a mass of magma below

volt   unit of potential difference equivalent to J/C

voltage drop   the electric potential difference across a resistor or other part of a circuit that consumes power

voltage source   source of electric power in an electric circuit that maintains a constant voltage supply to the circuit

volume   how much space something occupies

vulcanism   volcanic activity; the movement of magma

W

warm front   the front that forms when a warm air mass advances against a cool air mass

water table   the boundary below which the ground is saturated with water

watershed   the region or land area drained by a stream; a stream drainage basin

watt   metric unit for power; equivalent to J/s

wave   a disturbance or oscillation that moves through a medium

wave equation   the relationship of the velocity of a wave to the product of the wavelength and frequency of the wave

wave front   a region of maximum displacement in a wave; a condensation in a sound wave

wave height   the vertical distance of an ocean wave between the top of the wave crest and the bottom of the next trough

wave mechanics   alternate name for quantum mechanics derived from the wavelike properties of subatomic particles

wave period   the time required for two successive crests or other successive parts of the wave to pass a given point

wavelength   the horizontal distance between successive wave crests or other successive parts of the wave

weak acid   acids that only partially ionize because of an equilibrium reaction with water

weak base   a base only partially ionized because of an equilibrium reaction with water

weathering   slow changes that result in the breaking up, crumbling, and destruction of any kind of solid rock

white dwarf stars   one of two groups of stars on the Hertzsprung-Russell diagram that have a different set of properties than the main sequence stars; faint, white-hot stars that are very small and dense

wind   a horizontal movement of air that moves along or parallel to the ground, sometimes in currents or streams

wind abrasion   the natural sand-blasting process that occurs when wind particles break off small particles of rock and polish the rock they strike

wind chill factor   the cooling equivalent temperature that results from the wind making the air temperature seem much lower; the cooling power of wind

winter solstice   in the Northern Hemisphere, the time when the sun reaches its minimum altitude, which occurs on or about December 22 and identifies the beginning of the winter season

work   the magnitude of applied force times the distance through which the force acts; can be thought of as the process by which one form of energy is transformed to another

Z

zone of saturation   zone of sediments beneath the surface in which water has collected in all available spaces

Credits

Photographs

Chapter 18 Opener: © Jeff Foot/Tom Stack & Associates; **18.1:** © USGS Photo Library, Denver, CO; **18.7:** © NASA; **Page 455:** © 2000 by the Trustees of Princeton University; **Page 457:** © Sheila Davies.

Chapter 19 Opener: © Douglas Cheeseman/ Peter Arnold, Inc.; **19.1:** © A. Post, USGS Photo Library, Denver, CO; **19.2:** © Bill W. Tillery; **19.4A:** © Robert W. Northrop, Photographer/ Illustrator; **19.4B:** © Bill W. Tillery; **19.5:** C.C. Plummer; **19.6:** © John S. Shelton; **19.9A:** © National Park Service/Photo by Cecil W. Stoughton; **19.9B:** © D.E. Trimble, USGS Photo Library, Denver, CO; **19.10:** © Frank M. Hanna; **19.14:** © University of Colorado, Courtesy National Geophysical Data Center, Boulder, CO; **19.18:** © NASA; **19.21:** © Bill W. Tillery; **19.22:** © D.W. Peterson, USGS; **19.23:** © B. Amundson; **Page 478:** Edinburgh University Library.

Chapter 20 Opener: © G Ziesler/Peter Arnold, Inc; **20.1:** © W.R. Hansen, USGS Photo Library, Denver, CO; **20.2:** National Park Service, Photo by Wm. Belnap, Jr.; **20.4A:** © A. J. Copley/Visuals Unlimited; **20.4B:** © L. Linkhart/ Visuals Unlimited; **20.5:** © Ken Wagner/Visuals Unlimited; **20.6A-B:** © Bill W. Tillery; **20.8:** © William J. Weber/Visuals Unlimited; **20.9:** © Doug Sherman/ Geofile; **20.11:** © B. Amundson; **20.12A:** © D.A. Rahm, photo courtesy of Rahn Memorial Collections, Western Washington University; **20.13:** © C.C. Plummer; **20.16A-B:** © Bill W. Tillery; **Page 496:** U.S. Geological Survey.

Chapter 21 Opener: © Alfred Pasieka/SPL/ Photo Researchers, Inc; **21.1, 21.2:** © Robert W. Northrop, Photographer/Illustrator; **21.7:** © Bob Wallen; **21.10:** © Frank M. Hanna; **21.13:** U.S. Geological Survey; **Page 513:** Library of the Geological Survey of Austria.

Chapter 22 Opener: © Galen Rowell/ Mountain Light Photography; **22.1, 22.16, 22.18A-B, 22.19A-F:** © Bill W. Tillery.

Chapter 23 Opener: © Charles Mayer/ Science Source/Photo Researchers; **23.1:** © Peter Arnold/Peter Arnold, Inc; **23.5, 23.7:** © NOAA; **23.8:** © Rachel Epstein/PhotoEdit, Inc; **23.17:** © Telegraph Herald/Photo by Patti Carr; **23.18, 23.19:** © NOAA; **23.21:** © David Parker/SPL/ Photo Researchers; **23.22:** © Bill W. Tillery; **23.25:** © Dr. Charles Hogue, Curator of Entomology, Los Angeles County Museum of Natural History; **23.26:** © Bob Wallen; **23.27:** © Elizabeth Wallen; **Page 563:** Public Domain; **Page 564:** Courtesy of the National Portrait Gallery London.

Chapter 24 Opener: © James H. Karales/ Peter Arnold, Inc; **24.1:** Salt River Project; **24.9:** City of Tempe, AZ; **24.10:** Salt River Project; **24.11, 24.14, 24.15:** © Bill W. Tillery; **24.19:** © John S. Shelton; **Page 588:** © Bodleian Library, University of Oxford (MS. Eng. Misc. c. 1103, F2).

Line Art/Text

Chapter 3 **3.19:** Source: Energy Information Administration www.eia.doe.gov/emen/sep/us/ frame.html.

Chapter 9 **Box Figure 9.3:** Source: www.accessscience.com/server-ava/arknoid/ science/AS/Biographies/B (see Bunsen). Reprinted by permission of The McGraw-Hill Companies.

Chapter 16 **Table 16.1:** Source: Data from NASA.

Chapter 17 **17.23:** From Carla W. Montgomery, *Physical Geology*, 3rd edition. Copyright © 1993. Reprinted by permission of The McGraw-Hill Companies.

Chapter 18 **18.A-B:** From Charles C. Plummer and David McGeary, *Physical Geology*, 6th edition. Copyright © 1993. Reprinted by permission of The McGraw-Hill Companies. **18.5:** From Charles C. Plummer and David McGeary, *Physical Geology*, 6th edition. Copyright © 1993. Reprinted by permission of The McGraw-Hill Companies. **18.6:** From Charles C. Plummer and David McGeary, *Physical Geology*, 6th edition. Copyright © 1993. Reprinted by permission of The McGraw-Hill Companies. **18.11:** Pitman, W.C., III, Larson, R.L., and Herron, E.M., compilers, 1974, The age of the ocean basin: Boulder, Colorado, Geological Society of America Maps and Charts 6, 2 sheets. Reprinted by permission of The Geological Society of America, Boulder, CO. **18.12:** From Carla W. Montgomery, *Physical Geology*, 3rd edition. Copyright © 1993. Reprinted by permission of The McGraw-Hill Companies. **18.13:** Source: After W. Hamilton, U.S. Geological Survey. **18.18:** From Charles C. Plummer and David McGeary, *Physical Geology*, 6th edition. Copyright © 1993. Reprinted by permission of The McGraw-Hill Companies.

Chapter 19 **Table 19.1:** Source: National Oceanic and Atmospheric Administration. **19.3:** From Carla W. Montgomery, *Physical Geology*, 3rd edition. Copyright © 1993. Reprinted by permission of The McGraw-Hill Companies. **19.6A:** From David McGeary and Charles C. Plummer, *Physical Geology, Earth Revealed*, 2nd Edition. Copyright © 1994. Reprinted by permission of The McGraw-Hill Companies. **19.7:** From Carla W. Montgomery, *Physical Geology*, 3rd edition. Copyright © 1993. Reprinted by permission of The McGraw-Hill Companies. **19.8:** From Carla W. Montgomery, *Physical Geology*, 3rd edition. Copyright © 1993. Reprinted by permission of The McGraw-Hill Companies. **19.10A:** From David McGeary and Charles C. Plummer, *Physical Geology, Earth Revealed*, 2nd Edition. Copyright © 1994. Reprinted by permission of The McGraw-Hill Companies. **19.11A-C:** From Charles C. Plummer and David McGeary, *Physical Geology*, 6th edition. Copyright © 1993. Reprinted by permission of The McGraw-Hill Companies. **19.12A-B:** From Charles C. Plummer and David McGeary, *Physical Geology*, 6th edition. Copyright © 1993. Reprinted by permission of The McGraw-Hill Companies. **19.14A-C:** From Charles C. Plummer and David McGeary, *Physical Geology*, 6th edition. Copyright © 1993. Reprinted by permission of The McGraw-Hill Companies. **19.15:** From Carla W. Montgomery, *Physical Geology*, 3rd edition. Copyright © 1993. Reprinted by permission of The McGraw-Hill Companies. **19.17A-B:** From Carla W. Montgomery, *Physical Geology*, 3rd edition. Copyright ©1993. Reprinted by permission of The McGraw-Hill Companies. **19.20:** From Carla W. Montgomery, Physical Geology, 3rd edition. Copyright © 1993. Reprinted by permission of The McGraw-Hill Companies. **19.25:** From Charles C. Plummer and David McGeary, *Physical Geology*, 5th edition. Copyright © 1991. Reprinted by permission of The McGraw-Hill Companies.

Chapter 20 **20.3A-B:** From Carla W. Montgomery, *Physical Geology*, 3rd edition. Copyright © 1993. Reprinted by permission of The McGraw-Hill Companies. **20.10A-C:** From Charles C. Plummer and David McGeary, *Physical Geology*, 6th edition. Copyright © 1993. Reprinted by permission of The McGraw-Hill Companies. **20.14:** From Charles C. Plummer and David McGeary, *Physical Geology*, 5th edition. Copyright © 1991. Reprinted by permission of The McGraw-Hill Companies.

Chapter 21 **21.3:** From Carla W. Montgomery and David Dathe, *Earth: Then and Now*, 2nd Edition. Copyright © 1994. Reprinted by permission of The McGraw-Hill Companies. **21.5A-B:** From Carla W. Montgomery, *Physical Geology*, 3rd edition. Copyright ©1993. Reprinted by permission of The McGraw-Hill Companies. **21.6:** From Carla W. Montgomery, *Physical Geology*, 3rd edition. Copyright © 1993. Reprinted by permission of The McGraw-Hill Companies. **21.8:** From Carla W. Montgomery, *Physical Geology*, 3rd edition. Copyright © 1993. Reprinted by permission of The McGraw-Hill Companies. **21.9:** From Carla W. Montgomery, *Physical Geology*, 3rd edition. Copyright © 1993. Reprinted by permission of The McGraw-Hill Companies. **21.11:** From Carla W. Montgomery, *Physical Geology*, 3rd edition. Copyright © 1993. Reprinted by permission of The McGraw-Hill Companies. **21.12:** From Carla W. Montgomery, *Physical Geology*, 3rd edition. Copyright © 1993. Reprinted by permission of The McGraw-Hill Companies. **21.15:** Modified from "Decade of North American Geology," 1983 Geologic Time Scale—Geological Society of America.

Chapter 24 **Pages 575–577:** Drawings, and some text, from *How Wastewater Treatment Works . . . The Basics*, The Environmental Protection Agency, Office of Water, www.epa.gov/owmitnet/ basics.html.

Index

Augite, *429*
Autumnal equinox, 401
Average acceleration, 30
Average velocity, 27–28, 31–32
Avogadro, Amedeo, 266
Avogadro's number, 267, 271
Axis of Earth, 400–401

B

Babylonian astronomy, 384, 406
Background radiation, 335, 349
Bacteria, 576, 580
Bakelite, 319
Balancing chemical
  equations, 257–61
Balmer, J. J., 211
Balmer series, 211
Band of stability, 329–30, 349
Bandwidth, 199
Barium, 237, 337, *339*
Barometers, 521
Barrels of petroleum, 76
Barrow, Isaac, 54
Baryonic dark matter, 369
Basalts
  formation of, 435
  on Moon, 414
  in oceanic crust, 446, 451
  surface features created by, 474
Base names, 302
Bases, 287–90, 294
Basins, 465
Batholith, 475
Bay of Fundy, 417
Beats (sound), 127, *128*
Beaufort, Francis, 564
Becquerel, Henri, 326, *327*
Becquerels, 335
Bell Burnell, Jocelyn, 371
Benzene
  Faraday's isolation of, 173
  refraction index, 187
  ring symbol, 306, *307*
Bergen Geophysical Institute, 563
"Best fit" lines, 600
Beta particles
  defined, 327, 349
  emission during radioactive
    decay, 330, 331
Bicarbonates, 290–91
Bicycle racers, 36
Big bang theory, 346–47, 366–68
Big Dipper, *355, 356*
Bimetallic strips, 89
Binary system, 199
Binding energy, 336, 349
Biomass, 75, 78
Biotite, *429*
Bjerknes, Vilhelm Firman Koren, 563
Blackbody radiation, 180–81, *182,* 210
Black Hills, 473
Black holes, 363, 370
Blood pH, 290
Blue crabs, 584
Blueshifts, 134
Bode, 393

Bohr, Niels, 210
Bohr model of atom, 210–13, 225
Boiling point
  phase change at, 99
  of solutions, 284–86, 294
Bonding pairs, 239, 240
Bonds. *See* Chemical bonds
Boric acid, 286
Boron, 171
Boulders, 435
Boundaries between physical
    conditions, 125
Bowen's reaction series, 431, *432,* 433
Brachiopods, 509
Breakers, 582, *583*
Breccias, 414, 435
Brick, rate of conduction, 97
British thermal unit, 94
Bromine ions, 237
Brönsted, Johannes Nicolaus, 293
Buckminsterfullerene, 270
Buckyballs, 270
Bulldozing by glaciers, 491
Bunsen, Robert Wilhelm, 248
Bunsen burner, 248
Burning glasses, 189
Butane, 261, 303

C

Cadmium ions, 238
Calcium
  bonding with fluorine, 237
  in hard water, 290
  ionization, 235, 237
  outer orbital electrons, 222
Calcium bicarbonate, 260
Calcium carbide, 319
Calcium carbonate, 436
Calcium chloride, 244
Calcium hydroxide, 287
Calculus, 54
Calendars, 410
California Current, 585
Callisto, 383, *385*
Caloric theory of heat, 108
Calories, 93–94
Canceling (mathematics), 597
Canyons, submarine, 586
Carbohydrates
  generalized oxidation equation,
    261–62
  overview, 314–15, 321
Carbon. *See also* Hydrocarbons
  as basis of organic compounds, 300
  as reducing agent, 262
  valence, 247
Carbon-12, 209, 267
Carbon-60, 270
Carbonaceous chondrites, 392
Carbonated beverages, 290
Carbonates, 288
Carbonation, chemical weathering
    by, 484
Carbon dioxide
  cycle in atmosphere, 518
  regulation in atmosphere, 520

role in greenhouse effect, 522, 533
  in seawater, 579–80
  speed of sound in, 124
Carbonic acid
  basic features, 286
  chemical weathering by, 484
  in rain, 493
  in seawater, 580
Carboniferous age, 509
Carbon monoxide, 533–34
Carbon tetrachloride, 187, 279
Carboxylic acid, 312
Cascade Mountains, 473, 474, 477
Casts, fossil, 502
Catalysts, 269
Catalytic converters, 269
Catastrophism, 462
Cathode rays, 206–7, 225
CAT scans, 338, 449
Cave formation, 484, *486*
Cavendish, Henry, 51
Celestial equator, 354
Celestial meridian, 355
Celestial sphere, 354, *355*
Cellophane, 319
Cells, 313
Celluloid, 318
Cellulose, 315, *316*
Celsius scale, 91
Cementation of sediments, 436–37
Cenozoic era, 508–9, *510*
Centaurs, 387
Centimeters, 8
Centrifugal force, 50
Centripetal force, 50, 52–53, 56
Cepheid variables, 360
Cesium atoms, 7
Chain reactions, 340
Chalcopyrite, 432
Charge. *See* Electric charge
Charge-to-mass ratio, 207–8
Charles' law, 18, 20
Chemical activity of metals, 264
Chemical bonds
  basic functions, 232, 233, 234, 249
  covalent, 235, 238–41, 241–42,
    245–47
  ionic, 235, 236–38, 241–42, 243–45
  metallic, 235
  polarity, 241–42
Chemical elements. *See* Elements
Chemical energy, 72, 233–34, 249
Chemical equations
  balancing, 257–61
  for different reaction types, 262–65
  elements of, 233–34
  generalizing, 261–62
  obtaining information
    from, 265–71
Chemical reactions
  basic features, 233–34, 248–49
  creating hydrocarbon derivatives,
    309–13, 320
  defined, 232
  equations for, 257–62
  nuclear reactions versus, 326
  types, 262–65, 271

Chemical sediments, 435–36
Chemical weathering, 483, 484,
    493, 497
Chernobyl accident, 344–45
Chert, 432
Chesapeake Bay, 584
Chinook winds, 527
Chlorine
  ion form, 237
  in seawater, 579
  use in wastewater treatment, 576
Chlorofluorocarbons, 535
Chondrites, 390–91, 392
Chondrules, 390–91
Chromium, 238
Chrysotile, 434
Ciliary muscle, 189–90
Cinder cone volcanoes, 474
Circuit breakers, 154
Circuits, electric, 149–50, 153–55
Circular motion, 49–50, 52–53
Cirques, 491, *492*
Cities, local climates, 560–61, 562
Citric acid, 312
Cladding, 199
Clastic sediments, 435
Clausius, Rudolf, 88
Clay
  particle size range, 435
  porosity and permeability, 571, *572*
  in soil, 486–87
Claystone, 435
Clean Air Act of 1971, 534
Cleavage, 430, 438
Clementine spacecraft, 414
Climate, 556–61, 564
Clouds
  along fronts, 548–49
  formation of, 534–35, *536,* 542,
    543–44, 564
  seeding, 545
  on Venus, 378
Coal, 75, 76–78
Coalescence process of
    precipitation, 544
Cobalt ions, 238
Coefficients, 258
Coherent motion, 107
Cohesion of molecules, 87
Cold fronts, 548
Color
  causes of, 187–91
  effects of acids and bases on, 287
  of minerals, 428
  of stars, 358
  temperature and, 181
  vision receptors, 200
Colorado River watershed, 570,
    571, 573
Columbia Plateau, 474
Columbia River watershed, 570, *571*
Columnar jointing, 466
Comas of comets, 388
Combination reactions, 263
Combining volumes, law of, 266
Comet P/Wild 2, 389
Comets, 386–89, 395

Comet Shoemaker-Levy, 383, *386*
Common names of chemical
    compounds, 242
Compact discs, 198
Compaction of sediments, 436
Compasses, 159
Complex rock sequences, *506*
Composite volcanoes, 474–75
Compound motion, 38–40, 55–56
Compounds (chemical)
    basic features, 232–34, 248
    bond polarity, 241–42
    bond types, 235–41
    chemical formulas, 254–56
    composition, 242–47
    defined, 86
Compressive stress, 464, 467, *468*
Computed axial tomography
    (CAT), 338, 449
Concave lenses, 189, 190
Concave mirrors, 185
Concentration of solutions,
    280–83, 294
Concepts
    defining with equations, 12, 43
    elements of, 2–4, 20
Concrete, 95, 97
Condensation
    clouds, 534–35, *536*, 542, 543–44
    dew and frost, 531
    fog, 532–34
    overview, 103–4, 528–31, 537
Condensation nuclei, 531–32, 544, 564
Condensation point, 99
Condensations in waves, 120, 135
Conduction
    electrical, 145, 153
    overview, 96, *97*, 110
Conservation of energy, 75, 79
Conservation of mass, 258
Conservation of momentum,
    47–48, 56
Conservation of water, 276
Constructive interference, 127, 192
Contact time, 49
Continental air masses, 546
Continental climates, 560
Continental crust, 446
Continental divides, 570
Continental drift, 449, *450*, 458.
    See also Plate tectonics
Continental glaciers, 491
Continental shelf, 586, *587*
Continental slope, 586, *587*
Continent-continent plate
    convergence, 455
Contrails, 531
Control groups, 16, 21
Controlled experiments, 16, 21
Control rods, 341
Convection
    in earth's atmosphere, 524
    in earth's mantle, 456
    overview, 97–98, 110
    in stars, 357
Convection currents, 97, *98*
Conventional current, 150

Convergent plate boundaries, 452–55
Conversion of energy, 73–74, *75*, 93
Conversion of units, 14, 596–97
Conversion ratios, 597
Convex lenses, 189, 190
Convex mirrors, 185
Coordinated Universal Time, 7
Copper
    ionization, 225, 238
    mass density, 10
    rate of conduction, 97
    specific heat, 95
Core
    of earth, 447, 457
    of stars, 357
Coriolis effect, 403, 585
Corpuscular theory of light, 55
Corrective lenses, 190
Correlation of rock units, 505–7
Cosmic Background Explorer
    spacecraft, 366
Cosmological constant, 368
Cotton, 97
Coulombs, 145
Coulomb's law, 146, 174
Covalent bonds
    electronegativity and, 241–42
    formation of, 235, 238–41, 249
    in hydrocarbons, 301
Covalent compounds
    boiling point, 285
    defined, 240
    formulas, 247, 254
    naming, 245–47, 249
Covalent molecules, 242
Craters
    on Callisto, 383
    on Mercury, 377, *379*
    on Moon, 411–13
    of volcanoes, 474
Creep, 487
Crests of waves, 123, 581
Cretaceous extinctions, 511
Critical angle, 186, 201
Critical density of universe, 369
Critical mass, 340
Crosscutting relationships, 504
Crossover technique, 244–45, 259
Crude oil, 307–8
Crust (earth)
    deformation of, 463–67
    elements in, 424
    overview, 446, 457
Crystals
    formation in minerals, 431–32
    mineral structures, 425–26,
        427, 430
    in solid-state devices, 170–71
Cubits, 5
Cumulonimbus clouds, *536*
Cumulus clouds
    appearance, *536*
    in thunderstorms, 548, *549,*
        550–52
Curie, Marie, 348
Curie, Pierre, 348
Curies, 335, 349

Curl, Robert Floyd Jr., 270
Current. *See* Electric current
Currents (ocean)
    major, 584–85, *586*
    overview, 581, 589
    as regional climate factors, 559
    shoreline, 583
Cycles, of vibrations, 117, 135
Cyclists, 36, 42–43
Cycloalkanes, 306
Cyclones, 550
Cyclonic storms, 549–50, *551*

D

Dallas (TX) climate, 559
Dalton, John, 206, 225, 266
Dark energy, 367, 368
Dark matter, 369
Data, 9, 21
Data points, 600
Davis, William Morris, 496
Davy, Humphrey, 108, 172
Daylight saving time, 409
Days, measuring, 406–9, 419
De Broglie, Louis, 214, 224, 225
Decane, 303
Deceleration, 31, 39
Decibel scale, 128
Decimeters, 7
Declination, 159, *160*
Decomposition of rocks and
    minerals, 483
Decomposition reactions, 263–64
Deep-focus earthquakes, 470
Deflation, 492–93
Degradable pollutants, 575
Degradation of energy, 107
Deimos, 382
Deltas, 489, *490*
Delta symbol, 12, 259, 265, 600
Democritus, 86, 206
Denatured alcohol, 311
Density
    of atmosphere, 518–19
    of minerals, 431
    ratios, 10–11
Density currents, 584–85
Denver (CO) climate, 558–59
Destructive interference, 127, 192
Deuterium, 346, 574
Devil's Post Pile, *466*
Devil's Tower, *466*
Devonian period, 509
Dew formation, 531–32
Dew point temperature, 531
Diamond, 187
Diastrophism, 463–67
Diatomic molecules, 86, 232
2,2-dichloro-3-methyloctane, 305
Diesel fuel, 309
Diethylether, 311
Differential heating
    causing general atmospheric
        circulation, *527, 537, 545*
    causing thunderstorms, 550
Differentiation of earth's interior, 445

Diffusion
    of molecules, 88
    of reflected light, 182–83, *184*
Digital cameras, 191
Digital information, 199
Dikes, 476
2,2-dimethylbutane, 304
2,3-dimethylbutane, 304
Dinosaurs, 509
Dipoles, 242, 246
Dippy Bird, 106
Direct current, 151, 153, 166
Direction, 29, 31
Direct proportion, 12, 21
Direct solar gain, 101
Dirty-snowball cometary model, 386
Disaccharides, 315, 321
Disintegration of rocks and
    minerals, 483
Dispersion of light, 191, 201
Dissolving process, 279–80
Distance-versus-time graphs, *28*
Distillation of petroleum, 307–8
Divergent plate boundaries, 452, *453*
Divides, 570
Dolomite, 436
Domes, 465, 473
Doping, 171
Doppler, Johann Christian, 133
Doppler effect, 132–33
Doppler radar, 134
Dot matrix printers, 163–64
Double bonds, 240
Double rainbows, *19*
Downdrafts, *551, 552*
Drain cleaners, 287
Drift velocity of electrons, 151
Dry zones, 558, *559*
Dual nature theory of light, 197, 214
Dunes, 493
Dust, 520
Dust Bowl, 492–93
Dwarf galaxies, 363–65
Dynamite, 311

E

Ear anatomy, 121
Earth
    basic properties, 378
    coordinate systems, 403–6
    dating, 507–8
    distance from Sun, 393, 400
    distribution of elements, 424
    formation of, 392–93, 444–45, 512
    interactions with Moon, 415–19
    internal structure, 445–48, 457
    magnetic field, 158–59, 161,
        449–50
    motions, 398, *399,* 400–403
    shape and size, 398–400
    time standards, 406–11
Earthquakes
    causes, 468–69
    locating, 469–71
    measuring, 471–72
    seismic tomography from, 449

Fluorine, 237, 238, 239
Fluorite, 432
Fly ash, 78
Foci of earthquakes, 469–70
Fog, 532–34
Folding
  causes, 464–66, 478
  mountains created by, 472–73
Foliation, 438
Follow-through, 49
Food preservation, 337
Foot-pounds, 62–63
Footprints, fossil, 503
Footwalls, 467
Force arrows, 34
Force fields. *See* Fields of force
Forces
  on elastic materials, 116–18
  measuring, 43
  overview, 32–34
  relation to change of motion, 41
  relation to work, 62
Formaldehyde, 311
Formic acid, 286, 287, 312
Forms of energy, 71–73
Formulas
  covalent compounds, 240, 247
  empirical and molecular, 254–56
  ionic compounds, 237, 244–45
Formula weight, 255, 271
Fossil fuels, 76. *See also* Petroleum
Fossils
  dating rocks from, 505, *507*, 513
  defined, 500, 513
  early concepts of, 500–501
  incomplete record, 509
  interpreting, 511
  types, 501–3
Foucault, J. L., 195, 402
Foucault pendulum, 402
Fractions, 593
Fracture of minerals, 430–31
Free fall, 35, 36–37, 38
Freezing
  fossil preservation by, 502
  of seawater, 286, 579, 585
  weathering effects, 483
Freezing point, 99, 286, 294
Freon, 523, 524
Frequency. *See also* Spectra
  defined, 117, 135
  interference, 126–27
  motion and, 132–35
  natural, 129, 135
  of waves, 123, 180
Freshwater distribution, 568–74
Fresnel, A. J., 192, 200
Friction
  electrostatic charge from, 144
  heat from, 92
  resistance to moving objects from, 34, 36
  work against, 71
Fronts, 548–50, 563, 564
Frost formation, 531–32
Frost wedging, 483, *485*
Fructose, 314

Fuel rods, 341, 342, *345*
Fullerenes, 270
Full moon, 415
Functional groups, 310
Fundamental charge of electrons, 145, 174
Fundamental frequency, 131–32, 135
Fundamental properties, 6–7
Fuses, 154
Fusion, latent heat of, 100
Fusion (nuclear)
  overview, 345–46
  proposed research facility, 11
  in stars, 345, 357, 360

G

Gabbro, 435
Gaia hypothesis, 535
Galactic clusters, 363
Galaxies, 363–68, 372
Galena, 432
Galilean moons, 383, *385*
Galileo
  motion theory, 34, 35–37, 55
  use of scientific method, 15–16
Galileo spacecraft, 380
Gallium scans, 338
Galvanometers, 162–63
Gamma cameras, 338
Gamma photons, 328
Gamma radiation
  basic features, 330–31
  defined, 327, 349
  particle properties, 197
Ganymede, 383, *385*
Gases
  as compounds, 232
  convection in, 97
  evaporation and condensation, 103–4, 110
  line spectra, 210–11
  molecular arrangement, 88, 109
  nebulae, *353*, 356
  origins of stars from, 356–57
  sound wave transmission in, 123
Gasohol, *311*
Gasoline, 10, 308–9, *311*
Gastric solution, 290
Gay-Lussac, Joseph, 266
Geiger counters, 333, *334*
Generalization, 9–10, 16, 261–62
Generators, 166, 174
Geologic events, 503–7
Geologic time, 507–11
Geomagnetic time scale, 508
German Plankton Expedition, 588
Giant planets, 376
Giotto spacecraft, 388
Glaciers, 404, 490–91, *492*
Glass
  formation of, 431, 433, 434
  rate of conduction, 97
  refraction index, 187
  specific heat, 95
  speed of sound in, 124
Glass wool, 97

Glen Canyon Dam, *63*
Global Surveyor spacecraft, 381
Global wind patterns, 527–28
Globular clusters, 363
Glucose
  basic features, 314–15
  molecular formula, 254, 314
  molecular structure, 314
  oxidation equation, 261–62
Glycerol, 311
Glycogen, 315
Glycols, 311
Gneiss, 438
Gnomons, 407
Gold, 95, 238
Goldschmidt, Victor Moritz, 439
Grabens, 467
Gram-atomic weight, 268, 271
Gram-formula weight, 268, 271
Gram-molecular weight, 268, 271
Granite
  basic features, 434
  component minerals, 432, *433*
  weathering of, 484
Grapefruit pH, 290
Graphs, 599–600
Gravel, 435
Gravitational fields, 147
Gravitational forces
  Jupiter's influence on asteroid belt, 394
  Newton's law, 50–53, 54, 56
  role in star formation, 356–57, 360
  as source of nuclear energy, 347
  tides and, 417–19
Gravitational potential energy, 67
Gravitons, 148
Gravity. *See also* Acceleration due to gravity; Gravitational forces
  erosion due to, 487
  role in weight, 44, 53
  work against, 71
Grays, 335
Great Dark Spot (Neptune), *375*
Great Red Spot (Jupiter), 383, *384*
Greek letters, 12
Greenhouse effect
  on Earth, 522, 533
  in passive solar applications, 101
  on Venus, 379
Gregorian calendar, 410
Grit chambers, 575–76
Ground-fault interrupters, 155
Ground state of electrons, 212–13
Groundwater, 569, 570–74, 589
Gulf of Mexico, 417
Gulf Stream, 585
Guyots, 455
Gypsum, *430*
Gyres, 585

H

Hail, 552–53
Hair hygrometer, 530
Half-life, 333
Halley, Edmund, 54

Halley's comet, 390
Halogens, 220, 239, 309
Hanging walls, 467
Hardness of minerals, 429–30
Hard water, 290–91
Hardy, Alistair Clavering, 588–89
Hardy Plankton Continuous Recorder, 588–89
Harmonic instruments, 132
Hawaiian volcanoes, 474
Hawking, Stephen William, 370
Health
  effects of radiation on, 335–36
  nuclear medicine technology, 338–39
Hearing, 120–21
Heat. *See also* Temperature
  basic attributes, 92–93
  evaporation and condensation, 103–4, 110
  flow of, 96–99
  measuring, 93–94, 110
  mechanical equivalent, 79
  phase changes and, 99–103
  specific heat, 94–96
  thermodynamics, 104–9
Heat death of the universe, 109
Heat engines, 104–5
Heat islands, 561, 562
Heat of formation, 236
Heat pumps, 106, *107*
Heisenberg uncertainty principle, 216
Heliostats, 77
Helium
  fusion in stars, 361
  as nuclear fusion product, 345, 357, 360
  sound transmission in, 123, 124
Hematite, 429, 484
Henry, Joseph, 165, 173
Heptane, 303
Herodotus, 500
Hertz, 117
Hertzsprung-Russell diagram, 359–60, 372
Hess, Harry Hammond, 455
Hexane, 303
High latitudes, 556–57
High-level nuclear wastes, 347
High-performance alloys, 223
High-pressure centers, 550
High-quality energy, 107
Hipparchus, 357
Historical geology, 500
Hogbacks, 475, *476*
"Hole in the wall" sun calendar, 407
Horizontal motion, 34, 39–40
Hornblende, *429*
Horsepower, 66, 156
Horsts, 467
Household circuits, 154–55
Howard, Luke, 534
Hubble, Edwin, 360, 365, 366
Hubble's law, 360
Hubble Space Telescope, 363, 366
Huggins, William, 133
Humid climates, 560

Humidity, 529–30
Humus, 486
Hurricanes, 554–55
Hutton, James, 462, 478
Huygens, Christian, 191
Hydration, 279, 484
Hydrocarbons
   alkanes, 301–5, 320
   alkenes, 305–6, 320
   alkynes, 306, 320
   aromatic, 306
   cycloalkanes, 306
   derivatives, 309–13, 320
   generalizing chemical reactions, 261
   overview, 300–301, 320
   petroleum, 307–9, 320
   in smog, 269
Hydrochloric acid
   basic features, 286, 287
   formula, 284, 288
   pH, 290
   as strong acid, *289*
Hydrogen
   combustion equation, 265–66
   covalent bonds, 239
   energy level diagram, 212
   fusion in stars, 345, 357, 360
   ion forms, 237
   on Jupiter, 383
   line spectrum, 211
   sound transmission in, 123, 124
   valence, 247
Hydrogenation of oils, 316
Hydrogen bonding, 278, 293
Hydrologic cycle, 520, 543, 568–69
Hydronium ions, 284, 288, 289
Hydropower, 78–80
Hydrostatics, 520
Hydroxide ions, 243, 288
Hyperopia, 190
Hypotheses, 16–17, 21

I

Ice
   hydrogen bonding, 278
   in oceans, 286, 579, 585
   refraction index, 187
   in solar system origins, 392–93
   specific heat, 95
Ice ages, 404, 586
Ice-crystal process of precipitation, 545
Ice-forming nuclei, 545
Iceland, 452
Igneous rocks
   defined, 432, 433
   magnetized, 450
   overview, 433–35, 440
Immiscible fluids, 279
Impact theory of Moon's origins, 414–15
Impulse, 49, 56
Incandescence
   defined, 180
   line spectra and, 210–11, 248
Inches, 5
Incident rays, 184, 185

Inclined planes, 64–65
Incoherent motion, 107
Index fossils, 505, 508
Index of refraction, 187
Indian Ocean, 586
Indirect solar gain, 101
Induction
   electromagnetic, 164–68, 173, 175
   electrostatic charge from, 144
Inertia
   defined, 34, 55
   Newton's law, 41
   relation to mass, 7, 44
   work against, 70
Inertial confinement, 346
Infrared radiation
   from earth's surface, 521–22, 531
   as energy form, 72
   measuring body temperature
      from, 90
   resonant heating of water, 246
Infrasonic waves, 120–21
Inner transition elements, 223
Inorganic chemistry, 300
Instability in earth's atmosphere,
   543–44
Instantaneous speed, 27, 55
Insulators
   electrical, 145
   heat, 96, 110
Intangible concepts, 2–3
Intensity of sound waves, 128, 135
Interference
   of light, 191–92, *193*, 200, 201
   of sound waves, 126–27, 131
Intermediate-focus earthquakes, 470
Intermittent streams, 569
Internal energy
   defined, 92–93, 110
   in thermodynamics, 105, 107
Internal reflection, 186, 199, 201
International date line, 409, *411*
International System of Units, 6.
   See also Measurement
Intertropical convergence zones,
   527–28
Intrusive igneous rocks
   formation of, 433, 475–76
   principles of studying, 504, *506*
Inverse, defined, 593
Inverse proportion, 12, 21, 42–43
Inverse square relationships, 13, 52
Inversions, 522
Invert sugars, 315
Io, 383, *385*
Iodine, 237, 338
Ion–polar molecule force, 279
Ion exchange reactions, 264–65
Ionic bonds
   electronegativity and, 241–42
   formation of, 235, 236–38, 249
Ionic compounds
   boiling point, 285–86
   defined, 237
   formulas, 244–45, 254
   naming, 243, 249
   solubility in water, 279–80

Ionization
   overview, 143, 221–25
   from radiation exposure, 335–36
   in seawater, 579
   of sodium, 234–35
   in water, 284
Ionization counters, 333
Ionosphere, 524
Iris of eye, 189
Iron
   at earth's core, 447
   ion forms, 238
   on Mars, 393
   mass density, 10
   oxidation, 263, 484
   rate of conduction, 97
   in solenoids, 161–62
   specific heat, 95
Iron meteorites, 390–91
Iron ore, reduction with carbon, 262
Island arcs, 454–55
Isolated solar gain, *101,* 102
Isomers, 302–4, 320
Isotopes
   defined, 209, 225
   half-lives, 333
   identifying, 327–28
ISSE spacecraft, 388

J

Jackson, Shirley Ann, 349
Jet streams, 527–28
Joints in rock, 466
Joule, James Prescott, 79–80, 94
Joules, 62, 68, 155
Joule-Thomson effect, 80
Juan de Fuca plate, 477
Julian calendar, 410
Jupiter
   basic properties, 378
   Comet Shoemaker-Levy impact, *386*
   distance from Sun, 393
   Great Red Spot, 383, *384*
   magnetic field, 161
   overview, 382–83
   satellites, 383, *385*

K

Kelvin, Lord (William Thomson),
   79–80, 172, 174
Kelvin scale, 91
Kerosene, 309
Ketones, 310, 311
Kilocalories, 94
Kilograms, 7, 44–45
Kinetic energy
   conversion to potential energy, 73–74
   evaporation and, 103
   increases from work, 71
   overview, 69–70
   relation to temperature, 88,
      92–93, 109
Kinetic molecular theory, 86–88,
   99–104
Kuiper Belt, 387

L

Laccoliths, 476
Lactic acid, 286, 312
Lactose, 315
Lake Powell, *63*
Lakes, 570
Lamp oil, 307
Landform development, 494, *495,*
   512. *See also* Mountains;
   Weathering
Landslides, 487
La Niña, 562
Lanthanide series, 223
Laser Ranging Retro-reflector
   Experiment, 454
Lasers, 193, 198
Last quarter moon, 415–16
Latent heat, 99, 110
Latent heat of fusion, 100
Latent heat of vaporization
   causing hurricanes, 554
   in cloud formation, 544, 550
   overview, 100–102
   as unusual property of water, 277
Latitude, 405–6, 419, 556
Lattices, 170–71
Lava, 431, 474. *See also* Magma;
   Volcanoes
Laws of motion, 40–47, 54
Laws of science. *See* Scientific laws
Lead
   ion forms, 238
   mass density, 10
   rate of conduction, 97
   specific heat, 95
   speed of sound in, 124
Leap seconds, 7
Leap years, 7, 410
Lectiones Opticae (Newton), 54
Lemon battery, 163
Lemon juice, 290
Length, 4, 6
Lenses, 189–91
Levers, 64, 65
Light
   colors, 181, 187–91
   dual nature, 197, 214
   electromagnetic spectrum, 72, *73*
   Faraday's studies, 174
   interaction with matter, 182–83, 201
   inverse square relationship, 13
   Newton's studies, 54–55, 188, 191,
      193–94
   particle features, 195–97, 210
   reflection, 182–85, 201
   refraction, 185–87, 201
   sources, 180–82
   speed of, 6, 185–87
   theories of, 55, 197
   wave features, 191–95
Lightning, 552
Light ray model, 182
Light-year, 354
Limestone, 436, 484, 494
Linear model of radiation
      exposure, 336

Trickling filters, 576
Trilobites, 508, *509*
Triple bonds, 240–41
Tritium, 346
Trojan asteroids, 390
Trombe walls, 101
Tropical air masses, 546
Tropical climate zone, 557
Tropical depressions, 554
Tropical storms, 554
Tropical year, 409, 410
Tropic of Cancer, 406
Tropic of Capricorn, 406
Tropopause, 523
Troposphere, 522–23
Troughs, 123, 581
Tsunamis, 471, 583, 589
Tugboats, 32, *33*
Tungsten ions, 238
Turbidity currents, 585
Turbines, *74,* 341, *343*
Two New Sciences (Galileo), 34
Typhoons, 554

U

Ultrasonic waves, 120–21, 122
Ultraviolet radiation, 523–24
Umbra, 416
Unconformities, 504
Undertow, 583
Uniformitarianism, 462–63, 478, 503–4
Units. *See also* Measurement
 converting, 14, 15, 596–97
 defined, 4
 radiation, 335
 of work, 62–65
Universal constant (*G*), 51
Universal law of gravitation, 50–53, 54, 56
Universe, 346–47, 366–68
Unleaded gasoline production, 308–9
Unpolarized light, 194
Unsaturated hydrocarbons, 305–6, 316
Updrafts, 550, *551*
Uranium-235, 337, *339,* 340
Uranium-238, 328, 332–33, *334*
Uranus, 378, 385–86, 393
Urban heat islands, 561, 562
Urea, 300, 319
Ussher (Archbishop), 507

V

Valence, 247
Valence electrons, 234–35, 249. *See also* Compounds (chemical)
Validity, 17
Valley glaciers, 491
Vaporization, latent heat of, 100–102
Vapor pressure, 284–85
Vapors, molecular arrangement, 88. *See also* Gases

Variable-charge ions, 243, 244
Variables, 12, 21, 599–600
Variable stars, 360
Vega spacecrafts, 380, 388
Velocity
 defined, 29
 final velocity equation, 73
 kinetic energy and, 69
 momentum and, 47
 symbol for, 12
Venera spacecrafts, 380
Ventifacts, 492
Venus
 basic properties, 378
 distance from Sun, 393
 lack of magnetic field, 161, 379
 overview, 378–79
 surface appearance, *380*
Vernal equinox, 401
Vertical projectiles, 39
Vibrations. *See also* Sound
 basic attributes, 116–18
 basic wave features, 118–23
 from earthquakes, 471–72
 interference, 126–27, 131
 refraction and reflection, 124–26
Viking spacecrafts, 380, 381, *382*
Vine, Frederick John, 457
Vinegar, 288, 290
Virtual images, 185
Virtual particles, 370
Virtual photons, 148
Visible light
 colors in, 187–88
 in electromagnetic spectrum, 72, *73,* 180
Vitalist theory, 300
Volcanism, 463
Volcanoes
 atmospheric dust from, 404
 earthquakes with, 469, 474
 examples, *463*
 formation of, 473–77
 on Io, 383
 on Mars, *382*
 on Moon, 414
Voltage, 150, 164–68
Volts, 148
Volume
 defined, 9
 origin of units, 5–6
 standard metric unit, 8
 symbol for, 12
Voyager spacecraft, *375*

W

Walking styles, 29
Warm fronts, 548–49
Wastewater, 574, 575–77
Water
 climatic effects, 559
 discovery on Moon, 414
 in earth's atmosphere, 520, 528–35, *536,* 542–45

 as energy source, 78–80
 erosion due to, 487–89, *490*
 evaporation and condensation, 103–4
 freshwater distribution, 568–74
 household uses, 276
 law of combining volumes for, 266–67
 mass density, 10
 molecular structure, 277–79
 pH, 290
 pollution, 276, 277, 573, 584
 rate of conduction, 97
 refraction index, 187
 resonant heating, 246
 seawater (*see* Seawater)
 softening, 290–91
 solutions, 280–86
 solvent properties, 279–80
 specific heat, 95
 speed of light in, 186
 speed of sound in, 124
 unusual properties, 276–77
 weathering by, 483, 484, *485*
Watersheds, 570
Water tables, 572
Water vapor (atmospheric), 520, 528–35, *536,* 542–45
Waterwheels, 149
Watts, 66, 155–57
Wave equation, 123, 187, 224
Wave fronts, 124–25
Wavelength
 of colors in visible light, 188
 defined, 123, 135
 of ocean waves, 581, 582
 of tsunamis, 471
Wave mechanics, 215
Waves
 between air masses, 549–50, 554
 basic features, 118–23, 180
 light as, 191–95
 of oceans, 581–84, 589
 sound, 120–21, 123–27 (*see also* Sound)
Weakly interacting massive particles (WIMPs), 369
Weather. *See also* Atmosphere (Earth)
 air masses, 546–48, 564
 cloud formation, 534–35, *536,* 542, 543–44, 564
 describing, 8
 forecasting, 555–56
 fronts, 548–49, 564
 general atmospheric circulation, 545
 hydrologic cycle and, 542–43
 precipitation formation, 544–45, 564
 storms, 550–55, 564
 volcano effects on, 476
 waves and cyclones, 549–50
Weathering
 landforms created by, 494
 processes in, 482–84, 497
 soils from, 484–87

Weather radar, 134
Wedges, 65
Wedging effects, 483, *485*
Wegener, Alfred, 449
Weight, 7, 44, 56
Weightlessness, 53
Westinghouse, George, 166
Wet zones, 558, *559*
Wheel and axle, 65
White asbestos, 434
Whitecaps, 583
White dwarf stars, 360, 361, 367
Wilkinson Microwave Anisotropy Probe, 366
WIMPs (weakly interacting massive particles), 369
Wind
 atmospheric patterns, 524–28
 Beaufort scale, 564
 El Niño effects on, 562
 erosion due to, 491–94
Wind chill factor, 526
Wind energy, 77–78
Wind instruments, 132
Wine, oxidation of, 312
Winter solstice, 401, 419
Wöhler, Friedrich, 300, 319
Wood, rate of conduction in, 97
Word equations, 257
Work
 defined, 62, 80
 in electric circuits, 149–50, 153–55, 175
 in first law of thermodynamics, 106
 measuring, 62–65
 power and, 65–67
 relation to energy, 67, 70–71, 80, 93

X

X rays, 326, 338
x-variables, 599–600

Y

Yards (measurement), 5
Yearly time, 409–10, 419
Young, Thomas
 biography, 200
 theory of polarized light, 194
 wave model of light, 55, 191
Young's modulus, 200
Youthful mountains, 494
Youth stage of streams, 488–89
y-variables, 599–600

Z

Zero gravity, 53
Zinc ions, 238
Zinc sulfide, 334
Zodiac, 384
Zones of saturation, 572

Table of Atomic Weights (Based on Carbon-12)

Name	Symbol	Atomic Number	Atomic Weight	Name	Symbol	Atomic Number	Atomic Weight
Actinium	Ac	89	(227)	Meitnerium	Mt	109	(268)
Aluminum	Al	13	26.9815	Mendelevium	Md	101	258.10
Americium	Am	95	(243)	Mercury	Hg	80	200.59
Antimony	Sb	51	121.75	Molybdenum	Mo	42	95.94
Argon	Ar	18	39.948	Neodymium	Nd	60	144.24
Arsenic	As	33	74.922	Neon	Ne	10	20.179
Astatine	At	85	(210)	Neptunium	Np	93	(237)
Barium	Ba	56	137.34	Nickel	Ni	28	58.71
Berkelium	Bk	97	(247)	Niobium	Nb	41	92.906
Beryllium	Be	4	9.0122	Nitrogen	N	7	14.0067
Bismuth	Bi	83	208.980	Nobelium	No	102	259.101
Bohrium	Bh	107	(264)	Osmium	Os	76	190.2
Boron	B	5	10.811	Oxygen	O	8	15.9994
Bromine	Br	35	79.904	Palladium	Pd	46	106.4
Cadmium	Cd	48	112.40	Phosphorus	P	15	30.9738
Calcium	Ca	20	40.08	Platinum	Pt	78	195.09
Californium	Cf	98	242.058	Plutonium	Pu	94	244.064
Carbon	C	6	12.0112	Polonium	Po	84	(209)
Cerium	Ce	58	140.12	Potassium	K	19	39.098
Cesium	Cs	55	132.905	Praseodymium	Pr	59	140.907
Chlorine	Cl	17	35.453	Promethium	Pm	61	144.913
Chromium	Cr	24	51.996	Protactinium	Pa	91	(231)
Cobalt	Co	27	58.933	Radium	Ra	88	(226)
Copper	Cu	29	63.546	Radon	Rn	86	(222)
Curium	Cm	96	(247)	Rhenium	Re	75	186.2
Darmstadtium	Ds	110	(281)	Rhodium	Rh	45	102.905
Dubnium	Db	105	(262)	Rubidium	Rb	37	85.468
Dysprosium	Dy	66	162.50	Ruthenium	Ru	44	101.07
Einsteinium	Es	99	(254)	Rutherfordium	Rf	104	(261)
Erbium	Er	68	167.26	Samarium	Sm	62	150.35
Europium	Eu	63	151.96	Scandium	Sc	21	44.956
Fermium	Fm	100	257.095	Seaborgium	Sg	106	(266)
Fluorine	F	9	18.9984	Selenium	Se	34	78.96
Francium	Fr	87	(223)	Silicon	Si	14	28.086
Gadolinium	Gd	64	157.25	Silver	Ag	47	107.868
Gallium	Ga	31	69.723	Sodium	Na	11	22.989
Germanium	Ge	32	72.59	Strontium	Sr	38	87.62
Gold	Au	79	196.967	Sulfur	S	16	32.064
Hafnium	Hf	72	178.49	Tantalum	Ta	73	180.948
Hassium	Hs	108	(269)	Technetium	Tc	43	(99)
Helium	He	2	4.0026	Tellurium	Te	52	127.60
Holmium	Ho	67	164.930	Terbium	Tb	65	158.925
Hydrogen	H	1	1.0079	Thallium	Tl	81	204.37
Indium	In	49	114.82	Thorium	Th	90	232.038
Iodine	I	53	126.904	Thulium	Tm	69	168.934
Iridium	Ir	77	192.2	Tin	Sn	50	118.69
Iron	Fe	26	55.847	Titanium	Ti	22	47.90
Krypton	Kr	36	83.80	Tungsten	W	74	183.85
Lanthanum	La	57	138.91	Uranium	U	92	238.03
Lawrencium	Lr	103	260.105	Vanadium	V	23	50.942
Lead	Pb	82	207.19	Xenon	Xe	54	131.30
Lithium	Li	3	6.941	Ytterbium	Yb	70	173.04
Lutetium	Lu	71	174.97	Yttrium	Y	39	88.905
Magnesium	Mg	12	24.305	Zinc	Zn	30	65.38
Manganese	Mn	25	54.938	Zirconium	Zr	40	91.22

Note: A value given in parentheses denotes the number of the longest-lived or best-known isotope.